T0331820

Atlantis Studies in Mathematics for Engineering and Science

Volume 14

Series editor

Aims and scope of the series

The series 'Atlantis Studies in Mathematics for Engineering and Science' (AMES) publishes high quality monographs in applied mathematics, computational mathematics, and statistics that have the potential to make a significant impact on the advancement of engineering and science on the one hand, and economics and commerce on the other. We welcome submission of book proposals and manuscripts from mathematical scientists worldwide who share our vision of mathematics as the engine of progress in the disciplines mentioned above.

For more information on this series and our other book series, please visit our website at: www.atlantis-press.com/publications/books.

Atlantis Press
8, square des Bouleaux
75019 Paris, France

More information about this series at http://www.springer.com/series/10071

Thomas J. Buckholtz

Models for Physics of the Very Small and Very Large

Thomas J. Buckholtz
T.J. Buckholtz & Associates
Portola Valley, CA
USA

ISSN 1875-7642 ISSN 2467-9631 (electronic)
Atlantis Studies in Mathematics for Engineering and Science
ISBN 978-94-6239-165-9 ISBN 978-94-6239-166-6 (eBook)
DOI 10.2991/978-94-6239-166-6

Library of Congress Control Number: 2016937514

Printed on acid-free paper

To Helen Buckholtz.
In memory of Joel and Sylvia J. Buckholtz.
With appreciation for each of many people
who contributed to my being able
to attempt this work.

Preface

Welcome to *Models for Physics of the Very Small and Very Large.*

This is a monograph about math-based modeling.

However, let me start by describing a book I hope this monograph helps enable.

~ ~ ~

Ideally, a book would do the following. List all known elementary particles. List all elementary particles people have not found. Show properties for each particle. Describe interactions in which each particle partakes. Use that information to close gaps between known data and traditional theory. Close gaps regarding particle physics, astrophysics, and cosmology. Predict data people have yet to measure. Point to practical applications. Do all that, based on one model or theory.

That program faces difficulties. For example, suppose someone produced that book. Not enough data exists to verify some aspects of the book.

That program features the following question. To what extent can models correlate with elementary particles?

That is a useful question to explore.

Consider theory. 95 % of the inferred stuff in the observable universe is unknown. To what extent is that stuff made of elementary particles? What properties do those particles have?

Consider practice. Experimentalists look for new elementary particles. What candidate particles might people look for? How might people look for them?

Consider society. Knowing of more particles might lead to useful applications.

~ ~ ~

This monograph may not be that book.

Perhaps, this monograph demonstrates modeling and/or physics that that book might feature. Perhaps, this monograph provides steps toward that book.

~ ~ ~

Models for Physics of the Very Small and Very Large discusses modeling.

This monograph tries to add to the extent models can correlate with elementary particles.

This monograph provides a meta-model. And, uses the meta-model to produce models. And, shows that such models correlate with the list of all known elementary particles. And, shows that the models correlate with some known properties of the particles. And, shows that the models correlate with some known interactions in which the particles partake.

Perhaps, people can use such a meta-model and such models to predict elementary particles. And, predict properties for predicted particles. And, predict interactions in which predicted particles partake. And, use concepts about predicted particles to explain data for which traditional physics theory does not seem to have adequate explanations. Some of this data might pertain to cosmology. Some of this data might pertain to astrophysics.

~ ~ ~

I hope people find value regarding the science and art of modeling.

People might decide to work on meta-modeling. People might hone the meta-model this monograph discusses. People might develop other meta-models.

People might use meta-models—perhaps including the one this monograph discusses—to develop models.

People might use models—perhaps including some this monograph discusses—to gain new understanding regarding known data. People might use such models to make predictions. People might use such models to guide experimental or observational efforts. People might use such models to augment traditional theories.

~ ~ ~

Perhaps, people will find value regarding aspects of nature.

This monograph discusses a meta-model.

The meta-model outputs a list of known and candidate elementary particles. The meta-model outputs some properties for the particles.

This monograph shows examples of producing and using models. Some models output additional properties for particles. Some models output possible interactions in which particles partake.

Each of the next three paragraphs discusses a use of models. The three uses use the list of elementary particles that the meta-model outputs. The uses interpret one symmetry differently.

One use features the following. Interpret the symmetry as not correlating with reuse of particle sets. Add particles, beyond today's Standard Model particles. Perhaps, point to dark matter particles and to dark energy particles. Perhaps, explain the rate of expansion of the universe.

A second use features the following. Add some elementary particles, beyond today's Standard Model particles. Interpret the symmetry as correlating with reuse of particle sets. Reuse the particle set and Standard Model physics five times. Perhaps, explain much about dark matter. Add particles, beyond today's Standard Model particles. Perhaps, point to dark energy particles. Perhaps, explain the rate of expansion of the universe.

A third use features the following. Add some elementary particles, beyond today's Standard Model particles. Interpret the symmetry as correlating with reuse of particle sets. Reuse the particle set and Standard Model physics five times. Perhaps, explain much about dark matter. Add gravitons and some other particles to the particle set. Reuse the cumulative particle set. Perhaps, explain much about dark energy stuff. Add a particle related to photons and gravitons. Perhaps, explain the rate of expansion of the universe.

Perhaps, the first of the three uses dovetails best with some aspects of how people interpret data today.

Perhaps, the meta-model and various models will dovetail with new aspects of how people discuss nature in the future.

~ ~ ~

I hope people will use this work. I hope people will benefit from this work. I hope people will tell me of extensions to this work, shortcomings in the work, and developments to which the work contributes.

~ ~ ~

Finally, I especially acknowledge Keith Jones' efforts in coordinating the reviewing and editing processes for this monograph.

Portola Valley, CA, USA
March 2016

Thomas J. Buckholtz

Table of Contents

Chapter 1 Overview

Section 1.1 Some perspective

Section 1.1 discusses context for this monograph. We identify needs for models to predict aspects of nature. We review a past scenario in which modeling improved physics. We draw parallels between that scenario and today's situation. We discuss an opportunity to improve physics. We suggest an agenda for capturing that opportunity.

A need

We think physics needs ways to predict elementary particles.

People can use predictions to guide efforts to infer or find elementary particles.

People can use knowledge of elementary particles to explain presently unexplained facets of cosmology and of other aspects of nature.

An analogy

What is to elementary particles as the periodic table is to atoms?

A parallel

We think today's situation regarding elementary particles has parallels to past work regarding atoms.

By the nineteenth century, the notion of an element existed. Each chemical consists of one or more elements. An element correlates with a chemical that consists of just one type of basic unit.

By the nineteenth century, the notion of an atom was centuries old. Each element consists of one type of basic unit. That unit is an atom.

People knew that chemicals interact with each other. People measured atomic weights. People listed elements. People did not know how many elements nature includes or allows.

During the nineteenth century, people tried to catalog elements. Two principles proved useful. Consider chemical interactions. Consider atomic weights.

The following method proved useful. Create a two-dimensional table. In each column, try to list elements that interact similarly chemically. Within a column, list elements with higher atomic weights below elements with lower atomic weights.

The periodic table started as a model pertaining to what. What do people know about elements? What might people do depict such knowledge? What might people predict? People found gaps in tables. People predicted that gaps might correlate with elements.

T.J. Buckholtz, *Models for Physics of the Very Small and Very Large*,
Atlantis Studies in Mathematics for Engineering and Science 14,
DOI 10.2991/978-94-6239-166-6_1

People did not know much about what an atom is. People did not know much about how atoms interact chemically.

Later, people developed models pertaining to atoms and to how. How do atoms interact chemically? Those models feature electrons and electronic structures of atoms and molecules. How does atomic weight arise? Those models feature atomic nuclei and nuclear physics.

Some parallels

We think today's situation regarding elementary particles has parallels to past work regarding elements and atoms.

By the end of the twentieth century, people listed elementary particles. People knew of interactions involving elementary particles. People knew properties of elementary particles. People suggested specific possibilities for a few new elementary particles.

In the early twenty-first century people found the Higgs boson. A list of somewhat well-defined anticipated particles shrank to a list of one item. Perhaps, gravitons exist.

People posit that yet other particles exist. For example, how else might one explain dark matter?

We can summarize today's situation. People have a list of elementary particles. People do not have a catalog that includes blank spaces.

And, perhaps, people do not have an optimal definition of a basic unit. For example, to what extent is each of the three generations (electron, muon, and tauon) of charged leptons an independent basic unit? Or, to what extent is a combination of one quark and one color charge a basic unit? Or, people have tried to posit concepts for mechanisms underlying elementary particles.

An opportunity

We think people can develop ways to catalog elementary particles. We think people can make predictions about what. What particles might exist? What properties do the particles have? In what interactions do the particles partake?

Develop ways to catalog and predict particles. We think that is a useful pursuit.

We think the pursuit faces difficulties. People may not have enough data about particles to well judge candidate catalogs. A catalog should match particles, properties, and interactions of which people know. But, how might people use predictions (and not just extant data about particles) to home in on useful models?

We think a way forward exists.

An agenda

We propose the following agenda.

Pinpoint simple elementary-particle data that extant physics models use as inputs but do not produce as outputs. Examples of such data include spin and, for fermions, numbers of generations.

Develop math-based meta-models that produce models that can output some of the simple data and can output lists of elementary particles.

Develop models, based on parameters for which a meta-model allows choices.

Use the meta-model and models to produce a catalog of elementary particles. See how well the catalog correlates with known aspects of elementary particles. See how well results the meta-model and models produce correlate with simple elementary-particle data. See with what the catalog correlates regarding possible new particles, properties, and interactions. Then, turn from what to how. Explore extents to which the new particles might explain the how of various aspects of cosmology and other fields of physics.

Interrelate and iterate, regarding meta-models, models, catalogs, predictions, data, and explanations.

Then, perhaps, gain insight regarding the how of elementary particles, properties, and interactions.

Section 1.2 This monograph

Section 1.2 discusses relationships between this monograph and the agenda the previous section suggests. People might say that work in this monograph provides impetus to tackle the agenda. We think this monograph provides examples of trying to take steps the agenda features. We note that models we show produce candidates for physics predictions. We note that some of candidates may prove useful and that some may prove to be wrong. We suggest next steps people might want to take. We summarize, from a modeling perspective, chapters in this monograph. We summarize, from a physics perspective, types of insight and examples of predictions that following the agenda might produce.

Some perspective

We think this monograph shows some promise for the agenda.

We show aspects of the agenda. We show a meta-model. We show models, catalogs of particles, and possible predictions. We possibly provide insight regarding some cosmology and some other phenomena. We possibly show some insight regarding the how of elementary-particle physics.

We think this monograph shows a prototype for approaches to the agenda. The meta-model has parameters. People might feel they need more data about nature to judge the value of this one meta-model, the models we show, the catalogs we develop, the possible insight and predictions we extract from models, and so forth.

We hope people use work in this monograph. We hope people advance various facets of physics to be as well understood as atoms and elements. We hope those facets include aspects of elementary particles, astrophysics, and cosmology.

The concept

This monograph features math-based modeling.

We try to follow the agenda.

We try to show details regarding developing and using the meta-model and models.

We use known physics data to limit the scope the meta-model. Given, a lack of some physics data, we try to keep the meta-model broad. We use known physics data to help select parameters within the meta-model.

We think that the meta-model correlates with a prototype for a catalog of elementary particles.

We think that parameters allow for differing interpretations regarding the physics-relevance of items in the catalog.

We try to go far regarding possible predictions. We think doing so can help people judge the promise of the agenda, the meta-model, some models, and so forth. People may find some of the possible predictions to be useful. People may find some of the possible predictions to be wrong.

We try to go far regarding possibly explaining facets of cosmology and of other physics. We think doing so can help people judge the promise of the agenda, the meta-model, some models, and so forth. People may find some of the explanations to be useful. People may find some of the explanations to be wrong.

We try to suggest ways to improve or extend work in this monograph.

We try to suggest ways to integrate work in this monograph and results from traditional physics.

We try to suggest opportunities for future research.

Three examples

This monograph shows examples of using models.

Each of the next three paragraphs discusses a use of models. The three uses use the list of elementary particles that the meta-model outputs. The uses interpret one symmetry differently.

One use features the following. Interpret the symmetry as not correlating with reuse of particle sets. Add particles, beyond today's Standard Model particles. Perhaps, point to dark-matter particles and to dark-energy particles. Perhaps, explain the rate of expansion of the universe.

A second use features the following. Add some elementary particles, beyond today's Standard Model particles. Interpret the symmetry as correlating with reuse of particle sets. Reuse the particle set and Standard Model physics five times. Perhaps, explain much about dark matter. Add particles, beyond today's Standard Model particles. Perhaps, point to dark-energy particles. Perhaps, explain the rate of expansion of the universe.

A third use features the following. Add some elementary particles, beyond today's Standard Model particles. Interpret the symmetry as correlating with reuse of particle sets. Reuse the particle set and Standard Model physics five times. Perhaps, explain much about dark matter. Add gravitons and some other particles to the particle set. Reuse the cumulative particle set. Perhaps, explain much about dark-energy stuff. Add a particle related to photons and gravitons. Perhaps, explain the rate of expansion of the universe.

Perhaps, the first of the three uses dovetails best with some aspects of how people interpret data today.

Perhaps, at least one of these three uses will dovetail with new aspects of how people discuss nature in the future.

A convention

We start some sentences with the phrase people might say that. We do so to point to concepts people might want to ponder, strengthen, or rebut and ultimately might adopt or reject. We hope this convention adds usefulness to this monograph. We hope use of this convention hastens progress regarding the agenda.

The chapters

Generally, Chapter 2, Chapter 3, and Chapter 4 travel a series of steps. The steps provide a path from unexplained data to models for unexplained data. The first step starts from elementary-particle data that traditional models do not fully output. The first step develops a meta-model and, from the meta-model, models. The second step shows how to apply some such models. The second step provides a list of candidate elementary particles and some of their properties and interactions. The step applies the list to some elementary-particle data with which traditional models do not correlate well. The third step applies the list beyond the realm of elementary particles. The third step suggests models that could correlate with data with which traditional models do not correlate well. Some of the data is cosmology data. Some of the data is astrophysics data.

Generally, Chapter 5, Chapter 6, and Chapter 7 broaden perspectives from the perspectives previous chapters offer. Chapter 5 discusses correlations between work in this monograph and established models. This chapter provides concepts regarding harmonizing and integrating work in this monograph with traditional work. Chapter 6 discusses correlations between work in this monograph and established theories. This chapter provides concepts regarding harmonizing and integrating work in this monograph with traditional work. Chapter 7 discusses the extent to which models in this monograph might provide components for broader understanding of nature. This chapter suggests, especially to the extent people find merit in such models, opportunities people might want to pursue.

~ ~ ~

Specifically, the next paragraphs summarize the chapters.

Chapter 1 (Overview) discusses context for this monograph and discusses work this monograph presents. We discuss an agenda to which this work might contribute. We discuss scope of the work. We discuss possible uses for the work. We summarize chapters in the monograph. We summarize some of the mathematical physics that models this monograph shows may help advance.

Chapter 2 (From data to the MM1 meta-model and MM1MS1 models) discusses modeling. We discuss differences between inputs to models and outputs from models. We discuss the desirability of extending modeling practices to include models that produce as outputs information that traditional models use as inputs but do not produce as outputs. We point to some physics data that generally accepted physics models treat as inputs and generally do not produce as outputs. We develop a meta-model. We use the MM1 meta-model to produce MM1MS1 models. We correlate aspects of the models with aspects of traditional mathematical physics. We correlate aspects of the models with possible future mathematical physics.

Chapter 3 (From the MM1 meta-model to particles and properties) starts from solutions we identify in the previous chapter, shows possibilities for listing known and possible elementary particles, and shows possibilities for modeling properties of particles and for modeling interactions in which particles partake.

Chapter 4 (From particles to cosmology and astrophysics) starts with the known and candidate elementary and composite particles that the previous chapter discusses and shows possibilities for correlating with or anticipating data about cosmology or astrophysics. We show models that possibly correlate with various cosmological phenomena and with various astrophysical phenomena. People might say that some models point to how to close some gaps between physics data and results from traditional theory.

Chapter 5 (From MM1MS1 models to traditional models) starts with models this monograph features, discusses extents to which people might correlate or integrate our models with traditional physics models, and discusses extents people might use our models to extend or improve on traditional physics models.

Chapter 6 (From MM1MS1 models to traditional theories) starts with models this monograph features, discusses extents to which people might correlate or integrate our models with traditional physics theories, and discusses extents people might use our models to extend or improve on traditional physics theories.

Chapter 7 (From the MM1 meta-model to perspective) discusses the extent to which models in this monograph might provide components for broader understanding of nature. We show results from applying the models. We suggest possible opportunities for research regarding and applications of models. We suggest possible applications regarding results from models.

Chapter 8 (Appendices) shows some data and math this monograph uses as inputs.

Chapter 9 (Compendia) provides a list of acronyms this monograph uses and provides lists for use in finding items in this monograph.

Possible physics

People might say that some models we present correlate with aspects of nature or correlate with possibly useful insight or predictions. Statements below correlate with these notions.

People might say that results in Chapter 2 (From data to the MM1 meta-model and MM1MS1 models) might correlate with or provide insight about the following aspects of physics. Spins of elementary particles. Similarities and differences between boson elementary particles and fermion elementary particles. Fields and particles. Number of generations, for elementary fermions. Number of color charges, for relevant elementary fermions. Aspects related to interactions that preserve generation for elementary fermions. A length characterizing the weak interaction. Conservation laws. Conservation of charge. Possible bases for dark-energy stuff. Possible bases for dark matter. Quantum aspects related to magnetic dipole moments of elementary particles. Symmetries related to QCD (or, quantum chromodynamics). Aspects of kaon CP-violation. Aspects of neutral B meson flavor oscillation.

People might say that results in Chapter 3 (From the MM1 meta-model to particles and properties) might correlate with or provide insight about the following aspects of physics. A list of known elementary particles. A list of possible elementary particles. Some interactions in which the particles partake. Spins for known and possible elementary particles. Possible charges, for some possible elementary bosons. The weak mixing angle. The ratio of the mass of the Z boson to the mass of the Higgs boson. Possible approximate masses, for possible boson elementary particles. A possible mass, for the tauon, that could be more accurate than a mass determined from experimental results. Correlations between measurements of the tauon mass and the gravitational constant. Mechanisms correlating with neutrino oscillations. Possible new appearances of the fine-structure constant.

People might say that results in Chapter 4 (From particles to cosmology and astrophysics) might correlate with or provide insight about the following aspects of physics. Mechanisms governing the rate of expansion of the universe. Dark energy. Dark matter. Stuff other than ordinary matter, dark matter, and dark-energy stuff. Ratios of densities of dark matter to densities of ordinary matter. Clustering and anti-clustering mechanisms, regarding ordinary matter and dark matter. Objects, such as galaxies, that include both ordinary matter and dark matter. A mechanism possibly leading to imbalance between matter and antimatter. Mechanisms, correlating with quantum phenomena, that may correlate with phenomena people describe via general relativity. The galaxy rotation problem. Phenomena regarding the spacecraft flyby anomaly. Mechanisms leading to quasars. Cosmic microwave background cooling. Some phases, for quark-based plasmas.

People might say that results in Chapter 5 (From MM1MS1 models to traditional models) might correlate with or provide insight about the following aspects of physics.

Additions to the Standard Model. Details regarding aspects of the cosmology timeline. Additions to the cosmology timeline.

People might say that results in Chapter 6 (From MM1MS1 models to traditional theories) might correlate with or provide insight about the following aspects of physics. Similarities and differences regarding concepts that pertain regarding MM1MS1 models, traditional quantum physics, and classical physics. The Dirac equation. Relationships among quadratic operators, linear operators, and conservation laws. Relationships between models correlating with conservation laws and models correlating with internal properties of elementary particles. For the hydrogen atom, energy levels, fine-structure splitting, hyperfine splitting, and the Lamb shift. Possible alternatives to some traditional-physics correlations between some aspects of quantum chromodynamics and some applications of concepts correlating with special relativity. Facets of, uses of, and limits regarding models based on general relativity.

People might say that results in Chapter 7 (From the MM1 meta-model to perspective) might correlate with or provide insight about the following aspects of physics. Various aspects of physics, assuming specific choices of models that may close gaps between known data and traditional theory. Various aspects of physics, assuming a specific choice of models that may correlate with traditional theory via which people interpret data. A table of families of known and possible elementary particles and of some properties correlating with those particles. Possible opportunities for research, centric to aspects of the MM1 meta-model and MM1MS1 models that correlate with the MM1 meta-model. General opportunities, regarding modeling, for research and applications. General opportunities, regarding physics, for research and applications.

Chapter 2 From data to the MM1 meta-model and MM1MS1 models

Chapter 2 (From data to the MM1 meta-model and MM1MS1 models) discusses modeling. We discuss differences between inputs to models and outputs from models. We discuss the desirability of extending modeling practices to include models that produce as outputs information that traditional models use as inputs but do not produce as outputs. We point to some physics data that generally accepted physics models treat as inputs and generally do not produce as outputs. We develop a meta-model. We use the MM1 meta-model to produce MM1MS1 models. We correlate aspects of the models with aspects of traditional mathematical physics. We correlate aspects of the models with possible future mathematical physics.

~ ~ ~

People might say that results in Chapter 2 (From data to the MM1 meta-model and MM1MS1 models) might correlate with or provide insight about the following aspects of physics. Spins of elementary particles. Similarities and differences between boson elementary particles and fermion elementary particles. Fields and particles. Number of generations, for elementary fermions. Number of color charges, for relevant elementary fermions. Aspects related to interactions that preserve generation for elementary fermions. A length characterizing the weak interaction. Conservation laws. Conservation of charge. Possible bases for dark-energy stuff. Possible bases for dark matter. Quantum aspects related to magnetic dipole moments of elementary particles. Symmetries related to QCD (or, quantum chromodynamics). Aspects of kaon CP-violation. Aspects of neutral B meson flavor oscillation.

Section 2.1 Math-based models for quantum phenomena

Section 2.1 discusses modeling. We discuss differences between inputs to models and outputs from models. We discuss the desirability to extend modeling practices to include models that produce as outputs information that traditional models use as inputs but do not produce as outputs. We point to some physics data that generally accepted physics models treat as inputs and generally do not produce as outputs. We discuss goals for the MM1 meta-model and for MM1MS1 models. We note that trying

T.J. Buckholtz, *Models for Physics of the Very Small and Very Large*,
Atlantis Studies in Mathematics for Engineering and Science 14,
DOI 10.2991/978-94-6239-166-6_2

to develop models that use quantum harmonic oscillator math may be useful. We introduce terminology regarding classes of physics theories, classes of models this monograph shows, and some physics concepts.

~ ~ ~

This subsection provides perspective about this section.

One key to advancing science can be to pinpoint known data for which people do not think theories or models are adequate. We start this section by noting some such data.

Another key to advancing science can be to pinpoint models that seem to be more precise than are other models. Perhaps, such more-precise models provide promise for reuse. We note a seemingly precise model.

Another key can be to develop meta-models via which people can choose or develop models. This section discusses goals for such a meta-model and for models that correlate with the meta-model.

Another key can be to categorize models and theories into useful sets. This section discusses some categories we think can be useful for people's understanding, developing, and using some models and theories.

~ ~ ~

This subsection provides perspective about modeling - in general and specifically as pertains to this monograph.

Math-based models provide key aspects of science and of applied science. People use outputs from models to shape thinking and action. People use outputs to design research efforts. People use outputs to do engineering and to develop products and services.

Models are not nature. People design models to reflect thoughts about nature. Models incorporate people's thinking. Models incorporate or use assumptions and other inputs.

Sometimes inputs to models reflect interpretations of experiments or observations for which people think that the totality of generally accepted models does not produce as outputs those interpretations.

People might say that this monograph reflects attempts to develop models that produce, as outputs, information that traditional models do not produce as outputs. Generally, some such information correlates with results of experiments or observations. People might say that other such output information constitutes possible predictions.

People might say that we try to develop models that reflect and extend a trend. Some traditional classical-physics modeling features concepts for which people might correlate the term continuous. Over time, concepts for which people might correlate the term discreet gained prominence regarding physics and physics models. For example, models relevant to chemistry changed with the discovery of atoms. People had previously discussed concepts for a discreet unit of stuff. People's thinking embraced the discovery - the atom. The advent of quantum mechanics led to more concepts that correlate with the notion of discreet. People might say that today's quantum mechanics features some discreet concepts that correlate with inputs to, but not outputs from, generally accepted models. People might say that we try to develop models that produce as outputs some of those discreet numbers.

People might say that we develop models that do not significantly contradict known data. Room for debate exists. Some of that data reflects not only results of experiments or observations but also interpretations of such results via traditional models. So, some such debate could be significantly about traditional models as well as about the extent to which new models dovetail with known data.

We think that not enough data exists to validate completely today models we present.

We adopt a practice common in numerical simulation. We provide, in effect, a meta-model though which people can develop or specify models. Within the meta-model, people can choose from among various parameters and various modeling assumptions. People might say that the MM1 meta-model is adequately specific that people can efficiently make such choices and can effectively use the resulting specific MM1MS1 models.

People might say that at least one choice of parameters and assumptions correlates with producing as outputs information adequately matching known data that generally accepted traditional models do not produce as outputs.

People might say that at least one choice of parameters and assumptions correlates adequately well with mainstream traditional modeling regarding aspects of elementary-particle physics, astrophysics, and cosmology.

~ ~ ~

This subsection lists data for which the word discreet pertains and the word continuous does not pertain.

Table 2.1.1 features sets of discreet physics numbers. People use these numbers in physics models. People might say that, generally, people do not derive some of these numbers from physics models.

Table 2.1.1 Types of physics observations that yield discreet numbers

1. For each known elementary particle, the expression $j|q_e|/3$, with j being an integer, describes the charge of the particle.
   1.1. q_e denotes the charge of an electron.
2. For each elementary particle, the expression $S\hbar$, with 2S being a nonnegative integer, describes the spin of the particle.
   2.1. $\hbar$ denotes Planck's constant (reduced).

3. Harmonic-oscillator-math raising operators and lowering operators correlate with quantum descriptions of modes of the vector potential.
   3.1. For each photon, the spectrum of excitation states is discreet.
4. The number S(S + D* – 2), with D* = 3, pertains for the radial-component of some harmonic oscillator math-based models correlating with particles having spin/ħ of S.

$$S(S + D^* - 2) = S(S + 1), \text{ for } D^* = 3 \quad (2.1)$$

   4.1. For each known elementary particle, such an S(S + 1) is relevant to the physics of the particle.
5. For each known elementary particle, a nonnegative number, m', describes the rest mass of the particle.
   5.1. The spectrum of such rest masses is discreet.
6. The single speed c and the next expression pertain for all free-ranging elementary particles.

$$E^2 - c^2P^2 = (m')^2c^4 \quad (2.2)$$

   6.1. In the expression, E denotes energy, c denotes the speed of light (in a vacuum), P denotes momentum, and m' denotes the rest mass of the specific particle.
   6.2. Free-ranging elementary particles include electrons and photons.
   6.3. Free-ranging elementary particles do not include quarks.
7. For each known elementary fermion, the number of generations is 3.
   7.1. Examples of elementary fermions include the electron, neutrinos, and quarks.
   7.2. For neutrinos, people say that 3 flavors exist.

Table 2.1.2 points to a possible discreet physics number for which we use the term #ENS. Much traditional physics does not explore whether #ENS can be other than 1. Some MM1MS1 models correlate with #ENS = 48. For #ENS = 48, we discuss the possibility that 8 sets of ensembles exist in the universe. Each set includes 6 ensembles. Each set of 6 ensembles would have its own gravitons. One set includes one ensemble that correlates with ordinary matter and five ensembles that correlate with dark matter. Some MM1MS1 models correlate with the possibility that #ENS = 6. For #ENS = 6, we discuss the possibility that 1 set of 6 ensembles exists in the universe. The set includes one ensemble that correlates with ordinary matter and five ensembles that correlate with dark matter. Some MM1MS1 models correlate with the possibility that #ENS = 1.

Table 2.1.2 A discreet number (#ENS) for which much traditional physics assumes one value and for which MM1MS1 models explore more than one value

1. Below (See, Section 2.13.), ...
   1.1. We identify a symmetry we call INSSYM7.
   1.2. We correlate this symmetry with ...
      1.2.1. The (mathematics) group SU(7).
      1.2.2. The number 48, which is the number of generators of SU(7).
   1.3. We posit that (mathematically) each generator correlates with an ensemble.
   1.4. An ensemble includes ...
      1.4.1. A copy (or, instance) of the Standard Model (as of the year 2015) and, thereby, the following.
         1.4.1.1. A set of elementary particles including ...
            1.4.1.1.1. Leptons (the electron, the muon, the tauon, and neutrinos).
            1.4.1.1.2. Quarks.
            1.4.1.1.3. The weak-interaction bosons (the Z and W bosons).
            1.4.1.1.4. The Higgs boson.
            1.4.1.1.5. The photon.
            1.4.1.1.6. Gluons.
      1.4.2. Possibly, some other elementary particles that would correlate with the MM1 meta-model.
         1.4.2.1. Some possible relatives of the Higgs, Z, and W bosons provide examples.
   1.5. We assume that one ensemble correlates with ordinary matter (or, baryonic matter).
      1.5.1. That ensemble correlates with aspects of nature with which people are familiar on a daily basis.
   1.6. We develop MM1MS1 models that correlate with the following possibilities.
      1.6.1. #ENS = 1. Here, ...
         1.6.1.1. Models point to no forces that connect the ordinary-matter ensemble with other ensembles.
         1.6.1.2. People might say that ...
            1.6.1.2.1. The universe features 1 ensemble.
            1.6.1.2.2. The other 47 mathematically possible ensembles do not pertain to nature.
      1.6.2. #ENS = 6. Here, ...
         1.6.2.1. Models point to forces that connect the ordinary-matter ensemble with 5 other ensembles.
         1.6.2.2. People might say that ...
            1.6.2.2.1. The universe features 6 ensembles.

1.6.2.2.2. The other 42 mathematically possible ensembles do not pertain to nature.

1.6.3. #ENS = 48. Here, ...

1.6.3.1. Models point to forces that connect the ordinary-matter ensemble with 47 other ensembles.

1.6.3.2. People might say that ...

1.6.3.2.1. The universe features 48 ensembles.

1.7. We mention the possibility for MM1MS1 models that correlate with the following possibility. This monograph does not further explore this possibility.

1.7.1. #ENS = 2.

2. #ENS denotes the number of ensembles that nature exhibits.

2.1. For #ENS ≠ 1, an ensemble does not include some elementary particles that would correlate with the MM1 meta-model.

2.1.1. The graviton provides an example.

3. We use the following acronyms.

3.1. The acronym ENS48 pertains to models for which #ENS = 48.

3.2. The acronym ENS6 pertains to models for which #ENS = 6.

3.3. The acronym ENS1 pertains to models for which #ENS = 1.

Table 2.1.3 defines some notation that we use.

Table 2.1.3 Notation of the form #Z

1. Usually, ...

1.1. A symbol of the form #Z (in which Z can be a combination of letters, numbers, and punctuation marks) ...

1.1.1. Denotes a nonnegative integer.

2. Sometimes, ...

2.1. #Z correlates with the termination of one of the following series of sequential nonnegative integers.

2.1.1. #Z, #Z – 1, ..., 1, 0.

2.1.2. 0, 1, ..., #Z – 1, #Z.

2.2. In such cases, we may use the statement #Z = 'Ø to correlate with the empty set of integers.

2.2.1. In mathematics, the symbol Ø to denotes the empty (or, null) set.

3. For example, ...

3.1. In Table 2.3.1, we use the expression #'E = 'Ø.

3.1.1. This usage correlates with SIDE = Ø. (See Table 2.2.3.)

Table 2.1.4 attempts to categorize concepts from Table 2.1.1 and Table 2.1.2.

Table 2.1.4 Categories of some discreet physics numbers

1. Dimensionless integers that seem to be independent of physics constants.
    1.1. Integers correlating with excitation states of the vector potential.
    1.2. D*.
    1.3. Numbers of generations.
    1.4. #ENS.
2. Dimensionless integers that couple with physics constants.
    2.1. 3Q', which we define as charge/($|q_e|/3$).
    2.2. 2S, which equals spin/($\hbar/2$).
3. Other.
    3.1. $|q_e|/3$.
    3.2. $\hbar/2$.
    3.3. Values of m'.

~ ~ ~

This subsection notes successes and possible limitations of some traditional math-based models that people use regarding physics.

People try to develop and use math-based models that people can correlate with observations about nature.

People call one collection of such models the particle-physics Standard Model. (Here, we use the term particle-physics Standard Model to differentiate from uses of the terms cosmology standard model and cosmological standard model. We do not use the terms cosmology standard model and cosmological standard model further in this monograph.) The Standard Model correlates with some observations about elementary particles, composite particles, and interactions between particles.

People say that the Standard Model falls short regarding correlating with various observations and regarding predicting various possible future observations. People might say that, for example, numbers of generations, values of m', and a list of elementary particles can each be considered to be (mainly) inputs to the Standard Model and not (from a standpoint of theory) results of the Standard Model. People might say that, also, the Standard Model (as of 2015) did not adequately predict elementary particles that had yet to be discovered.

~ ~ ~

This subsection discusses goals we have regarding the MM1 meta-model and physics models people can produce from the MM1 meta-model.

Table 2.1.5 provides attributes for which we strive, regarding the MM1 meta-model we present and regarding models we develop from the MM1 meta-model.

Table 2.1.5 Goals regarding a meta-model and regarding physics models

1. Show a math-based meta-model, such that ...

- 1.1. People can use the meta-model to develop and explore physics models.
- 1.2. At least one model ...
  - 1.2.1. Seems to correlate with data for which traditional physics models do not adequately correlate.
  - 1.2.2. Seems not to significantly require possible assumptions or conclusions that contradict relevant known data.
  - 1.2.3. Seems adequately harmonious with extant successful physics models.
- 1.3. At least one model ...
  - 1.3.1. Seems to correlate adequately with traditional physics models that successfully correlate with some data.
- 1.4. People can use the meta-model to explore how to unify new and traditional models.

~ ~ ~

This subsection discusses traditional physics uses of quantum harmonic oscillator math-based models.

People might say that the item in Table 2.1.1 about harmonic-oscillator-math raising operators and lowering operators seems to pertain exactly, at least to the extent of some aspects of models for known photonics.

Other uses of quantum harmonic oscillator math-based models in traditional physics represent attempts to layer quantum models on top of models for non-quantum physics. Such attempts feature approximations to physics phenomena. For example, attempts to quantize some aspects of classical physics exactly via harmonic oscillator math would require modeling an infinitely large potential energy.

~ ~ ~

This subsection notes reasons for considering trying to find new physics uses for quantum harmonic oscillator math-based models.

Table 2.1.6 pertains. Much of the work in this monograph uses mathematics for quantum harmonic oscillators. (Compare with the first item in Table 2.1.4. Below, we show work that correlates with the second item and the third item in Table 2.1.6.)

Table 2.1.6 Correlations between quantum harmonic oscillator math-based models and discreet physics numbers

1. Math-based models for harmonic oscillators provide a seemingly exact model for aspects of quantum photonics phenomena.
2. We find, in quantum harmonic oscillator math-based models, a key role for D*.
3. We find that quantum harmonic oscillator math-based models can correlate with generations.

4. We find that quantum harmonic oscillator math-based models may correlate with masses of zero-mass elementary particles and with masses of some non-zero-mass elementary particles.

We note other concepts that may dovetail with attractiveness for trying to use quantum harmonic oscillator math-based models. People might say that, for quantum harmonic oscillators, the kinetic energy and potential energy contribute equally. People might say that such equal contributions correlate with a stationary point or other feature regarding the expression T – V in action (or, Lagrangian) mathematics. Here, T correlates with kinetic energy and V correlates with potential energy. Much traditional modeling related to elementary particles correlates with action-based math.

~ ~ ~

This subsection discusses relationships between math, models, and physics that this monograph addresses.

We show and discuss solutions for math describing isotropic pairs of isotropic quantum harmonic oscillators. The set of such solutions has an infinite number of ground-state members. This chapter contains subsections, paragraphs, sentences, and phrases prefaced by the symbol [Physics:]. Work in this chapter uses some of those remarks to limit the ground states this monograph features. After such uses, we phase out using the symbol [Physics:].

[Physics:] Correlations between nature and models this monograph features may span aspects of elementary-particle physics and cosmology. For example, some solutions correlate with all known elementary particles. For some aspects of elementary-particle physics and of cosmology, data about nature are not well-developed.

Work in this chapter allows people to make current and future selections regarding which math solutions to try to correlate with aspects of nature.

[Physics:] For example, some solutions may correlate with yet-to-be-discovered elementary particles. Should future data rule out some such particles, people can, when thinking about particle physics, abandon use of those solutions and of work in this monograph based on those solutions.

Thus, this monograph provides a meta-model (or, platform) for developing models that might match data regarding nature. (See Table 2.1.5.)

~ ~ ~

This subsection provides vocabulary for describing some classes of physics theories and models.

Table 2.1.7 pertains.

Table 2.1.7 Terminology for classes of some physics theories and models

1. Classical physics.
    1.1. Pre-relativity classical physics.
    1.2. Special-relativistic classical physics.
        1.2.1. This includes (non-quantum mechanical) relativistic electromagnetism.
    1.3. General-relativistic classical physics.
2. Quantum physics.
    2.1. Some models pertaining to objects (such as solids, fluids, molecules, atoms, and atomic nuclei) that people would say contain elementary particles and composite particles.
    2.2. Some models pertaining to elementary particles and composite particles.

People might say that, regarding quantum physics pertaining to elementary particles, Table 2.1.8 pertains. The classes of models the table shows are not necessarily completely thorough or rigorous. We think that the classes are defined adequately for purposes of this monograph.

Table 2.1.8 Terminology (including QMUSPR, QMPRPR, MM1, and MM1MS1) for classes of some quantum physics models regarding elementary particles

1. We use the term QMUSPR to abbreviate the phrase quantum model that uses (or quantum models that use) assumed elementary-particle properties.
    1.1. Much of traditional quantum mechanics ...
        1.1.1. Assumes a list of elementary particles.
        1.1.2. Uses the list of elementary particles as inputs to models.
        1.1.3. Assumes that elementary particles have properties (such as spin, mass, or charge).
        1.1.4. Uses the properties as inputs to models.
        1.1.5. Models interactions between elementary particles and/or objects that include elementary particles.
2. We use the term QMPRPR to abbreviate the phrase quantum model that provides (or quantum models that provide) elementary-particle properties.
    2.1. Some of the work this monograph features ...
        2.1.1. Assumes that models can provide lists of known elementary particles and candidates for elementary particles.
        2.1.2. Assumes that, for elementary particles, properties (such as spin, mass, or charge) can (at least, somewhat) correlate with outputs from models.
        2.1.3. Models properties of elementary particles.

        2.1.4. Models interactions between elementary particles.
3. We use the term MM1 to abbreviate the phrase meta-model 1.
    3.1. The term MM1 correlates with the meta-model this monograph discusses.
4. We use the term MM1MS1 to abbreviate the phrase meta-model 1 model set 1.
    4.1. The term MM1MS1 correlates with models that people can develop from the MM1 meta-model.
5. People might say that MM1MS1 models correlate with attempts to develop QMPRPR.

Much of physics collates with describing objects and their relationships with the environments in which the objects exist. For example, some physics might model the motion of a planet within a solar system. Here, there might be many objects. Here, a model might correlate the motion of the planet with an environment characterized by the gravitation fields that the model associates with the other objects in the solar system. Or, for example, some physics might model motions of solids and liquids that make up the planet. A model of earth might consider oceans to be part of the planet. A model of tides might consider gravitational effects correlating with objects (such as the moon and the sun) not associated with the planet and effects (gravitational and otherwise) that the model would correlate with the planet.

Some work, regarding elementary particles, in this monograph features characteristics of the particles that people might characterize as internal properties. Examples of such properties include spin, mass, and charge. Some work, regarding elementary particles, in this monograph features relationships between elementary particles and their environments. Examples correlating with such relationships might include forces that act on particles, perceived energy of particles, and perceived momentum of particles. People might say that making a complete separation between internal and environmental may be inappropriate in some models. Paralleling the example of a planet, its tides, and its solar system, we might agree. However, for purposes of some work in this monograph, we think that Table 2.1.9 provides useful (though not necessarily completely rigorous) distinctions.

Table 2.1.9 The terms internal (INTERN), fermion-transformational (FERTRA), extended internal (EXTINT), and environmental (ENVIRO), regarding models regarding elementary particles

1. Internal.
    1.1. This term correlates with modeling of aspects (of an elementary particle) that a model treats as being more associated with the particle than with the environment in which the particle exists.
    1.2. Sometimes we use the acronym INTERN.
        1.2.1. INTERN abbreviates the word internal.
2. Extended internal.

- 2.1. This term also correlates with modeling of aspects (of an elementary particle) that a model treats as being more associated with the particle than with the environment in which the particle exists.
  - 2.1.1. People might say that extended internal models ...
    - 2.1.1.1. May support theoretically studying correlations between types of elementary-particle properties.
    - 2.1.1.2. May not be needed for other purposes.
  - 2.1.2. Sometimes we use the acronym EXTINT.
  - 2.1.3. EXTINT abbreviates the phrase extended internal.

3. Fermion-transformational.
   - 3.1. This term correlates with modeling the extent to which an elementary fermion can change into another elementary fermion, based on interactions with non-zero-mass bosons.
   - 3.2. Sometimes we use the acronym FERTRA.
     - 3.2.1. FERTRA abbreviates the phrase fermion transformational.
4. Environmental.
   - 4.1. People might say that this term correlates with modeling of aspects (of an elementary particle) that a model treats as being more associated with the environment in which the particle exists than with the particle.
   - 4.2. Relevant environments can include the following. (See Table 2.1.10)
     - 4.2.1. SPATIM.
     - 4.2.2. FRERAN.
     - 4.2.3. COMPAR.
     - 4.2.4. ATOMOL.
   - 4.3. Sometimes we use the acronym ENVIRO.
     - 4.3.1. ENVIRO abbreviates the word environmental.
5. This monograph links concepts regarding INTERN models, concepts regarding FERTRA models, and concepts regarding ENVIRO models.

[Physics:] Some aspects of physics distinguish, for a single type of elementary particle, between various environments. For example, for electrons, some aspects of physics pertain for electrons bound in molecules or atoms. Some aspects of physics pertain for electrons that roam over distances inside metals or semiconductors. Some aspects of physics pertain for electrons that travel in a near vacuum. Some aspects of physics regarding electrons span all such environments.

We discuss more than one environment for elementary particles. Table 2.1.10 pertains.

Table 2.1.10 The terms SPATIM, FRERAN, COMPAR, and ATOMOL regarding ENVIRO models regarding elementary particles

1. SPATIM.

1.1. This term abbreviates the phrase space-time coordinate or the phrase space-time coordinate symmetries.
1.2. People might say that SPATIM ENVIRO models correlate with particles existing in a universe for which space-time coordinates pertain for models.
1.3. We show SPATIM models for all elementary particles.
  1.3.1. People might say that some of the elementary particles for which we develop SPATIM models have not been observed to exhibit behavior compatible with some traditional uses of the term free-ranging.
    1.3.1.1. Examples of such elementary particles include quarks and gluons.
1.4. We use this term to abbreviate ...
  1.4.1. SPATIM ENVIRO.

2. FRERAN.
  2.1. This term abbreviates the phrase free-ranging.
  2.2. People might say that FRERAN models correlate with aspects of traditional quantum physics models regarding free-ranging elementary particles.
  2.3. People might say that FRERAN models correlate with aspects (such as some aspects correlating with Feynman diagrams) of traditional quantum physics models that people use regarding aspects of behavior of non-free-ranging elementary particles. (Compare with COMPAR models, to which this table alludes below.)
  2.4. We use this term to abbreviate ...
    2.4.1. FRERAN SPATIM.
    2.4.2. FRERAN SPATIM ENVIRO.
3. COMPAR.
  3.1. This term abbreviates the phrase composite particles.
  3.2. People might say that COMPAR models correlate with aspects (regarding elementary particles) that correlate with composite particles.
    3.2.1. Examples of composite particles include pions and the proton.
    3.2.2. Examples of elementary particles bound in pions or protons include quarks and gluons.
      3.2.2.1. People consider quarks and gluons to be non-free-ranging elementary particles.
  3.3. We use this term to abbreviate ...
    3.3.1. COMPAR SPATIM.
    3.3.2. COMPAR SPATIM ENVIRO.
4. This monograph shows, for some elementary particles, both FRERAN and COMPAR models.

- 4.1. [Physics:] For the W boson, FRERAN models differ from COMPAR models.
  - 4.1.1. People might say that FRERAN models pertain to interactions between leptons and W bosons.
  - 4.1.2. People might say that FRERAN models might pertain to some interactions between quarks and W bosons.
  - 4.1.3. People might say that COMPAR models pertain to some interactions between quarks and W bosons.

5. ATOMOL.
   - 5.1. This term abbreviates the phrase atoms and molecules.
   - 5.2. People might say that ATOMOL models correlate with aspects (of elementary particles and atomic nuclei) that correlate with atoms and molecules.
     - 5.2.1. Examples of atoms include the hydrogen atom.
   - 5.3. We use this term to abbreviate ...
     - 5.3.1. ATOMOL SPATIM.
     - 5.3.2. ATOMOL SPATIM ENVIRO.

~ ~ ~

This subsection notes that this monograph uses two types of representations for quantum harmonic oscillators.

This monograph features models based on math correlating with isotropic quantum harmonic oscillators. Traditionally, people use at least two math-based approaches regarding quantum harmonic oscillators. For some cases, the two approaches yield similar results. (For example, see Table 2.2.14 and Table 2.5.12.) For some cases, the two approaches yield different results. (For example, see Table 2.2.15 and Table 2.5.13.)

One approach features partial differential equations and functions ([Physics:] wave functions) of continuous coordinates ([Physics:] often, spatial coordinates). Here, for multidimensional harmonic oscillators, people may have a choice regarding coordinate systems to use. One choice features using one linear coordinate for each dimension. For an isotropic harmonic oscillator, another choice features using radial coordinates.

This monograph sometimes uses approaches that feature partial differential equations. For multidimensional aspects, we use radial coordinates. For single-dimensional aspects we use a linear coordinate or a projection of a radial coordinate.

We correlate the term DIFEQU with such uses. DIFEQU provides an acronym for (usually) radial-coordinate (generally) multidimensional partial differential equation.

Another approach does not feature spatial coordinates. Such approaches start from, for each dimension, a characterization that features a number that represents the number of times the one-dimensional harmonic oscillator correlating with that dimension is excited. The models can feature raising-operators and lowering-operators that people express in a way that includes a numerical factor that depends on the original number of excitations. People sometimes use the term ladder operator to characterize raising operators and lowering operators. For an isotropic harmonic oscillator, people can develop models that feature a number that represents a total of excitations, with the sum running across the one-dimensional harmonic oscillators that correlate with the various dimensions.

This monograph sometimes uses approaches that feature ladder operators and do not feature partial differential equations. For multidimensional aspects, we generally use representations that feature excitation numbers and ladder operators for the individual component oscillators.

We correlate the term LADDER with such uses. LADDER provides an acronym for ladder operators.

Other approaches pertain regarding harmonic oscillators. For example, for multidimensional isotropic harmonic oscillators expressed in radial coordinates, people use approaches that feature ladder operators. This monograph does not use such approaches.

Table 2.1.11 pertains. Treatments of LADDER representations feature math pertaining to sets of lone (or, one-dimensional) oscillators. Treatments of DIFEQU representations feature radial coordinates and/or linear coordinates and functions ([Physics:] wave functions) expressed in terms of radial and/or linear coordinates.

Table 2.1.11 Models, for harmonic oscillators, that this monograph uses

1. Models based on LADDER approaches.
2. Models based on DIFEQU approaches.

~ ~ ~

This subsection provides definitions of the terms boson and fermion.

Table 2.1.12 shows definitions of boson and fermion. These definitions correlate with traditional physics uses of the two terms.

Table 2.1.12 Definitions of S, boson, and fermion

1. [Physics:] This monograph uses S to denote spin/$\hbar$.
2. [Physics:] The term boson pertains to any object for which (in math-based models) 2S equals an even nonnegative integer.
3. [Physics:] The term fermion pertains to any object for which (in math-based models) 2S equals an odd positive integer.

~ ~ ~

This subsection lists categories of LADDER solutions this monograph uses and categories of models (based on solutions) this monograph uses.

Table 2.1.13 lists categories of LADDER solutions this monograph uses. [Physics:] Here, regarding the term elementary particles, we deemphasize distinctions people might try to make regarding the term field and the term particle.

Table 2.1.13 CORMAT and CORPHY categories of LADDER solutions

1. CORMAT solutions ...
   1.1. Correlate with the simplest LADDER solutions (to equations involving isotropic pairs of isotropic harmonic oscillators) that we think correlate with known elementary particles or with possible elementary particles.
   1.2. Take their acronym (of CORMAT) from the phrase core relevant mathematical solutions.
2. CORPHY solutions ...
   2.1. Correlate with CORMAT solutions in ways such that ...
      2.1.1. For boson-related solutions, CORPHY solutions equal CORMAT solutions.
      2.1.2. For fermion-related solutions, CORPHY solutions represent extensions (based on adding two oscillators) to CORMAT solutions.
   2.2. Take their acronym (CORPHY) from the phrase core solutions relevant to elementary-particle physics.
   2.3. Provide a basis for solutions this monograph shows for the following categories of solutions. (See, for example, Table 2.1.14.)
      2.3.1. INTERN LADDER.
      2.3.2. EXTINT LADDER.
      2.3.3. FERTRA LADDER.
      2.3.4. SPATIM LADDER.
      2.3.5. FRERAN LADDER.
      2.3.6. COMPAR LADDER.

~ ~ ~

[Physics:] This subsection previews applications for categories of solutions this monograph uses and for categories of models (based on solutions) this monograph uses.

Table 2.1.14 summarizes some applications for various categories of models or solutions. (See Table 2.2.9, Table 2.5.6, and Section 2.6 for discussions about the term solution. See Table 2.3.4 for a list of families of elementary particles.)

Table 2.1.14 Some applications of various categories of solutions

1. INTERN LADDER solutions correlate with ...
   1.1. Families of elementary particles.
   1.2. Elementary particles within a family.

   - 1.3. Spin.
   - 1.4. Excitation states, for bosons.
   - 1.5. Charge.
   - 1.6. Some aspects regarding approximate masses, for elementary bosons.
   - 1.7. Zero size for some aspects related to elementary particles.
   - 1.8. Number of color charges, for non-lepton elementary fermions.
   - 1.9. The extent to which the strengths of interactions mediated by G-family particles vary, when the interactions involve fermions, by fermion generation. (See, for example, Table 2.8.2.)
2. DIFEQU solutions correlate with ...
   - 2.1. Families of elementary particles.
   - 2.2. Elementary particles within a family.
   - 2.3. Spin.
   - 2.4. Some aspects regarding approximate masses, for elementary bosons.
   - 2.5. Non-zero size for the extent to which fields pertain.
   - 2.6. Zero size for some aspects related to elementary particles.
   - 2.7. The number of generations for each non-zero-mass elementary fermion.
3. FERTRA LADDER solutions correlate with ...
   - 3.1. Abilities of fermion elementary particles to transform into other fermion elementary particles via interactions with non-zero-mass elementary bosons.
4. FRERAN LADDER solutions correlate with ...
   - 4.1. Symmetries related to special relativity symmetries (and the Poincare group).
   - 4.2. Number of instances of fields for each particle.
     - 4.2.1. Possible explanations for dark-energy stuff and dark matter.
   - 4.3. The notion that, in FRERAN environments, interactions mediated by W-family particles do not change the generations of fermions. (See, for example, remarks preceding Table 2.7.1.)
   - 4.4. For G-family particles, the number of channels pertaining to interactions that the particles intermediate. (See Section 2.13.)
   - 4.5. Aspects related to quarks (and possibly other somewhat similar particles) and gluons (and possibly other somewhat similar particles) bound into composite particles.
5. COMPAR LADDER solutions correlate with ...
   - 5.1. Aspects related to quarks (and possibly other somewhat similar particles) and gluons (and possibly other somewhat similar particles) bound into composite particles.

~ ~ ~

[Physics:] This subsection discusses our using different terminology for some aspects this monograph treats than terminology some people use for somewhat similar aspects of traditional physics.

Table 2.1.15 pertains.

Table 2.1.15 Terminology (MM1MS1-...) that differentiates work in this monograph from some traditional concepts

1. G-family solutions that we correlate with photonics may not exactly correlate with some traditional uses for models for photons. (See Table 2.8.3.)
    1.1. The following differences may pertain.
        1.1.1. Polarization.
            1.1.1.1. We emphasize circular-polarization modes.
                1.1.1.1.1. In effect, each polarization (that is, left circular polarization or right circular polarization) correlates with an anti-mode (in the sense of antiparticle) to the other mode.
            1.1.1.2. Much of traditional physics seems to emphasize linear polarization modes.
                1.1.1.2.1. Linear polarization modes do not function as each other's anti-modes.
        1.1.2. Electromagnetism and magnetic dipole moments that are not generated by motions of charges.
            1.1.2.1. We treat effects of magnetic dipole moments that correlate with (possibly stationary) elementary particles as correlating with a boson other than the boson we correlate with electromagnetism pertaining to stationary and moving charges.
            1.1.2.2. Traditional physics may consider that photons interact with each of ...
                1.1.2.2.1. Stationary and moving charges.
                1.1.2.2.2. Magnetic dipole moments not generated by the motion of charges.
2. This monograph uses the term MM1MS1-photon to denote some concepts we propose and use.
3. People might say that the following statements characterize some possible differences between MM1MS1-photons and photons (or, QMUSPR-photons).
    3.1. MM1MS1-photons may not interact with elementary-particle magnetic dipole moments.
        3.1.1. In this monograph, ...

            - 3.1.1.1. MM1MS1-photons interact with charges and charge-based currents.
            - 3.1.1.2. A different spin-1 boson interacts with elementary-particle magnetic dipole moments.
    - 3.2. In traditional physics models, photons interact with charges, charged-based currents, and elementary-particle magnetic dipole moments.
4. This monograph ...
    - 4.1. Discusses a MM1MS1 boson that interacts with elementary-particle magnetic dipole moments.
    - 4.2. Notes the possibility for correlating this monograph's models for a (conceptual) combination of this MM1MS1 boson and MM1MS1-photons with traditional models for photons.
5. We are less thorough regarding QMUSPR-centric and MM1MS1-centric labelling regarding other elementary particles that correlate with other G-family solutions. (Regarding the G-family, see Table 2.3.4, Section 3.2, and Section 3.3.)
    - 5.1. For example, traditional physics does not yet encompass gravitons.
        - 5.1.1. People might say that detection of gravitational waves does not necessarily correlate with detection of individual gravitons.
            - 5.1.1.1. People might say that this notion has parallels regarding light. People discussed observations correlating with light waves before people discussed detection of photons.
    - 5.2. Given the lack of a graviton in traditional physics models, ...
        - 5.2.1. We do not try to distinguish between the terms graviton and MM1MS1-graviton.
        - 5.2.2. In this monograph, we do not further use the term MM1MS1-graviton.
6. We are not certain as to the extent to which MM1MS1-neutrinos differ from neutrinos (or, QMUSPR-neutrinos).
    - 6.1. MM1MS1-neutrinos have zero mass.
    - 6.2. People say that the existence of neutrino oscillations (or, neutrino flavor mixing) implies that at least one neutrino has non-zero mass.
        - 6.2.1. People attribute neutrino oscillations to interactions between neutrinos and gravity.
    - 6.3. Our models correlate with the notions that ...
        - 6.3.1. MM1MS1-neutrinos can change flavor based on interactions other than interactions with gravity.
        - 6.3.2. MM1MS1-neutrinos interact with gravity.
        - 6.3.3. Interactions between MM1MS1-neutrinos and gravity may not lead to neutrino-oscillations.

7. We use MM1MS1 notation for MM1MS1-neutrinos.
8. Generally, various not well-defined terms (such as dark energy) exist.
    8.1. Generally, we do not try to use MM1MS1 notation to distinguish between our uses of such terms and traditional uses of such terms.

Section 2.2 LADDER models for isotropic quantum harmonic oscillators

Section 2.2 discusses LADDER models for isotropic pairs of isotropic quantum harmonic oscillators. We discuss aspects regarding solutions to equations. We discuss discreet-math models pertaining to 1-dimensional harmonic oscillators. We discuss discreet-math models pertaining to multi-dimensional isotropic quantum harmonic oscillators. We define a constraint, Œ = 0, that each MM1MS1 LADDER model features. We define, in the context of LADDER models, the term solution. We show notation for labeling columns in tables that show aspects of LADDER models. People might say that we show a ground-state solution that traditional physics models may have underutilized.

~ ~ ~

This subsection provides perspective about mathematics that features solutions to equations in general and about mathematics that pertains to isotropic pairs of isotropic quantum harmonic oscillators specifically.

Some branches of mathematics feature solving equations. Some equations have characteristics that people might characterize as isolation and constraints. For example, consider the algebraic equation y = ax + b. Here, people might consider a and b to be constants and x to be an independent variable. Here, people might consider y to be a dependent variable. Or, people might consider that y exists (as an independent variable) in isolation from ax + b and that the equation provides a constraint linking y and x.

People might state that such isolation characterized accounting until around the beginning of the fourteenth century. In effect, people could add numbers and characterize assets. In effect, people could add numbers and characterize liabilities. People might not need to link the two calculations. Starting around the year 1300, people developed and used the practice of double-entry bookkeeping. In double-entry bookkeeping, 0 = assets – liabilities.

Today, much work regarding physics applications of math that pertains to quantum harmonic oscillators exhibits parallels to single-entry practices (or, isolated solutions). People specify some constants and (for one-dimensional quantum harmonic oscillators) a coordinate. (For multi-dimensional quantum harmonic oscillators, there is more than one coordinate.) People solve an equation. For a one-dimensional quantum harmonic oscillator or for an isotropic multi-dimensional quantum harmonic oscillator, a solution features a ground-state energy and an incremental energy that characterizes the differences in energy between any two nearest (but unequal) energy levels.

Work in this section features math for quantum harmonic oscillators. Some work in this section shows isolated solutions. This section makes a transition. Much of the section features an approach that parallels double-entry bookkeeping. People might say that, here, (regarding an equation we show above) each one of y and ax + b correlates with a quantum harmonic oscillator. Here, an analog of a double-entry balance that nets to zero proves useful. People might say that, by analogy, each of y and x is somewhat an independent variable and that $0 = y - (ax + b)$ provides a constraint.

We make use of such a double-entry-like approach to define the math from which we develop physics models.

[Physics:] People might say that the double-entry-like net balance of zero points to how to avoid some possible problems that people state regarding some traditional physics models. One such problem is the seemingly infinite energy that people correlate with a sum of photon ground-state energies. Another such problem correlates with the possibility that the universe includes a seemingly infinite amount of energy.

~ ~ ~

This subsection shows notation for LADDER models for lone quantum harmonic oscillators.

Table 2.2.1 shows well-known information about 1-dimensional harmonic oscillators. Below, this monograph finds uses for negative values of N.

Table 2.2.1 Notation and math pertaining to a lone quantum harmonic oscillator (LADDER models)

1. People specify the state of a 1-dimensional harmonic oscillator via a linear combination of base states.
2. Each base state features an integer that characterizes that base state.
    - 2.1. We use the notation $| N >$ for a base state.
        - 2.1.1. N denotes the integer.
3. Raising operators correlate with increasing the value of N.
    - 3.1. The symbol a^+ denotes a raising operator.
    - 3.2. The next equation pertains.

$$a^+ | N > = (1 + N)^{1/2} | N + 1 > \quad (2.3)$$

4. Lowering operators correlate with decreasing the value of N.
    - 4.1. The symbol a^- denotes a lowering operator.

   4.2. The next equation pertains.

$$a^- \mid N > = N^{1/2} \mid N - 1 > \tag{2.4}$$

5. Traditional treatments of lone harmonic oscillators limit the range of N to nonnegative integers.
   5.1. People might say that the following expressions disconnect possible applicability of negative N from applicability of nonnegative N.

$$a^- \mid 0 > = 0 \mid -1 > \tag{2.5}$$

$$a^+ \mid -1 > = 0 \mid 0 > \tag{2.6}$$

6. People may use the terminology ground state to denote the base state $\mid 0 >$.
7. [Physics:] People may associate an energy with each base state.
   7.1. The energy is proportional to the next result.

$$a^+ a^- + 1/2 = N + 1/2 \tag{2.7}$$

8. [Physics:] In physics, the state of a quantum harmonic oscillator can be described as a linear combination of base states ...
   8.1. $\Sigma_N b_N \mid N >$, with ...
      8.1.1. Each b_N being a complex number.
      8.1.2. $\Sigma_N |b_N|^2 = 1$.

One concern that people might raise regarding negative values of N is a choice of sign for the $(1 + N)^{1/2}$ and $N^{1/2}$ factors associated, respectively, with raising operators and lowering operators. However, a similar choice of sign can pertain for these factors when N is a nonnegative integer. We think that choice of signs is not an issue for work in this monograph.

~ ~ ~

This subsection shows notation for LADDER representations for multi-dimensional isotropic quantum harmonic oscillators.

Table 2.2.2 shows information about multi-dimensional isotropic quantum harmonic oscillators. Below, this monograph finds uses for isotropic pairs of multi-dimensional isotropic quantum harmonic oscillators.

Table 2.2.2 Notation and math pertaining to LADDER representations for a multi-dimensional isotropic quantum harmonic oscillator

1. For a (#D + 1)-dimensional isotropic quantum harmonic oscillator, ...
   1.1. #D denotes a nonnegative integer.
   1.2. The set { Pj | j = 0 , 1 , ... , #D − 1 , or #D } provides an index to the lone quantum harmonic oscillators that make up the (#D + 1)-dimensional isotropic quantum harmonic oscillator.

1.3. Numbers N(P0), N(P1), ... correlate with base states of the respective lone harmonic oscillators.
    1.3.1. The lone harmonic oscillators correlate with the (#D + 1)-dimensional isotropic quantum harmonic oscillator.
    1.3.2. Each number N(..) provides part of a description of a base state for the (#D + 1)-dimensional isotropic quantum harmonic oscillator.
2. People specify the state of a (#D + 1)-dimensional isotropic quantum harmonic oscillator via a linear combination of base states for the (#D + 1)-dimensional harmonic oscillator.
3. Traditional treatments of isotropic quantum harmonic oscillators limit the range of each N(..) to nonnegative integers.
4. Below, this monograph uses negative values of N(..).
5. [Physics:] People may associate an energy with each base state of the (#D + 1)-dimensional isotropic quantum harmonic oscillator.
    5.1. The energy is proportional to the sum over the Pj of terms of the following form.

$$(a_{Pj})^{+} (a_{Pj})^{-} + 1/2 = N(Pj) + 1/2 \quad (2.8)$$

    5.2. The term isotropic correlates with the sum's giving equal weight to each of the N(Pj) + 1/2.

~ ~ ~

This subsection shows notation relevant to LADDER representations for isotropic pairs of isotropic quantum harmonic oscillators.

Table 2.2.3 describes an isotropic pair of isotropic quantum harmonic oscillators. The table shows a relationship between the two paired isotropic quantum harmonic oscillators. Each of the symbols #EMAX and #PMAX comports with Table 2.1.3. #EMAX can denote 'Ø or a nonnegative integer. #PMAX always denotes a nonnegative integer. #EMAX correlates with the term QE-like. #PMAX correlates with the term QP-like. ([Physics:] Think of energy-momentum space. QE-like matches energy-like. The E in QE correlates with use of the symbol E to denote energy. QP-like matches momentum-like. The P in QP correlates with use of the symbol P to denote momentum.)

Table 2.2.3 Notation and relationships pertaining to an isotropic pair of isotropic quantum harmonic oscillators (LADDER models)

1. Each of #EMAX and #PMAX comports with Table 2.1.3.
    1.1. If #EMAX is an even integer, #PMAX is an even integer.
    1.2. If #EMAX is 'Ø, #PMAX is an even integer.
    1.3. If #EMAX is an odd integer, #PMAX is an odd integer.
2. We define two sets. Each set consists of indices associated with a sequence of consecutive nonnegative integers.

$$\text{SIDE} = \{ \text{Ej} \mid \text{j} = \#\text{EMAX}, \#\text{EMAX} - 1, \ldots, 1, \text{or } 0 \} \quad (2.9)$$

$$\text{SIDP} = \{ \text{Pj} \mid \text{j} = 0, 1, \ldots, \#\text{PMAX} - 1, \text{or } \#\text{PMAX} \} \quad (2.10)$$

3. We provide notation for the union of the two sets. (If #EMAX = 'Ø, SID = SIDP.)

$$\text{SID} = \text{SIDE} \cup \text{SIDP} \quad (2.11)$$

4. N(n) denotes the quantum number for the lone harmonic oscillator n.
   - 4.1. Each n has one of forms Ej or Pj.
   - 4.2. Each N(n) + 1/2 term below matches a result (from traditional quantum harmonic oscillator math) for a 1-dimensional harmonic oscillator.
5. We define two sums. (If #EMAX = 'Ø, SIDE = Ø and $Æ_{QE} = 0$.)

$$Æ_{QE} = \Sigma_{n \in SIDE} (N(n) + 1/2) \quad (2.12)$$

$$Æ_{QP} = \Sigma_{n \in SIDP} (N(n) + 1/2) \quad (2.13)$$

6. The term isotropic pair of isotropic quantum harmonic oscillators correlates with sums giving equal weight to the magnitude of each of the N(Ej) + 1/2 and to the magnitude of each of the N(Pj) + 1/2.
7. We provide two equivalent versions of a constraint that correlates with the pair of isotropic oscillators being isotropic, with each of the two isotropic oscillators being isotropic, and with conditions this monograph imposes. Here, $\pm_n = +1$, for n ∈ SIDE. Here, $\pm_n = -1$, for n ∈ SIDP.

$$0 = Œ = Æ_{QE} - Æ_{QP} \quad (2.14)$$

$$0 = Œ = \Sigma_{n \in SID} \pm_n (N(n) + 1/2) \quad (2.15)$$

~ ~ ~

This subsection shows a constraint we use to limit models (based on isotropic pairs of isotropic quantum harmonic oscillators) we consider further.

Table 2.2.4 repeats a key feature from Table 2.2.3. This constraint pertains throughout this monograph.

Table 2.2.4 A constraint on Œ

$$Œ = 0 \quad (2.16)$$

We show LADDER examples of this constraint. People might say that this monograph deemphasizes explicitly pointing out DIFEQU examples of this constraint.

~ ~ ~

This subsection defines the concepts of open oscillator pair and closed oscillator pair.

Table 2.2.5 explains some notation.

Table 2.2.5 The notation E[j] and P[j]

1. This monograph uses notations E[j] and P[j] in which j can be an arithmetic expression that evaluates to an integer.
   1.1. For example, we might use the notations P[#PMAX −1] and P[#PMAX].

Table 2.2.6 defines the term local-Œ. In the rightmost column, each +1 comes from two instances of +1/2. Each instance of +1/2 correlates with one of the two oscillators in the pair. Here, the term even denotes an even positive integer.

Table 2.2.6 Definition of local-Œ for a base state correlating with an oscillator pair

For an oscillator pair ...	... for a base state defined by ...	... local-Œ equals ...
E[even]-and-E[even − 1]	N(E[even]) and N(E[even − 1])	N(E[even]) + N(E[even − 1]) + 1.
P[even − 1]-and-P[even]	N(P[even − 1]) and N(P[even])	N(P[even − 1]) + N(P[even]) + 1.

Table 2.2.7 shows examples in which local-Œ = 0. Here, each of j and k is a complex number.

Table 2.2.7 Examples in which local-Œ = 0

1. This item exhibits, for an oscillator pair E[even]-and-E[even − 1], amplitudes such that local-Œ = 0.
   1.1. Amplitude ...

   j × | N(E[even]) = 0 and N(E[even − 1]) = −1 >

   \+

   k × | N(E[even]) = −1 and N(E[even − 1]) = 0 >

   1.2. Such that ...

   $|j|^2 + |k|^2 = 1.$
2. This item exhibits, for an oscillator pair P[even − 1]-and-P[even], amplitudes such that local-Œ = 0.
   2.1. Amplitude ...

   j × | N(P[even − 1]) = 0 and N(P[even]) = −1 >

   \+

   k × | N(P[even − 1]) = −1 and N(P[even]) = 0 >

   2.2. Such that ...

   $|j|^2 + |k|^2 = 1.$

Table 2.2.8 exhibits two ways for an oscillator pair to have local-Œ = 0. (Table 2.2.7 discusses j and k.)

Table 2.2.8 Definitions of open oscillator pair and closed oscillator pair

For an oscillator pair ...	... for state defined (with $\|j\|^2 + \|k\|^2 = 1$) by ...	... the following term pertains to the oscillator pair
E[even]-and-E[even − 1]	j = 0 or k = 0	open
E[even]-and-E[even − 1]	j and k are indeterminate	closed
P[even − 1]-and-P[even]	j = 0 or k = 0	open
P[even − 1]-and-P[even]	j and k are indeterminate	closed

Regarding a table such as Table 2.3.11, people can consider to be a closed oscillator pair any E[even]-and-E[even − 1] oscillator pair for which the table shows two blank entries or any P[even − 1]-and-P[even] oscillator pair for which the table shows two blank entries. For example, for 2W in Table 2.3.11, people can consider that the P4L-and-P4R oscillator pair is closed. Also, people can consider to be closed each of the (infinite number of) E[even]-and-E[even − 1] oscillator pairs for which the table does not show columns and each of the (infinite number of) P[even − 1]-and-P[even] oscillator pairs for which the table does not show columns.

~ ~ ~

This subsection defines, for LADDER models, the term solution and discusses possible ways to order lists of solutions for isotropic pairs of isotropic quantum harmonic oscillators.

Table 2.2.9 pertains. This monograph also makes other uses of the word solution. (See, for example, Table 2.5.6.) We think that people can differentiate appropriately among various uses of the term solution.

Table 2.2.9 Defining the term solution (regarding LADDER model uses for which Œ = 0) and introducing a means for ordering solutions

1. We define a solution by ...
   1.1. #EMAX or a similar QE-like limit.
   1.2. #PMAX or a similar QP-like limit.
   1.3. A set of N(n) that satisfies the following constraint.
   $$0 = Œ = Æ_{QE} - Æ_{QP} \tag{2.17}$$
      1.3.1. People might associate the term solution (as in solution to an algebraic equation) with the satisfying of this equality.
2. The number of solutions is infinite.
   2.1. For example, there is an infinity of choices for #EMAX.
3. [Physics:] For this monograph, ...
   3.1. The following constraints pertain.
   $$\#EMAX \leq 16 \text{ or } \#EMAX = '\emptyset \tag{2.18}$$
   $$\#PMAX \leq 16 \tag{2.19}$$

   3.2. The number of relevant pairs of #EMAX and #PMAX is finite.
4. For a choice of #EMAX and #PMAX, the number of solutions is infinite.
5. [Physics:] Generally (but not always), we exhibit, for isotropic pairs of isotropic quantum harmonic oscillators, solutions that could correlate with ground states that we think are relevant to the physics we address.
   5.1. For a choice of #EMAX (or other QE-like limit) and #PMAX (or other QP-like limit), the number of ground-state solutions this monograph considers is finite.
6. When displaying (in a table) ground-state solutions, this monograph may use (but does not always use) an ordering based on ...
   6.1. First, increasing value of #PMAX.
   6.2. Then, other considerations.

~ ~ ~

This subsection introduces notation for displaying LADDER solutions.

Table 2.2.10 shows how some tables use vertical nomenclature to describe names of the lone oscillators that are parts of isotropic pairs of isotropic quantum harmonic oscillators. In Table 2.2.10, the first row shows oscillator names. The next two rows show how we use two rows to show (in tables below) oscillator names. To form a name, append the number from the last row in Table 2.2.10 to the letter from the next-to-last row of Table 2.2.10. We call these names linear-numbering names. Table 2.2.11 extends this work to include an alternative set of names (that is, polarization-centric names) for oscillators.

Table 2.2.10 Correlations between some linear-numbering names for lone oscillators and the display in tables of column headings correlating with those names

E6	E5	E4	E3	E2	E1	E0	P0	P1	P2	P3	P4	P5	P6	P7	P8	Oscillator names
E	E	E	E	E	E	E	P	P	P	P	P	P	P	P	P	Linear-numbering
6	5	4	3	2	1	0	0	1	2	3	4	5	6	7	8	names

Table 2.2.11 shows alternative names for oscillators. (Compare with Table 2.2.10. [Physics:] This monograph sometimes uses polarization-centric names.) For polarization-centric names, form the n in the N(n) correlating with a column by appending the last row in the column to the next-to-last row in the column. L denotes left (as in left-circularly polarized). R denotes right (as in right-circularly polarized). For example, the name P8L correlates with the oscillator that this monograph also names P7.

Table 2.2.11 Polarization-centric names for oscillators

E6	E5	E4	E3	E2	E1	E0	P0	P1	P2	P3	P4	P5	P6	P7	P8	Oscillator names
E	E	E	E	E	E	E	P	P	P	P	P	P	P	P	P	Polarization-centric
6R	6L	4R	4L	2R	2L	0	0	2L	2R	4L	4R	6L	6R	8L	8R	names

Table 2.2.12 shows how this monograph extends linear numbering beyond E9 and beyond P9. Here, A denotes 10, B denotes 11, ..., and G denotes 16. Here, except for columns for E0 and P0, the table shows two oscillators per column. For example, the P9A column correlates with oscillators P9 and PA. This monograph uses notations E[j] and P[j] in which j can be an arithmetic expression that evaluates to one of 0, 1, ..., or G. For example, P[10] = PA. (See Table 2.2.5.)

Table 2.2.12 Extended linear numbering for oscillators and for oscillator pairs

E	E	E	E	E	E	E	E	E	P	P	P	P	P	P	P	P	P
GF	ED	CB	A9	87	65	43	21	0	0	12	34	56	78	9A	BC	DE	FG

~ ~ ~

This subsection shows a well-known solution and another ([Physics:] non-traditional) solution.

Table 2.2.13 shows a well-known solution.

Table 2.2.13 The traditional-physics ground state for 3-dimensional QP-like isotropic quantum harmonic oscillators

1. The following equations characterize the ground state for this solution.

$$\#'E = 0 \tag{2.20}$$

$$\#'P = 2 \tag{2.21}$$

$$N(E0) = 1 \tag{2.22}$$

$$N(P0) = N(P1) = N(P2) = 0 \tag{2.23}$$

2. Œ = 0 results from ...
   2.1. A contribution of +3/2 correlating with N(E0) = 1.
   2.2. A contribution of −3/2 correlating with N(P0) = N(P1) = N(P2) = 0.
3. [Physics:] For a traditional ground state of an isotropic quantum harmonic oscillator with 3 spatial (or, QP-like) dimensions, people would correlate the energy with N(E0) + 1/2 = 3/2.

Table 2.2.14 depicts the solution Table 2.2.13 describes.

Table 2.2.14 Depiction of the ([Physics:] traditional) ground state for 3-dimensional QP-like isotropic quantum harmonic oscillators

E0	P0	P1	P2
1	0	0	0

Table 2.2.15 shows a solution for which N(E0) + 1/2 = 0 + 1/2 = 1/2. ([Physics:] People might say that the energy correlates with N(E0) + 1/2 = 1/2. People might dispute this solution. Work below shows another solution for which N(E0) = 0. (See Table 2.5.13.) There, the work uses radial coordinates. People might not dispute that solution. People might apply the term non-traditional to each of these two solutions for which N(E0) = 0.) For the state Table 2.2.15 shows, applying a raising operator to the P0 oscillator produces a state with 0 amplitude. Specifically, a_{P0}^{+} | N(P0) = −1 > = (1 + N(P0))$^{(1/2)}$ | N(P0) = 0 > = 0 | N(P0) = 0 >, because 1 + N(P0) = 0. ([Physics:] Below, we correlate this solution with MM1MS1-photons. The P0 oscillator correlates with longitudinal polarization. This application is consistent with the concept that each of a photon and a MM1MS1-photon has zero longitudinal polarization.)

Table 2.2.15 A ([Physics:] non-traditional) ground state for 3-dimensional QP-like isotropic quantum harmonic oscillators

E0	P0	P1	P2
0	−1	0	0

Section 2.3 INTERN applications of LADDER models

Section 2.3 discusses relevant solutions within models for isotropic pairs of isotropic quantum harmonic oscillators. We feature subsets of the solutions. We focus on INTERN LADDER subsets for which the solutions might correlate with data about elementary particles. We provide names for families of solutions. We characterize solutions pertaining to various families. People might say that we show or point to solutions that correlate with all known and some possible elementary particles. People might say that those solutions correlate with properties that include at least spin and whether rest mass is zero or non-zero.

~ ~ ~

This subsection introduces relevant solutions.

Table 2.3.1 and Table 2.3.3 show some ground-state solutions. For these tables, #'E = #EMAX and #'P = PMAX. For Table 2.3.1, a parameter σ equals +1. For Table 2.3.3, the parameter σ equals −1. [Physics:] Later, we indicate that the subfamilies these tables show correlate with all the elementary particles to which MM1MS1 models point as possibly existing. Later, we correlate σ = +1 with the term free-ranging. Later, we correlate σ = −1 with the term non-free-ranging. Later, we correlate elements of the subfamily names with more-familiar terminology people use regarding elementary particles. Later, we correlate spin/ħ with a parameter S and we indicate that, for subfamilies for which the letter G does not appear in the subfamily name, 2S = #'P.

Table 2.3.1 shows some ground-state solutions for which σ = +1. [Physics:] Here, we show mathematically possible 3C- and 4W-subfamilies. The table shows two 3C rows. We think that, for 3C, the table illustrates a condition we call inappropriate redundancy. (See Table 2.15.8.) We think that, for 3C, a condition we call lack of an appropriate D pertains. (See Table 2.15.8.) Based on either (or both) of these two conditions, we deemphasize 3C solutions. We think that, for 4W, the condition lack of appropriate D pertains. We think that, for 4W, a condition we call inappropriate square of rest mass pertains. (See Table 2.15.8.) Based on at least the first of these two conditions, we deemphasize 4W solutions. Later, we discuss constraints regarding excitations for subfamilies for which N(P0) ≤ −2. Later, we correlate inappropriate redundancy with appropriateness for deemphasizing G-family solutions for which #'P > 8. (See Table 2.15.8. Regarding the term G-family, see Table 2.3.4.)

Table 2.3.1 CORMAT σ = +1 ground-state solutions and subfamily names, based on #'E and #'P limits on harmonic oscillators

σ	#'E	E 6	E 5	E 4	E 3	E 2	E 1	E 0	P 0	P 1	P 2	P 3	P 4	P 5	P 6	P 7	P 8	#'P	Solution subfamily
+1	0							0	0									0	0H
+1	'Ø								0	−1								1	1C
+1	'Ø								−1	0								1	1N
+1	2					0	0	0	0	0	0							2	2W
+1	0							0	−1	0	0							2	022G
~~+1~~	~~'Ø~~								~~0~~	~~−1~~	~~0~~	~~−1~~						~~3~~	~~3C~~
~~+1~~	~~'Ø~~								~~0~~	~~0~~	~~−1~~	~~−1~~						~~3~~	~~3C~~
+1	'Ø								−1	0	0	−1						3	3N
~~+1~~	4			~~0~~	~~0~~	~~0~~	~~0~~	~~0~~	~~0~~	~~0~~	~~0~~	~~0~~	~~0~~					4	~~4W~~
+1	2					0	0	0	−1	0	0	0	0					4	24..G
+1	0							0	−2	0	0	0	0					4	042G
+1	4			0	0	0	0	0	−1	0	0	0	0	0	0			6	46..G
+1	2					0	0	0	−2	0	0	0	0	0	0			6	26..G
+1	0							0	−3	0	0	0	0	0	0			6	064G
+1	6	0	0	0	0	0	0	0	−1	0	0	0	0	0	0	0	0	8	68..G
+1	4			0	0	0	0	0	−2	0	0	0	0	0	0	0	0	8	48..G
+1	2					0	0	0	−3	0	0	0	0	0	0	0	0	8	28..G

σ	#'E	E 6	E 5	E 4	E 3	E 2	E 1	E 0	P 0	P 1	P 2	P 3	P 4	P 5	P 6	P 7	P 8	#'P	Solution subfamily
+1	0							0	−4	0	0	0	0	0	0	0	0	8	084G

Table 2.3.2 defines the term lepton-related solution.

Table 2.3.2 Definition of lepton-related solution

1. [Physics:] The term lepton pertains to an elementary particle if and only if ...
    1.1. The elementary particle correlates with a solution Table 2.3.1 shows.
    1.2. #'E = 'Ø for that solution.
    1.3. Nature exhibits the elementary particle.
        1.3.1. Known leptons include ...
            1.3.1.1. Neutrinos, which correlate with the 1N-subfamily.
            1.3.1.2. Electrons, muons, and tauons, which correlate with the 1C-subfamily.
        1.3.2. As yet, people have not found or ruled out elementary particles correlating with the 3N-subfamily.

Table 2.3.3 shows some ground-state solutions for which σ = −1. Mathematically, the 0Y solution cannot be excited. (See Table 2.15.8.) We deemphasize the 0Y-subfamily. [Physics:] For each of the cases 1Q, 1R, 3Q, 3I, 3R, and 3D, it matters that N(E[#'E]) = N(P[#'P]), but it does not matter which one of N(E[#'E]) = N(P[#'P]) = 0 or N(E[#'E]) = N(P[#'P]) = −1 we show in this table. (See Table 2.3.14.) Later, we discuss specific properties and constraints regarding values of N(..) for the QE-like oscillators for the 2Y- and 4Y-subfamilies. (See Section 3.6. See, especially, Table 3.6.1 and Table 3.6.8.)

Table 2.3.3 CORMAT σ = −1 ground-state solutions and subfamily names, based on #'E and #'P limits on harmonic oscillators

σ	#'E	E 6	E 5	E 4	E 3	E 2	E 1	E 0	P 0	P 1	P 2	P 3	P 4	P 5	P 6	P 7	P 8	#'P	Solution subfamily
−1	0							0	0									0	0O
~~−1~~	~~0~~							~~−1~~	~~−1~~									~~0~~	~~0Y~~
−1	1						0	0	0	0								1	1Q
−1	1						0	−1	−1	0								1	1R
−1	2					0	0	0	0	0	0							2	2O
−1	2					..	..	..	−1	−1	−1							2	2Y
−1	3				−1	0	0	0	0	0	0	−1						3	3Q
−1	3				−1	−1	−1	0	0	−1	−1	−1						3	3I
−1	3				−1	0	0	−1	−1	0	0	−1						3	3R
−1	3				−1	−1	−1	−1	−1	−1	−1	−1						3	3D

σ	#'E	E 6	E 5	E 4	E 3	E 2	E 1	E 0	P 0	P 1	P 2	P 3	P 4	P 5	P 6	P 7	P 8	#'P	Solution subfamily
−1	4			0	0	0	0	0	0	0	0	0	0					4	4O
−1	4			..	..	..	..	..	−1	−1	−1	−1	−1					4	4Y

Each row in Table 2.3.1 and Table 2.3.3 shows a solution subfamily for which the name includes one letter. Each letter denotes the name of one family of solutions.

~ ~ ~

This subsection lists names for families of solutions that this monograph considers further.

Table 2.3.4 discusses names for families of solutions. Technically, it could be permissible to consider the H-family to be a subset of the W-family. ([Physics:] This monograph lists the H- and W-families separately.) People use the term gamma rays to describe some photons. [Physics:] This monograph correlates some solutions from each family, except the I-, R-, D-, and O-families, with known elementary particles. Possibly, the I-family correlates with particles that interact directly with gravity and do not interact directly with electromagnetism. (We use the word directly so as to allow for the possibility of indirect interactions via clouds of virtual particles that would accompany the elementary particles. See Table 2.10.1.) We choose the letter I to correlate with the word invisible. We choose the letter R to correlate with the letter alphabetically next after the letter Q. Possibly, the D-family correlates with particles that do not interact directly with either gravity or electromagnetism. We choose the letter D to correlate with the word dark. We do not much discuss the extent to which O-family particles correlate with people's concepts for leptoquarks.

Table 2.3.4 Names for families of solutions

σ	Name	[Physics:] Elementary-particle motivation for the name
+1	H-family	Higgs boson.
+1	C-family	Charged leptons.
+1	N-family	Neutrinos.
+1	W-family	Weak-interaction bosons.
+1	G-family	Gamma ray and graviton (a hypothetic type of particle).
−1	Q-family	Quark.
−1	I-family	(No known-particle motivation.)
−1	R-family	(No known-particle motivation.)
−1	D-family	(No known-particle motivation.)
−1	Y-family	The shape of the letter suggests a type of gluon vertex.
−1	O-family	O, as in leptoquark (a hypothetical type of particle).

~ ~ ~

This subsection introduces notation useful for situations in which families or subfamilies exhibit similarities.

Table 2.3.5 pertains.

Table 2.3.5 Notation regarding similar $\sigma = +1$ families and subfamilies and regarding similar $\sigma = -1$ families and subfamilies

1. Regarding some solutions for which $\sigma = +1$, ...
   1.1. CN denotes either of the C- or N-families.
   1.2. WHO denotes any one of the W-, H-, or O-families.
   1.3. WO denotes either of the W- or O-families.
   1.4. 2WO denotes either of the 2W- or 2O-subfamilies.
   1.5. 4WO denotes either of the 4W- or 4O-subfamilies.
2. Regarding some solutions for which $\sigma = -1$, ...
   2.1. QIRD denotes any one of the Q-, I-, R-, or D- families.
   2.2. QR denotes either of the Q- or R-families.
   2.3. ID denotes either of the I- or D-families.
   2.4. YO denotes either of the Y- or O-families.
   2.5. 1QR denotes either of the 1Q- or 1R-subfamilies.
   2.6. 3QIRD denotes any one of the 3Q-, 3I-, 3R-, or 3D-subfamilies.
   2.7. 2YO denotes either of the 2Y- or 2O-subfamilies.
   2.8. 4YO denotes either of the 4Y- or 4O-subfamilies.
3. Sometimes, we use the above notations to mean ...
   3.1. Some, instead of any one.
   3.2. All, instead of any one.
   3.3. Both, instead of either of.

~ ~ ~

This subsection discusses two uses for family names, for subfamily names, and for names of solutions.

Table 2.3.6 pertains.

Table 2.3.6 Dual use of family names, subfamily names, and solution names

1. Sometimes, we use family names (such as Q-family) or subfamily names (such as 1Q) to designate sets of solutions.
2. [Physics:] Sometimes, we use (the same) family names or (the same) subfamily names to designate sets of known or possible elementary particles that this monograph correlates with a (respective) set of solutions.
   2.1. Here, the term elementary particle includes the notion of field (associated with the elementary particle).

3. Dual use pertains also for some solution names.
    3.1. [Physics:] However, ...
        3.1.1. For Y-family solutions, ...
            3.1.1.1. Pairs of 2Y solutions correlate with components of gluons.
            3.1.1.2. Pairs of 4Y solutions correlate with components of possible other particles.
        3.1.2. For G-family solutions and particles, ...
            3.1.2.1. For all cases, each solution correlates with two polarization modes.
            3.1.2.2. For some cases, direct correlations between single solutions and single elementary particles do not pertain.
                3.1.2.2.1. For example, direct correlation does not pertain for some cases that feature the SOMMUL-set models. (See Table 2.15.8 and Table 3.2.1.)

~ ~ ~

This subsection discusses names for subfamilies of solutions that this monograph considers further.

For each of the N-, Q-, and R-families, solutions exist for which 2S = 1 and solutions exist for which 2S = 3. This monograph uses the term subfamily to call attention to this concept. For the N-family, the subfamilies are 1N and 3N, respectively. For the Q-family, the subfamilies are 1Q and 3Q, respectively. For the R-family, the subfamilies are 1R and 3R, respectively. In each of these cases, for notation of the form nX, with X denoting a family, the equation n = 2S pertains.

For each of the I- and D-families, solutions exist for which 2S = 3. Here, for notation of the form nX, with X denoting a family, the equation n = 2S = 3 pertains.

For the Y-family, solutions exist for which 2S = 2 and solutions exist for which 2S = 4. For the Y-family, the subfamilies are 2Y and 4Y respectively. ([Physics:] A 0Y solution exists mathematically, but cannot be excited. Therefore, this work deemphasizes the 0Y solution. See Table 2.3.3.)

For the O-family, solutions exist for which 2S = 0, solutions exist for which 2S = 2, and solutions exist for which 2S = 4. For the O-family, the subfamilies are 0O, 2O, and 4O, respectively.

For the G-family, symbols of the form epnG denote subfamilies. Here, e = #'E, p = #'P, and n = 2S. Here, ep..G denotes a subfamily for which more than one solution may pertain and more than one S may pertain. An ep..G-subfamily may include more than one epnG-subfamily.

The C family has one subfamily, 1C. Here, 2S = 1. The W-family has one subfamily, 2W. Here, 2S = 2. The H-family has one solution and, thereby, one subfamily, 0H. Here, 2S = 0.

~ ~ ~

This subsection discusses some aspects of and distinctions between the MM1 meta-model and MM1MS1 models.

Table 2.3.7 pertains. (For more details, see Section 2.15.)

Table 2.3.7 Some aspects of and distinctions between the MM1 meta-metal and MM1MS1 models

1. The MM1 meta-model ...
   1.1. Embraces as correlating with candidate elementary particles ...
      1.1.1. Each of the families (of solutions) to which Table 2.3.4 alludes.
      1.1.2. Each of the subfamilies (of solutions) to which Table 2.3.1 alludes, except ...
         1.1.2.1. Those subfamilies the table shows as struck out.
      1.1.3. Each of the subfamilies (of solutions) to which Table 2.3.3 alludes, except ...
         1.1.3.1. Those subfamilies the table shows as struck out.
   1.2. Does not embrace as correlating with candidate elementary particles ...
      1.2.1. Any other constructs.
2. Each MM1MS1 model comports with the previous item in this table.
3. MM1MS1 models may vary based, for example, on ...
   3.1. A subset (of the overall set of solutions that correlate with candidate particles) to consider as correlating with possible particles.
   3.2. The number of fields a model correlates with a solution.
      3.2.1. [Physics:] This difference can lead to possibilities (that can differ by model) regarding ...
         3.2.1.1. The nature of dark-energy stuff. (See, for example, Section 4.3.)
         3.2.1.2. The nature of dark matter. (See, for example, Section 4.3.)
   3.3. [Physics:] Modeling and interpretation of models regarding numbers of generations for spin-3/2 elementary fermions. (See Table 2.6.4.)
   3.4. Other considerations. (See Section 2.15.)

~ ~ ~

This subsection discusses #E and #P, which are some limits that we place on oscillators for INTERN LADDER solutions.

Table 2.3.8 discusses reasons why this monograph features solutions for which there are an odd number of QE-like oscillators and an odd number of QP-like oscillators. CORPHY solutions correlate with solutions with an odd number of QE-like oscillators and an odd number of QP-like oscillators.

Table 2.3.8 Reasoning leading to featuring INTERN LADDER solutions with an odd number of QE-like oscillators and an odd number of QP-like oscillators, plus definitions of #E and #P (CORPHY solutions)

1. [Physics:] For elementary particles, ...
   - 1.1. Possibly the following should pertain.
     - 1.1.1. For boson elementary particles, ...
       - 1.1.1.1. For each non-G-family boson elementary particle, 2S = #'P.
       - 1.1.1.2. For each G-family boson elementary particle, 2S = 2 or 4.
     - 1.1.2. For each fermion elementary particle, 2S = #'P.
   - 1.2. For each fermion elementary particle, a solution correlating with the particle's antiparticle would use oscillator #'P + 1 and would not use oscillator #'P.
   - 1.3. We limit our choices of numbers of QP-like oscillators to odd numbers.
     - 1.3.1. We deemphasize the notation #'P.
     - 1.3.2. We use the notation #P, with the understanding that #P is a nonnegative even integer.
       - 1.3.2.1. Because oscillator P0 pertains, the number of QP-like oscillators is odd.
   - 1.4. For boson elementary particles, ...
     - 1.4.1. For each non-G-family boson elementary particle, ...

$$\#P = \#'P \tag{2.24}$$

$$2S = \#P = \#'P \tag{2.25}$$

     - 1.4.2. For each G-family boson elementary particle, ...

$$\#P = \#'P \tag{2.26}$$

$$2S = 2 \text{ or } 4 \tag{2.27}$$

   - 1.5. For each fermion elementary particle, ...

$$\#P = \#'P + 1 \tag{2.28}$$

$$2S = \#P - 1 = \#'P \tag{2.29}$$

   - 1.6. To maintain Œ = 0, ...
     - 1.6.1. We deemphasize the notation #'E.
     - 1.6.2. We use the notation #E, with the understanding that #E is a nonnegative even integer.
     - 1.6.3. For elementary bosons, ...

$$\#E = \#'E \tag{2.30}$$

     - 1.6.4. For elementary fermions, ...

$$\#E = 0, \text{ if } \#'E = '\emptyset \tag{2.31}$$

$$\#E = \#'E + 1, \text{ otherwise} \tag{2.32}$$

~ ~ ~

This subsection notes that, regarding LADDER solutions, the remainder of this monograph features CORPHY solutions and deemphasizes CORMAT solutions.

Until Table 2.3.8, regarding LADDER solutions, we feature CORMAT solutions. For the remainder of this monograph, regarding LADDER solutions, we feature CORPHY solutions.

For example, regarding Table 2.3.8, CORPHY solutions correlate with use of #EMAX = #E and #PMAX = #P. CORMAT solutions correlate with use of #EMAX = #'E and #PMAX = #'P.

~ ~ ~

This subsection lists ground-state INTERN LADDER solutions that this monograph considers further.

This subsection transforms results in Table 2.3.1 and in Table 2.3.3 from notation based on #'E and #'P to results that correlate with notation based on #E and #P. This correlates with the transition to, for LADDER solutions, featuring CORPHY solutions and deemphasizing CORMAT solutions.

Table 2.3.9 pertains.

Table 2.3.9 Algorithm for selecting ground-state INTERN LADDER solutions

1. [Physics:] Select solutions that we think are relevant.
   1.1. For example, limit selections to solutions for which #P ≤ 8.
   1.2. For example, do not include solutions that would correlate with a lack of physics-relevance, per Table 2.15.8.
2. Work above and below defines this algorithm.
   2.1. Table 2.3.11, Table 2.3.13, and Table 2.3.14 show results.

We discuss ground-state LADDER solutions for which $\sigma = +1$ and then ground-state LADDER solutions for which $\sigma = -1$. Table 2.3.10 provides information common to those two sets of ground-state LADDER solutions.

Table 2.3.10 Information pertaining to CORPHY ground-state LADDER solutions

1. [Physics:] N(P0) correlates with aspects related to rest mass.
   1.1. For INTERN LADDER solutions and for ENVIRO LADDER solutions, ...

$$N(P0) = 0 \text{ correlates with } m' \neq 0 \quad (2.33)$$
$$N(P0) < 0 \text{ correlates with } m' = 0$$

   1.2. For FERTRA LADDER solutions, ...

$$N(P0) = -1 \text{ correlates with } m' \neq 0 \quad (2.34)$$
$$N(P0) = -2 \text{ correlates with } m' = 0$$

2. For fermion INTERN LADDER solutions, ...
   2.1. For P[j], with j = #P − 1 or #P, ...
      2.1.1.1. People might say that N(P[j]) = −1 correlates with shutting down oscillator P[j].
3. The following limits pertain to S.

$$0 \le S \le 2 \tag{2.35}$$

   3.1. [Physics:] The upper limit for elementary particles comes from field theory. (Note Table 2.15.8.)

Table 2.3.11 shows CORPHY ground-state INTERN LADDER solutions for which σ = +1. Table 2.3.4 motivates the names of families. The notation ∼ ∼ denotes that (for that row in the table), for the two relevant oscillators, one of the two respective N(..) = 0 and the other N(..) = −1. There are two ways such can happen. Each way correlates with a solution. [Physics:] For that row, people might correlate one solution with the term particle and the other solution with the term antiparticle.

Table 2.3.11 CORPHY ground-state INTERN LADDER solutions for which σ = +1

E	E	E	E	E	E	E	P	P	P	P	P	P	P	P	P	(σ = +1)
6R	6L	4R	4L	2R	2L	0	0	2L	2R	4L	4R	6L	6R	8L	8R	Subfamily
						0	0									0H0
						0	0	∼	∼							1C
						−1	−1	∼	∼							1N
				0	0	0	0	0	0							2W
						0	−1	0	0							022G2&
						0	−1	0	0	∼	∼					3N
				0	0	0	−1	0	0	0	0					24..G
						0	−2	0	0	0	0					042G24&
		0	0	0	0	0	−1	0	0	0	0	0	0			46..G
				0	0	0	−2	0	0	0	0	0	0			26..G
						0	−3	0	0	0	0	0	0			064G246&
0	0	0	0	0	0	0	−1	0	0	0	0	0	0	0	0	68..G
		0	0	0	0	0	−2	0	0	0	0	0	0	0	0	48..G
				0	0	0	−3	0	0	0	0	0	0	0	0	28..G
						0	−4	0	0	0	0	0	0	0	0	084G2468&

Table 2.3.12 provides perspective regarding subfamily names for subfamilies for which only one ground-state solution exists.

Table 2.3.12 Perspective regarding subfamily names for subfamilies for which only one ground-state INTERN LADDER solution exists

1. For each of the following, the subfamily consists of one solution and Table 2.3.11 shows the name for the solution.
   1.1. 0H0.
      1.1.1. Here, the symbol 0H denotes the subfamily.
   1.2. 022G2&.
   1.3. 042G24&.
   1.4. 064G246&.
   1.5. 084G2468&.

Table 2.3.13 provides perspective about ground-state INTERN LADDER solutions for the G-family. The table provides a bound on which G-family ground-state solutions are physics-relevant. The table provides a means for computing spins for G-family solutions. [Physics:] We use models regarding circular polarization to interpret and use G-family solutions. For example, for a specific MM1MS1-photon (including its observed momentum), the MM1MS1-photon left-circular polarization mode can be multiply excited, the right-circular polarization mode can be multiply excited, but (at any one time) no more than one of the right-circular polarization mode and the left-circular polarization can be excited.

Table 2.3.13 Perspective regarding spin and other aspects of G-family ground-state INTERN LADDER solutions

1. The following statements describe an algorithm for computing S for G-family solutions.
   1.1. The symbol (0, ..., 0) denotes one of ...
      1.1.1. (0, 0);
      1.1.2. (0, 0, 0, 0);
      1.1.3. (0, 0, 0, 0, 0, 0); or
      1.1.4. (0, 0, 0, 0, 0, 0, 0, 0).
   1.2. Here, the symbol (0, ..., 0) correlates with the set that contains all relevant QP-like oscillators except P0.
   1.3. For a version of (0, ..., 0), use of the list of even integers specified as follows ...
      1.3.1. List = 2; for (0, 0).
      1.3.2. List = 2, 4; for (0, 0, 0, 0).
      1.3.3. List = 2, 4, 6; for (0, 0, 0, 0, 0, 0).
      1.3.4. List = 2, 4, 6, 8; for (0, 0, 0, 0, 0, 0, 0, 0).
   1.4. Select a sub-list from the list.
      1.4.1. For each of the cases 022G2&, 042G24&, 064G246&, and 084G2468&, the sub-list equals the list.

1.4.2. For each of the other cases, except for the following sub-lists, any sub-list pertains.
 1.4.2.1. 2, 8.
 1.4.2.2. 6.
 1.4.2.3. 8.
1.5. For the sub-list, find a combination of arithmetic operations, selected from plus and minus, such that arithmetic combining of the items in the sub-list sums to either 2 or 4.
 1.5.1. For example, + 2 = 2.
 1.5.2. For example, − 2 + 4 = 2.
 1.5.3. For example, − 2 + 6 = 4.
 1.5.4. For example, + 2 − 4 + 6 = 4.
 1.5.5. For example, − 2 + 4 − 6 + 8 = 4.
 1.5.6. Note that the following possibilities would not work because each correlates with S > 2 (See the next item.); hence, the exceptions above.
  1.5.6.1. − 2 + 8 = 6.
  1.5.6.2. + 6 = 6.
  1.5.6.3. + 8 = 8.
1.6. Set 2S equal to the result of the specified arithmetic combining of items in the sub-list.
 1.6.1. [Physics:] Here, 2S is formed based on a combination of circular polarizations.
  1.6.1.1. A circular polarization carries a spin/ħ equal to the appropriate one of 2/2, 4/2, 6/2, and 8/2.
   1.6.1.1.1. Here, the 2, 4, 6, and/or 8 (in the respective numerators) correlate with elements in the sub-list.
  1.6.1.2. The combination of circular polarizations is one circular polarization for a G-family particle.
  1.6.1.3. The negative of the combination of circular polarizations is the other circular polarization for the G-family particle.
 1.6.2. For a symbol of the form epnG..&, ...
  1.6.2.1. Note that e = #E.
  1.6.2.2. Note that p = #P.
  1.6.2.3. Note that n = 2S.
 1.6.3. For the symbol of the form epnG, ...
  1.6.3.1. Set e = #E.
  1.6.3.2. Set p = #P.
  1.6.3.3. Set n = 2S.
2. The number of G-family ground-state solutions for which S = 1 or S = 2 is unbounded (or, infinite).

3. Starting with #P = A = [10], multiple solutions can pertain for a given base state.
   3.1. For example, for 0A2G2468A&, the sub-list 2, 4, 6, 8, and A pertains and each of the following calculations leads to S = 1.
   3.2. 2S = (− 2 − 4 + 6) + (− 8 + 10)
   3.3. 2S = (+ 2 + 4 − 6) + (− 8 + 10)
4. We assume that, for #P ≥ A = [10], G-family solutions are not physics-relevant.
   4.1. We use the inappropriate redundancy criterion. (See Table 2.15.8.)
5. Work in here and elsewhere dovetails with the following statement.
   5.1. [Physics:] For each G-family solution that correlates with an elementary particle, #P ≤ 8.
6. The previous items (that discuss limits on #P for the G-family) dovetail with the following statement.
   6.1. [Physics:] For each G-family solution that correlates with an elementary particle, no other G-family solution has the same combination of #E, #P, 2S, ground-state N(P0), and sub-list.
7. Below, we sometimes denote sub-list by the symbol %.
   7.1. For example, we might use a symbol of the form epnG%& or a symbol of the form nG%&.

Table 2.3.14 characterizes CORPHY ground-state INTERN LADDER solutions for which σ = −1. Table 2.3.4 motivates the names of families. Here, considerations (similar to those regarding Table 2.3.11) for ~ ~ apply for adjacent oscillators. Each row for which ~ ~ applies has two applications of ~ ~ and, therefore correlates with four solutions. People might say that each such row correlates with two particles, and, for each of the two particles, one antiparticle. (See, for example, Table 3.5.3.) For Y-family solutions, Section 3.6 discusses values for N(..) for QE-like oscillators. (See Table 3.6.1 and Table 3.6.8.)

Table 2.3.14 CORPHY ground-state INTERN LADDER solutions for which σ = −1

E 6R	E 6L	E 4R	E 4L	E 2R	E 2L	E 0	P 0	P 2L	P 2R	P 4L	P 4R	P 6L	P 6R	P 8L	P 8R	(σ = −1) Subfamily
						0	0									0O
				~	~	0	0	~	~							1Q
				~	~	−1	−1	~	~							1R
				..	..	..	−1	−1	−1							2Y
				0	0	0	0	0	0							2O
		~	~	0	0	0	0	0	0	~	~					3Q
		~	~	−1	−1	0	0	−1	−1	~	~					3I
		~	~	0	0	−1	−1	0	0	~	~					3R
		~	~	−1	−1	−1	−1	−1	−1	~	~					3D
		..	..	..	..	..	−1	−1	−1	−1	−1					4Y

E	E	E	E	E	E	E	P	P	P	P	P	P	P	P	P	($\sigma = -1$)
6R	6L	4R	4L	2R	2L	0	0	2L	2R	4L	4R	6L	6R	8L	8R	Subfamily
		0	0	0	0	0	0	0	0	0	0					40

~ ~ ~

This subsection discusses generations and color charges.

People might say that the concept of generation correlates with an INTERN property of elementary fermions. People might say that the concept of color charge correlates with an INTERN property of, at least, quarks.

Table 2.6.4 discusses numbers of generations for elementary fermions. Section 2.14 provides possible insight regarding generations, numbers of color charges, and possible EXTINT LADDER models that show aspects regarding generations and color charges.

Section 2.4 One MM1 meta-model and various MM1MS1 models

Section 2.4 discusses relationships between the MM1 meta-model and MM1MS1 models.

~ ~ ~

This subsection discusses relationships between the MM1 meta-model and MM1MS1 models.

Table 2.4.1 pertains.

Table 2.4.1 Relationships between the MM1 meta-model and MM1MS1 models

1. The MM1 meta-model ...

 1.1. Features CORPHY ground-state INTERN LADDER solutions.

 1.2. Outputs ...

  1.2.1. Solutions that might correlate with elementary particles.

  1.2.2. Families and subfamilies (of solutions) that might correlate with elementary particles.

  1.2.3. The following numbers, which may correlate with properties of elementary particles.

   1.2.3.1. σ.

    1.2.3.1.1. +1 or −1.

   1.2.3.2. S.

1.2.3.2.1. 0, 1/2, 1, 3/2, or 2.
1.2.3.3. m'.
1.2.3.3.1. = 0 or ≠ 0.
2. Each MM1MS1 model ...
2.1. Comports with the MM1 meta-model.
2.2. Likely, accepts, as inputs, outputs from the MM1 meta-model.
2.2.1. To the extent the MM1MS1 model uses ground-state INTERN LADDER solutions, ...
2.2.1.1. The MM1MS1 model uses CORPHY ground-state INTERN LADDER solutions.
2.2.2. For each of some CORPHY ground-state INTERN LADDER solutions, the MM1MS1 model ...
2.2.2.1. May ignore the solution.
2.2.2.2. May assume the solution does not correlate with an elementary particle.
2.2.2.3. May assume the solution might (or does) correlate with an elementary particle.
2.3. Possibly, accepts, as inputs, results that other MM1MS1 models produce.
2.4. Likely, applies models other than INTERN LADDER models.
2.5. Possibly, produces, as outputs, ...
2.5.1. Candidate values for particle properties beyond properties the MM1 meta-model produces.
2.5.2. Candidates for interactions in which particles partake.
2.5.3. Interpretations of and/or candidate results regarding aspects of ...
2.5.3.1. Elementary-particle physics.
2.5.3.2. Cosmology.
2.5.3.3. Astrophysics.

Section 2.5 DIFEQU models for isotropic quantum harmonic oscillators

Section 2.5 discusses DIFEQU models for isotropic quantum harmonic oscillators. We discuss continuous-math models pertaining to lone harmonic oscillators. We discuss continuous-math models pertaining to multi-dimensional isotropic quantum harmonic oscillators. We discuss projecting representations into smaller numbers of dimensions than originally pertain to the representations. We discuss, in the context of DIFEQU models, the notion of solution. We note that the topic of normalization has physics-relevance. We show two types of solutions that normalize. People might say that we show solutions that traditional physics models may have underutilized. We discuss ranges of parameters that pertain for each of traditional solutions and non-traditional solutions. People might say that solutions we characterize correlate with known elementary particles and with possible elementary particles. People might say that we show integers that we later correlate with masses of elementary bosons.

~ ~ ~

This subsection provides notation for DIFEQU models for lone quantum harmonic oscillators.

Table 2.5.1 shows well-known information about 1-dimensional harmonic oscillators.

Table 2.5.1 Notation and math pertaining to a lone quantum harmonic oscillator (DIFEQU solutions)

1. The symbol x0 denotes a coordinate.
   1.1. The domain for x0 is $-\infty < x0 < \infty$.
2. The symbol $\Psi(x0)$ denotes a function of x0.
   2.1. The range of Ψ is the set of complex numbers.
3. The following equation correlates with the lone quantum harmonic oscillator.

$$\xi\,\Psi(x0) = (\xi'/2)\,(\,-\eta^2\,\partial^2/\partial(x0)^2 + \eta^{-2}\,(x0)^2\,)\,\Psi(x0) \qquad (2.36)$$

4. ξ and $\xi'/2$ denote numbers.
5. [Physics:] The following may pertain.
   5.1. $\Psi(x0)$ denotes a wave function.
   5.2. The variable x0 has dimensions of length.
   5.3. η denotes a parameter with dimensions of length.
   5.4. $V = (\xi'/2)\,\eta^{-2}\,(x0)^2$.
      5.4.1. People call this the potential.
   5.5. Generally, $\eta^2(\xi'/2)$ and $\eta^{-2}(\xi'/2)$ are constants that are not directly related to each other.
   5.6. Sometimes, people correlate ξ and $\xi'/2$ with numbers having dimensions of energy.
      5.6.1. People correlate $\xi'/2$ with the term ground-state energy.
6. People characterize solutions by ...
   6.1. Ψ has the form of a Hermite polynomial (in the variable x0) multiplied by $\exp(\,-(x0)^2/(2\eta^2)\,)$.

~ ~ ~

This subsection discusses an aspect of a difference between some uses of harmonic oscillator models in traditional physics and some uses of harmonic oscillator models in this monograph.

Table 2.5.2 pertains. People might say that Table 6.1.5 and Table 6.1.6 provide perspective about and harmonization regarding the possibly significant difference to which Table 2.5.2 points.

Table 2.5.2 Differences in units for ξ (and ξ') between some aspects of traditional physics and some aspects of work in this monograph

1. For some uses of harmonic oscillator models in traditional physics, ...
   1.1. ξ and $\xi'/2$ have dimensions of energy. (Note, for example, Table 2.5.1.)
2. For some uses of harmonic oscillator models in this monograph, ...
   2.1. ξ and $\xi'/2$ have dimensions of energy squared. (See, for example, (2.58) in Table 2.9.4.)

~ ~ ~

This subsection shows notation for DIFEQU solutions for multi-dimensional isotropic quantum harmonic oscillators.

Table 2.5.3 shows information about multi-dimensional isotropic quantum harmonic oscillators.

Table 2.5.3 Notation and math pertaining to a DIFEQU solution for a multi-dimensional isotropic quantum harmonic oscillator

1. For a (#D + 1)-dimensional isotropic quantum harmonic oscillator, ...
   1.1. #D denotes a nonnegative integer.
   1.2. The set { j | j = 0 , 1 , ... , #D − 1 , or #D } provides ...
      1.2.1. An index to the lone quantum harmonic oscillators that make up the (#D + 1)-dimensional isotropic quantum harmonic oscillator.
   1.3. The set { xj | j = 0 , 1 , ... , #D − 1 , or #D } provides ...
      1.3.1. A set of #D + 1 coordinates, with each coordinate pertaining to the respective lone quantum harmonic oscillator.
   1.4. Numbers N(0), N(1), ..., and N(#D) describe base states of the lone harmonic oscillators.
      1.4.1. The lone harmonic oscillators correlate with the (#D + 1)-dimensional isotropic quantum harmonic oscillator.

1.4.2. Each number N(..) provides part of a description of a base state for the (#D + 1)-dimensional isotropic quantum harmonic oscillator.

2. People specify the state of a (#D + 1)-dimensional isotropic quantum harmonic oscillator via a linear combination of base states for the (#D + 1)-dimensional harmonic oscillator.
3. [Physics:] People may associate an energy with each base state of the (#D + 1)-dimensional isotropic quantum harmonic oscillator.
   3.1. The energy is proportional to the sum over the relevant j of terms of following form.

$$(a_j)^+ (a_j)^- + 1/2 = N(j) + 1/2 \qquad (2.37)$$

   3.2. The term isotropic correlates with the sum's giving equal weight to each of the N(j) + 1/2.

For such a (#D + 1)-dimensional isotropic quantum harmonic oscillator, people can express representations for which sets of component oscillator are grouped into subsets (of the entire set of relevant harmonic oscillators) and for which radial coordinates can pertain. Traditionally, radial coordinates can pertain within each subset that includes at least two harmonic oscillators. We find uses for radial coordinates for subsets that include just one harmonic oscillator. (See, for example, Table 2.5.5.)

Table 2.5.4 pertains.

Table 2.5.4 Process for creating and preparing to use a partly linear-coordinate, partly radial-coordinate representation for a DIFEQU isotropic quantum harmonic oscillator

1. Subdivide the set { j | j = 0 , 1 , ... , #D − 1 , or #D } (See Table 2.5.3.) into subsets such that ...
   1.1. Each j appears in exactly one subset.
   1.2. No subset is the empty set.
2. For each subset that has exactly 1 member, ...
   2.1. Anticipate using math that Table 2.5.1 and Table 2.5.7 describe or using math that Table 2.5.6 describes.
3. For each subset that has more than 1 member, ...
   3.1. Anticipate using math that Table 2.5.6 describes.
      3.1.1. For example, the expression r^2 equals the sum of the squares of the relevant xj. (See, for example, (2.38) in Table 2.5.6.)
   3.2. Anticipate (for purposes of this monograph) not necessarily trying to correlate LADDER solutions with DIFEQU solutions.

~ ~ ~

This subsection anticipates uses of partly linear-coordinate, partly radial-coordinate representations for DIFEQU isotropic quantum harmonic oscillators.

Table 2.5.5 pertains. (Regarding D*, see, for example, Table 2.1.1.)

Table 2.5.5 Possible applications for models correlating with D* equaling each of 3, 2, and 1

1. We anticipate using subsets (of the set { j | j = 0 , 1 , ... , #D − 1 , or #D }) that have 3 elements.
   1.1. [Physics:] For example, some of these applications correlate with notions of at least one of the following.
      1.1.1. D* = 3.
      1.1.2. Projections of Ψ from D > 3 dimensions into D* = 3 dimensions.
      1.1.3. Three QP-like dimensions correlating with people's notions of space-time coordinates (or, possibly, notions of space-time).
2. We anticipate using subsets that have 2 elements.
   2.1. [Physics:] For example, some of these applications correlate with notions of at least one of the following.
      2.1.1. Projections of Ψ from D* = 3 dimensions into D* = 2 dimensions.
      2.1.2. Projections of Ψ from D ≥ 3 dimensions into D* = 2 dimensions.
      2.1.3. Integers that may correlate approximately with squares of masses for WHO-family particles.
3. We anticipate using subsets that have 1 element.
   3.1. [Physics:] For example, some of these applications correlate with notions of at least one of the following.
      3.1.1. Projections of Ψ from D* = 3 dimensions into D* = 1 dimension.
      3.1.2. Projections of Ψ from D ≥ 3 dimensions into D* = 1 dimension.
      3.1.3. Integers that may correlate approximately with squares of masses for WHO-family particles. (People might say that this correlation is somewhat more indirect than the corresponding correlation for subsets that have 2 elements.)

~ ~ ~

This subsection discusses radial-coordinate representations for isotropic quantum harmonic oscillators.

This subsection discusses radial-coordinate math-based models for QP-like isotropic quantum harmonic oscillators. This subsection does not consider normalization. (Later, we discuss normalization.) The use of QP-like notation does not imply a lack of applicability of similar math-based models to QE-like aspects of models. Some remarks in this subsection pertain to QE-like notation. For possible QE-like applications, see Table 2.6.6.

The math pertains to solutions this monograph denotes by $\Psi(r)$. Here, r is the radial coordinate. Ψ can be a function also of coordinates (that is, angular coordinates) other than r. (Regarding multiple definitions of the term solution, see remarks preceding Table 2.2.9.)

Table 2.5.6 pertains. The formulation is equivalent to traditional formulations in which $\eta^2(\xi'/2)$ and $\eta^{-2}(\xi'/2)$ are constants that are not directly related to each other.

Table 2.5.6 Math and notation for radial-coordinate solutions for QP-like isotropic quantum harmonic oscillators

1. The following equation pertains.

$$\xi\,\Psi(r) = (\xi'/2)\,(\,-\eta^2\,\nabla^2 + \eta^{-2}\,r^2\,)\,\Psi(r) \tag{2.38}$$

2. People call the following part of the above equation the Laplacian operator for D dimensions.

$$\nabla^2 = r^{-(D-1)}(\partial/\partial r)(r^{D-1})(\partial/\partial r) - \Omega r^{-2} \tag{2.39}$$

3. ξ and $\xi'/2$ denote numbers.
   3.1. [Physics:] In some applications, ξ and ξ' have dimensions of energy.
4. $\Psi(r)$ denotes a function ([Physics:] or, wave function).
5. r denotes a variable ([Physics:] with dimensions of length).
6. η denotes a parameter ([Physics:] with dimensions of length).
7. Ω denotes a number.
8. D denotes a positive integer.
9. $V = (\xi'/2)\,\eta^{-2}\,r^2$.
   9.1. People call this the potential.

Table 2.5.7 pertains.

Table 2.5.7 Solutions, based on Hermite polynomials, for quantum harmonic oscillators for which D = 1

1. For D = 1 (in Table 2.5.6), some solutions feature the following.
   1.1. $\Omega = 0$.
   1.2. The range $-\infty < r < \infty$ pertains.
   1.3. Ψ has the form of a Hermite polynomial (in the variable r) multiplied by $\exp(\,-r^2/(2\eta^2)\,)$.

Work below tends not to emphasize solutions people associate with Table 2.5.7. Below, the range $0 \le r < \infty$ pertains for traditional solutions. ([Physics:] People might dispute work below pertaining to D = 1. However, such non-traditional work regarding D = 1 possibly matches some physics notions. For example, see Table 2.5.5. Table 2.6.6

lists possible applications for QE-like solutions for which D = 1. For QE-like D = 1 applications, we might use the variable t' (for time) in place of r. And, we would use a symbol (with units of time) other than η in place of η.)

Table 2.5.8 describes solutions other than solutions people traditionally associate with Table 2.5.7.

Table 2.5.8 Some radial-coordinate solutions for isotropic quantum harmonic oscillator math

1. The following relationship characterizes these solutions.

$$\Psi(r) \propto r^{\nu} \exp(-r^2 / (2\eta^2))$$

2. For these solutions, ...
   2.1. The following algebraic equations pertain.

$$\xi = (D + 2\nu)\,(\xi'/2) \qquad (2.40)$$

$$\Omega = \nu(\nu + D - 2) \qquad (2.41)$$

   2.2. The parameter η does not appear in the two equations.
3. The following pertain to some ([Physics:] traditional) solutions.
   3.1. ν is nonnegative.
   3.2. ν is an integer.
   3.3. Ω is nonnegative.
4. Each of the following items points to other ([Physics:] non-traditional) solutions.
   4.1. ν can be negative.
   4.2. ν can be other than an integer.
5. For $D > 2$, the condition $\nu < 0$ is necessary (but not sufficient) for the next item to pertain.
   5.1. Ω can be negative.
6. For D = 2, ...
   6.1. $\Omega = \nu^2 \geq 0$.

~ ~ ~

This subsection discusses normalization.

[Physics:] Traditional physics does not include wave functions that do not normalize.

People say that a function such as Ψ normalizes if, and only if, integration (throughout the domain of the function) of the expression consisting of the product of the function and its complex conjugate produces a finite result.

For DIFEQU solutions, this monograph does not consider to be physics-relevant any wave function that does not normalize.

~ ~ ~

This subsection shows that solutions for which $D + 2\nu > 0$ normalize.

Table 2.5.9 pertains.

Table 2.5.9 Normalization for solutions for which $D + 2\nu > 0$

1. The following relationships characterize behavior of the r-related normalization integrand near $r = 0$.

$$(\Psi(r))^* \Psi(r)\, r^{D-1} \propto r^{D-1+2\nu} \exp(-2r^2 2^{-1}\eta^{-2}) \rightarrow r^{D-1+2\nu}, \text{ as } r \rightarrow 0 \quad (2.42)$$

   1.1. Here, * denotes complex conjugate.
   1.2. Here, the term r^{D-1} comes from the expression $\int \ldots r^{D-1}dr$.
2. The integral is finite ...
   2.1. If $-1 < D - 1 + 2\nu$.
   2.2. Or, equivalently, if $D + 2\nu > 0$.
3. Ψ normalizes if (but not only if) $D + 2\nu > 0$.

~ ~ ~

This subsection shows that solutions for which $D + 2\nu = 0$ normalize. Table 2.5.10 pertains.

Table 2.5.10 Normalization for solutions for which $D + 2\nu = 0$

1. Wolfram Alpha (2014a) provides the following definition of the Dirac delta function.

$$\delta(r) = \lim_{\varepsilon\rightarrow 0+} (1/(2(\pi\varepsilon)^{1/2})) \exp(\, -r^2/(4\varepsilon)\,) \quad (2.43)$$

2. This work makes the following association.

$$4\varepsilon = \eta^2 \quad (2.44)$$

3. The next item supplements a result in Table 2.5.9.
4. Ψ normalizes if (but not only if) $D + 2\nu = 0$.

~ ~ ~

This subsection provides terminology for characterizing some DIFEQU solutions.

Table 2.5.11 pertains. Here, we anticipate discussing a specific solution (for which the exponent ν pertains) in the contexts of more than one number of dimensions. Here, the terms inside and edge correlate with normalization pertaining for the solution in a context of D dimensions. Here, the term outside correlates with normalization not being possible in a context of D dimensions.

Table 2.5.11 Definitions, regarding a solution and a number of dimensions, of inside, edge, and outside (DIFEQU solutions)

1. For a solution characterized by the exponent ν, ...
   1.1. For a number of dimensions D, ...
      1.1.1. Inside denotes $D + 2\nu > 0$.
      1.1.2. Edge denotes $D + 2\nu = 0$.
      1.1.3. Outside denotes $D + 2\nu < 0$.

People might think of the terms inside, edge, and outside as correlating, for a value of ν, with a domain of D for which normalization pertains. People might think of the terms inside, edge, and outside as correlating, for a value of D, with a domain of ν for which normalization pertains.

~ ~ ~

This subsection shows some solutions.

This subsection focuses on math for which D* = 3. [Physics:] D* = 3 matches key elementary-particle data about spin. (See, (2.1) in Table 2.1.1.) Here, as elsewhere, S correlates with spin/ħ.

Table 2.5.12 shows some ([Physics:] traditional) solutions.

Table 2.5.12 Some ([Physics:] traditional) solutions (DIFEQU solutions)

1. The next items describe some ([Physics:] traditional) solutions.
   1.1. D* = 3.
   1.2. D = 3.
   1.3. S = ν, for some nonnegative integer ν.
   1.4. $\Omega = \nu(\nu + D - 2) = S(S + D^* - 2) = S(S + 1)$.
   1.5. 2S+1 angular solutions pertain.
2. For example, the next items describe a solution that correlates with the solution Table 2.2.13 shows. ([Physics:] Here, people would state that $3\times(\xi'/2)$ equals the energy.)
   2.1. D* = 3.
   2.2. D = 3.
   2.3. S = ν = 0.
   2.4. $\Omega = \nu(\nu + D - 2) = 0$.
   2.5. 1 angular solution pertains.
3. The next items describe the ground state and the excited states for the previous example.
   3.1. $\xi = (D + 2\nu)\,(\xi'/2) = (D^*/2 + \nu)\,\xi' = (3/2 + S)\,\xi'$.
   3.2. S is a nonnegative integer.
   3.3. $\Omega = S(S + 1)$.
   3.4. $\Psi(r) \propto r^S \exp(-r^2 / (2\eta^2))$.
   3.5. For the ground state, S = 0.

Table 2.5.13 shows math that allows for other ([Physics:] non-traditional) solutions for which D* = 3. In particular, solutions allow D ≠ D* (with D a positive integer) and solutions allow 2S to be an odd nonnegative integer.

Table 2.5.13 Some ([Physics:] non-traditional) solutions, including an example of a non-traditional ground state for 3-dimensional isotropic quantum harmonic oscillators (DIFEQU solutions)

1. The next items allow for some non-traditional solutions.
   1.1. $D^* = 3$.
   1.2. $D^* + 2\nu > 0$, with 2ν being an integer.
   1.3. $\Omega = \nu(\nu + D - 2)$, for some positive integer D.
   1.4. $|\Omega| = S(S + D^* - 2) = S(S + 1)$, for some S with 2S being a nonnegative integer.
   1.5. $2S + 1$ angular solutions pertain.
2. For example, the next items show a solution for which $S \neq \nu$. For r near 0, the normalization integral behaves like a non-zero factor multiplied by $\int r^0 dr$. This monograph considers this non-traditional solution to correlate with a ground state. ([Physics:] People might say that $1 \times (\xi'/2)$ equals the ground-state energy. People might say that the energy matches the energy of the state Table 2.2.15 shows. We find distinctly different applications for the two solutions.)
   2.1. $D^* = 3$.
   2.2. $\nu = -1$.
   2.3. $\Psi(r) \propto r^{-1} \exp(-r^2 / (2\eta^2))$.
   2.4. $D = 3$.
   2.5. $\Omega = \nu(\nu + D - 2) = -1(0) = 0$.
   2.6. $S = 0$.
   2.7. $\Omega = S(S + D^* - 2) = S(S + 1) = 0(1) = 0$.
   2.8. $\xi = (1/2)\,\xi'$.
   2.9. 1 angular solution pertains.

~ ~ ~

This subsection shows ranges of parameters for solutions based on radial coordinates.

Table 2.5.14 lists parameters that characterize solutions. Here, this monograph does not discuss traditional solutions for which $D = 1$. For non-traditional solutions that this monograph considers, this monograph limits 2ν to be an integer, limits D to be a positive integer, and (for $D > 1$) limits $D + 2\nu$ to be nonnegative. For each non-traditional solution, for one of $\pm = +$ and $\pm = -$, Ω must satisfy both of the equations the last row in the table shows. [Physics:] Applications for which $\xi' < 0$ may pertain regarding some aspects of models correlating with masses of non-zero-mass elementary bosons for which $S = 1$ or $S = 2$. (See Section 5.7 and Section 5.8.)

Table 2.5.14 Parameters for solutions (based on radial coordinates) that can be normalized (DIFEQU solutions)

Parameter	Value range ([Physics:] correlating with traditional solutions)	Value range ([Physics:] including non-traditional solutions)
ξ'	> 0	$\neq 0$
η	$\neq 0$	$\neq 0$
D*	Positive integer	Positive integer
D	= D*	Positive integer
2S	Even nonnegative integer	Nonnegative integer
2ν	= 2S	Integer, with $2\nu \geq -D$
Ω	$= S(S + D^* - 2)$	$= \nu(\nu + D - 2)$ $= \pm S(S + D^* - 2)$

Table 2.5.15 discusses the symbol σ. The symbol correlates with the sign of Ω. (Compare with Table 2.3.1 and Table 2.3.3.) [Physics:] Examples of free-ranging elementary particles include electrons and photons. Examples of non-free-ranging elementary particles include quarks and gluons.

Table 2.5.15 The number σ

1. $\Omega = \sigma S(S + 1)$, with ...
   1.1. [Physics:] For free-ranging particles, ...
      1.1.1. $\sigma = +1$, for solutions for which $S > 0$.
      1.1.2. $\sigma = +1$, for solutions for which $S = 0$.
   1.2. [Physics:] For particles that do not range freely, ...
      1.2.1. $\sigma = -1$, for solutions for which $S > 0$.
      1.2.2. $\sigma = -1$, for solutions for which $S = 0$.

In some aspects of this monograph, uses of σ correlate with $D^* = 3$ and, hence, with $\Omega = \sigma S(S + 1)$. For other purposes, people might define σ to correlate with $\Omega = \sigma S(S + D^* - 2)$.

~ ~ ~

This subsection lists non-traditional solutions, including solutions for which $S \leq 2$ and $D^* = 3$, 2, or 1.

Table 2.5.16 shows some cases of relationships between parameters. The table shows examples for which $D^* = 3$. Here, the work ignores the parameter ξ'. For $\nu = -1/2$ and for $\nu = -1$, any non-zero η can pertain. For $\nu = -3/2$, physics-relevant solutions pertain only with respect to one of the limits $\eta \to 0^+$ and $\eta \to 0^-$. Here, physics-relevance correlates with a solution's being able to be normalized. For each of three values of ν, the table shows how to compute D, given Ω. For each of those three values of ν, the table describes each candidate solution for which $S \leq 2$. ([Physics:] We use the term candidate solution to describe a solution that may correlate with an elementary particle.) For each of those values of ν, no mathematical upper limit exists for −Ω. For each of those values of ν, no mathematical upper limit exists for S.

Table 2.5.16 Relationships between some parameters, for some solutions for which D* = 3 (DIFEQU solutions)

D*	ν	D* + 2ν	D	S	Ω	σ	D	D + 2ν	2S + 1
3	−1/2	2	(5 − 4Ω)/2	1/2	3/4	+1	1	0	2
"	"	"	"	1/2	−3/4	−1	4	3	2
"	"	"	"	3/2	−15/4	−1	10	9	4
"	"	"	"	...					
"	−1	1	3 − Ω	1	2	+1	1	−1	3
"	"	"	"	0	0	+1	3	1	1
"	"	"	"	0	0	−1	3	1	1
"	"	"	"	1	−2	−1	5	3	3
"	"	"	"	2	−6	−1	9	7	5
"	"	"	"	...					
"	−3/2	0	(21 − 4Ω)/6	1/2	3/4	+1	3	0	2
"	"	"	"	1/2	−3/4	−1	4	1	2
"	"	"	"	3/2	−15/4	−1	6	3	4
"	"	"	"	...					

Table 2.5.17 shows some cases of relationships between parameters. The table shows examples having D* = 2. We do not anticipate using these solutions directly to identify candidate elementary particles. We anticipate using the solutions in ways that pertain to candidate elementary particles we identify via other means. For some candidate boson particles, we anticipate projecting solutions for which D* = 3 and ν = −1 into functions that correlate with D* = 2 and ν = −1. We anticipate that such projections correlate with masses for zero-mass elementary bosons and with approximate masses for non-zero-mass elementary bosons. (See, for example, Section 3.7.) For these applications, needs for normalization correlate with the respective solutions for which D* = 3 and do not necessarily correlate with the solutions for which D* = 2. For some of these applications, the limit S ≤ 2 does not pertain. For some of these applications, we think there is no need, for the S = 0 solution, to assign a value for σ. Here, for ν = −1, $\Omega = \pm S^2$.

Table 2.5.17 Relationships between some parameters, for some solutions for which D* = 2 (DIFEQU solutions)

D*	ν	D* + 2ν	D	S	Ω	σ	D	D + 2ν	2S + 1
2	−1/2	1	(5 − 4Ω)/2	1/2	1/4	+1	2	1	2
"	"	"	"	1/2	−1/4	−1	3	2	2
"	"	"	"	3/2	−9/4	−1	7	6	4
"	"	"	"	5/2	−25/4	−1	15	14	6

D*	ν	D* + 2ν	D	S	Ω	σ	D	D + 2ν	2S + 1
"	"	"	"	7/2	-49/4	+1	27	26	8
"	"	"	"	9/2	-81/4	-1	43	42	10
"	"	"	"	...					
"	-1	0	3 – Ω	1	1	+1	2	0	3
"	"	"	"	0	0		3	1	1
"	"	"	"	1	-1	-1	4	2	3
"	"	"	"	2	-4	-1	7	5	5
"	"	"	"	3	-9	-1	12	10	7
"	"	"	"	4	-16	-1	19	17	9
"	"	"	"	5	-25	-1	28	26	11
"	"	"	"	6	-36	-1	39	37	13
"	"	"	"	7	-49	-1	52	50	15
"	"	"	"	8	-64	-1	67	65	17
"	"	"	"	...					

Table 2.5.18 shows a formula pertaining to values of S that could arise for a choice of D* = 1. For some purposes of this monograph, we think we do not need to consider issues related to normalization. (See, for example, Table 2.6.2 and Table 2.10.8.)

Table 2.5.18 Values of S that could arise for the choice D* = 1

1. Ω = ±S(S + D* – 2) = ±S(S – 1), with ...
   - 1.1. D* = 1.
   - 1.2. S(S – 1) = 0, 0, 2, 6, ..., respectively, for S = 0, 1, 2, 3, ...
   - 1.3. S(S – 1) would be –1/4 for S = 1/2.
   - 1.4. S(S – 1) = 3/4, 15/4 ..., respectively, for S = 3/2, 5/2, ...
2. To the extent one considers just algebraic relationships (and not necessarily physics and not necessarily DIFEQU considerations), ...
   - 2.1. S(S – 1) = ..., 6, 2, respectively, for S = ..., –2, –1.
3. Thus, for the sequence S = .., –2, –1, 0, 1, 2, 3, .., (See Table 2.10.8.) ...
   - 3.1. The sequence of Ω is .., 6, 2, 0, 0, 2, 6, .. .
   - 3.2. Each of these Ω is nonnegative.
   - 3.3. Each of these Ω correlates with solutions for which D* = 3 and S (regarding D* = 3) is a nonnegative integer.

~ ~ ~

Reference 1 Wolfram Alpha (2014a)

Wolfram Alpha, computational knowledge engine, Wolfram Alpha LLC, (2014). http://mathworld.wolfram.com/DeltaFunction.html.

Section 2.6 Applications of DIFEQU models

Section 2.6 discusses, in the context of DIFEQU models, solutions. We focus on DIFEQU subsets for which the solutions might correlate with data about elementary particles. We correlate some subsets with the physics terms fermion and boson. We correlate some subsets with the physics terms fields and particles. We discuss aspects of projecting mathematical representations into smaller than the original numbers of dimensions. We show factors contributing to numbers of physics-relevant solutions. We list, for some DIFEQU subsets, numbers of relevant solutions. People might say that we discuss possible time-like uses of radial coordinates.

~ ~ ~

[Physics:] This subsection discusses concepts related to particles and fields.

People might say that physics uses of the term field include notions of constructs that extend broadly (with respect to space-time coordinates), that correlate with forces that impact elementary particles, and/or that correlate with the creation and destruction of elementary particles. Physics includes other uses of the term field.

People might say that physics uses of the term particle include notions of constructs that have properties (such as charge or mass) and/or that have localized existence (at particular times, with respect to space-time coordinates).

People might say that, for elementary particles, no internal component particles should exist. People might say that, for an elementary particle, internal properties should not vary. Perhaps, such non-variance of internal properties correlates with a matter of terminology. For example, people might debate the extent to which color charge is an internal property of quarks.

We prefer, absent math-based models, to deemphasize some detailed discussions of concepts of particles and fields.

People might say that work in this section provides working definitions of concepts related to the term particle and of concepts related to the term field. (See, for example, Table 2.6.1.)

~ ~ ~

This subsection discusses QE-like and QP-like aspects of DIFEQU solutions that correlate with MM1MS1 models.

People might say that Section 2.5 features QP-like notation, coordinates, and models.

[Physics:] In traditional physics, wave functions are functions of both QE-like coordinates and QP-like coordinates.

For MM1MS1 solutions that do not explicitly state QE-list aspects (such as coordinates), the monograph assumes QE-like aspects pertain. (See, for example, Table 2.6.2 and Table 2.6.6.)

Sometimes, we explicitly include QE-like coordinates. (See, for example, Section 2.10.)

~ ~ ~

[Physics:] This subsection previews some physics uses of DIFEQU solutions for which D* = 3 or D* = 2.

People use models to discuss elementary particles. People include, in some such discussions, math-based models for fields with which people say particles dovetail.

For this monograph's work, Table 2.6.1 pertains. For this monograph's work, some QP-like radial-coordinate representations feature D* = 3 or D* = 2. (For perspective regarding uses of DIFEQU representations, see Table 2.7.5.)

Table 2.6.1 Correlations among ν, η, particles, and fields, for solutions based on DIFEQU representations (for D* = 3 or D* = 2)

1. For D* = 3, ...
   1.1. $\nu = -1/2$ correlates with fermion fields.
   1.2. $\nu = -1$ correlates with boson fields.
   1.3. $\nu = -3/2$ correlates with fermion particles.
2. For D* = 2, ...
   2.1. $\nu = -1$ correlates with boson particles.
3. For fields, one of the following ranges pertains.
$$0 < \eta < \infty \tag{2.45}$$
$$0 < -\eta < \infty \tag{2.46}$$
4. For particles, the following limit pertains.
$$\eta^2 \to 0 \tag{2.47}$$
5. People might say that each of the following correlates with traditional-physics concepts of spatial extent.
   5.1. A range $0 < |\eta| < \infty$ for fields.
   5.2. A size $\eta = 0$ for particles.
6. Regarding sizes for elementary particles, ...
   6.1. People might say that the uncertainty principle correlates with non-zero sizes for elementary particles.
   6.2. The uncertainty principle correlates with non-zero-sized localization correlating with models that feature supposition of base-state wave functions.
   6.3. People might say that much traditional QED (or, quantum electrodynamics) correlates with point-like (not non-zero-volume-like) interaction vertices.
   6.4. Work in this monograph ...

6.4.1. Features MM1MS1 base states.
6.4.2. Tends not to discuss superposition.
6.4.3. Discusses point-like interaction vertices. (See Section 2.10.)

Table 2.6.2 pertains. (See Section 2.10.)

Table 2.6.2 Correlations between particles and dimensions, for boson solutions based on radial coordinates (for D* = 3)

1. [Physics:] For a DIFEQU solution for which D* = 3 and ν = −1 (that is, for a boson field), the following discussion pertains.
   1.1. Projecting a radial-coordinate solution into a ν = −1, D* = 2 solution correlates with aspects pertaining to the boson as a particle.
      1.1.1. People might say that ...
         1.1.1.1. The projection correlates with an interaction vertex.
         1.1.1.2. The projection maps into the 2 QP-like (or, spatial) dimensions perpendicular to the particle's direction of motion. Here, that direction pertains to motion either before the vertex (which would correlate with a vertex that correlates with an interaction that destroys the boson) or after the vertex (which would correlate with a vertex that correlates with an interaction that creates the boson).
         1.1.1.3. For the projection, D* + 2ν = 0, with D* = 2.
            1.1.1.3.1. This aspect correlates with some integer arithmetic pertaining to masses for non-zero-mass elementary bosons. (See Table 2.5.17, Section 2.10, Section 3.7, and Section 3.8.)
   1.2. People might expect that projecting the solution into a remaining 1 (= 3 − 2) dimension might pertain.
      1.2.1. Possibly, projection into a remaining 1 dimension dovetails with a projection of QE-like aspects of the wave function into 1 QE-like dimension.
      1.2.2. People might say that ...
         1.2.2.1. The QP-like projection maps into the dimension parallel to the particle's direction of motion.
         1.2.2.2. In effect, projecting the radial-coordinate solution into a point-like construct correlating with a D* = 1 construct correlates with aspects pertaining to the boson as a particle.

1.2.2.2.1. For this projection to correlate with $D^* + 2\nu = 0$, D^* would need to equal 2.
1.2.2.2.2. Such a projection could correlate with oscillators E0 and P0 in LADDER models.
1.2.2.2.3. A QE-like analog of ν might equal -1.
1.2.2.3. For the combined (QE-like and QP-like) projection, a $D^* + 2\nu = 0$ construct correlates with some integer arithmetic pertaining to masses for non-zero-mass elementary bosons. (See Table 2.5.17, Section 2.10, Section 3.7, and Section 3.8.)
1.2.2.3.1. Here, the choice of S (as is $\{D + 2\nu\}(2,S,\Omega)$) need not match S for the particle. (See Table 2.10.8.)

~ ~ ~

This subsection shows factors contributing to numbers of candidate solutions for $D^* = 3$ and $\nu = -1/2$ and for $D^* = 3$ and $\nu = -3/2$.

Each of $\nu = -1/2$ and $\nu = -3/2$ correlates with 2S equaling an odd integer (or, equivalently, S equaling a half-integer). Here, for $\Psi(r)$, r^ν has two possible values. The expressions $|r^\nu|$ and $-|r^\nu|$ denote the possible values. (This contrasts with $\nu = -1$ cases for which r^ν has just one value.) Recall that η can have a positive value or a negative value. We think that the relative signs of r^ν and η have meaning. ([Physics:] In terms of physics models, for $\nu = -3/2$, this work considers the case of Dirac elementary fermions. Here, one relative sign correlates with a particle and the other relative sign correlates with the particle's antiparticle. In terms of physics models, for $\nu = -1/2$, this work considers the case of fermion fields. Here, we think that one linear combination of terms correlates with a particle-and-antiparticle pair creation operator and that an orthogonal linear combination of terms correlates with a particle-and-antiparticle pair destruction operator.)

Table 2.6.3 provides a definition. Below, we use applications for which j and k are nonnegative integers.

Table 2.6.3 Definition of min(j, k)

1. We define min(j, k) by the following.

$$\min(j, k) = j, \text{ if } j \leq k \tag{2.48}$$

$$\min(j, k) = k, \text{ if } j \geq k \tag{2.49}$$

Table 2.6.4 pertains. Each item can provide a factor contributing to a calculation of a maximum number of solutions. This work excludes the case D = 1 because the math for D = 1 resembles math correlating with Table 2.5.7. To the extent the subgroup relationship $SU(4) \supset SU(2) \times SU(2) \times U(1)$ is relevant, #GEN(3/2) could be other than 15, 9, or 6.

Table 2.6.4 Factors contributing toward maximum numbers of solutions (for $D^* = 3$ and $\nu = -1/2$ and for $D^* = 3$ and $\nu = -3/2$)

1. For cases for which $D^* = 3$ and $D > 1$, ...
   - 1.1. For cases for which $\nu = -1/2$ or $\nu = -3/2$, ...
     - 1.1.1. $\min(D, 2(2S + 1))$ provides a factor for calculating the maximum number of candidate solutions.
   - 1.2. [Physics:] People might say that this factor correlates (in some sense) with correlating numbers of particle-INTERN states and numbers of particle spin states.
2. For cases in which $D^* = 3$ and $\nu = -3/2$, ...
   - 2.1. For solutions corresponding to one sign of η, ...
     - 2.1.1. A function of $2S + 1$ provides another factor for calculating the number of solutions.
     - 2.1.2. Correlating with models pertaining for $|\eta| > 0$, the number of spin states is $2S + 1$.
     - 2.1.3. Perhaps, the following modeling pertains.
       - 2.1.3.1. In the limit $\eta^2 \to 0$, considerations related to $2S + 1$ spin states correlate with considerations related to the number of generators of $SU(2S + 1)$.
     - 2.1.4. For spin-1/2 elementary particles, ...
       - 2.1.4.1. The factor (for use in calculating the number of solutions) correlates with the number of generations (or, for MM1MS1-neutrinos, with the number of flavors).
       - 2.1.4.2. [Physics:] The number of generations is 3.
       - 2.1.4.3. $2S + 1 = 2$.
       - 2.1.4.4. The number of generators of $SU(2S + 1)$ (or, of $SU(2)$) is 3.
     - 2.1.5. For spin-3/2 elementary particles, we use the parameter #GEN(3/2) to denote the factor (for use in calculating the number of solutions), which equals the number of generations. Possibly, one of the following two cases pertains.
       - 2.1.5.1. For the case we designate as GEN(3/2)15, ...
         - 2.1.5.1.1. We use concepts this table states above.
         - 2.1.5.1.2. $2S + 1 = 4$.
         - 2.1.5.1.3. The function is the number of generators of $SU(4)$.
         - 2.1.5.1.4. The number of generators of $SU(4)$ is 15.
         - 2.1.5.1.5. #GEN(3/2) = 15.
       - 2.1.5.2. For the case we designate as GEN(3/2)≠15, ...
         - 2.1.5.2.1. $2S + 1 = 4$.
         - 2.1.5.2.2. Possibly, the function correlates with the relevance of 2 instances of $SU(2)$.

2.1.5.2.2.1. People might say that choices for numbers of generators correlate with ...

2.1.5.2.2.1.1. #GEN(3/2) = 9, with 9 being the square of the number of generators of SU(2).

2.1.5.2.2.1.2. #GEN(3/2) = 6, with 6 being the sum, over 2 instances of SU(2), of the number of generators for each instance.

2.1.5.2.2.1.3. #GEN(3/2) possibly being other than 15, 9, or 6.

~ ~ ~

This subsection discusses numbers of candidate solutions for D* = 3.

This work extends Table 2.5.16 to show numbers of solutions.

Table 2.6.5 pertains. Here, #GEN denotes the number of generations. For the row for which ν = −1/2 and D = 1, math correlating with Table 2.5.7 may pertain. For this row, Table 2.6.5 shows possible numbers of candidate solutions, based on a minimum number of 1 (1 = D) and a maximum number of 4 (4 = 2(2S + 1)). For the row for which ν = −1 and D = 1, math correlating with Table 2.5.7 may pertain. For this row, Table 2.6.5 shows possible numbers of candidate solutions, based on a minimum number of 1 (1 = D) and a maximum number of 3 (3 = 2S+1). Otherwise, for boson particles and fields, the number of candidate solutions equals 2S + 1. ([Physics:] The number of spin states for a non-zero-mass elementary particle is 2S + 1.) For each of the three ν = −3/2 rows Table 2.6.5 shows, min(D, 2(2S + 1)) = D. [Physics:] People might say that, regarding ν = −3/2, to the extent nature exhibits spin-3/2 fermions, nature might exhibit fewer than 6×#GEN spin-3/2 fermions. (See Section 3.5.)

Table 2.6.5 Numbers of candidate solutions, for some solutions based on radial coordinates (for D* = 3)

D*	ν	D	Ω	σ	S	2S + 1	Generations (#GEN)	Number of candidate solutions
3	−1/2	1	3/4	+1	1/2	2		1, 2, 3, or 4
"	"	4	−3/4	+1	1/2	2		4
"	"	10	−15/4	−1	3/2	4		8
"	"				...			
"	−1	1	2	+1	1	3		1, 2, or 3
"	"	3	0	+1	0	1		1
"	"	3	0	−1	0	1		1
"	"	5	−2	−1	1	3		3
"	"	9	−6	−1	2	5		5

D*	ν	D	Ω	σ	S	2S + 1	Generations (#GEN)	Number of candidate solutions
"	"				…			
"	−3/2	3	3/4	+1	1/2	2	3	9 = 3×#GEN
"	"	4	−3/4	−1	1/2	2	3	12 = 4×#GEN
"	"	6	−15/4	−1	3/2	4	#GEN(3/2)	6×#GEN
"	"				…			

~ ~ ~

This subsection discusses possibly useful concepts about generations.

Assume, for this subsection, that the number of generations equals the number of generators of SU(2S+1). The concept of generations pertains only for $\nu = -3/2$ solutions. These solutions correlate with the limit $\eta^2 \to 0$. Generally, away from $\eta^2 \to 0$ or for solutions that correlate with fields, the number 2S + 1 provides a relevant factor regarding number of solutions. People might want to consider the extent to which implications exist for math-based models that correlate a number (2S + 1) of solutions that extend significantly away from $r \approx 0$ with a number (the number of generators of SU(2S + 1)) of solutions that, in some sense, pertain for $r = 0$ only.

[Physics:] People might say that, perhaps, generations correlate with a particle-INTERN counterpart to aspects of particle-ENVIRO wave functions. (To compare with EXTINT LADDER models, see Section 2.14.)

~ ~ ~

This subsection suggests reasons for considering QE-like uses of radial-coordinate math.

Table 2.6.6 pertains.

Table 2.6.6 Perspective regarding possible QE-like uses of DIFEQU models

1. Possible uses with models for the big bang or with models involving consequences of the big bang.
   1.1. [Physics:] The existence of the big bang correlates with a lower limit for the value of a coordinate people correlate with time.
   1.2. Use of radial coordinates for QE-like math correlates with models that can be compatible with the big bang.
2. Possible uses with models involving interaction vertices.
   2.1. [Physics:] People might say that some of our work regarding interaction vertices correlates with projecting QP-like solutions for which D* = 3 into functions for which D* = 1 in a way that also involves projecting QE-like solutions into functions for which D* = 1. (See, for example, Table 2.6.2.)

2.2. [Physics:] Some of our work regarding sizes of interaction vertices suggests that (for some vertices) a QE-like (or, temporal) size is zero. (See Section 2.10.) Perhaps such work correlates with models for which considering a lower limit (for particle creation) or upper limit (for particle destruction) regarding a temporal coordinate (or, a QE-like radial coordinate) is appropriate.

3. Possible uses regarding physics correlating with $\sigma = -1$.

3.1. [Physics:] Possibly, some considerations (correlating with physics for which $\sigma = -1$) correlate with more than one QE-like coordinate and, possibly therefore, with uses (within models) of QE-like radial coordinates.

3.1.1. Some of our work regarding $\sigma = -1$ phenomena correlates with symmetries we denote by (1;3)<. (See, for example, Table 2.13.33.)

3.1.2. These symmetries have similarities to the (1;3)> symmetries we correlate with various $\sigma = +1$ phenomena. (See, for example, Table 2.13.6.)

3.1.3. For $\sigma = +1$ phenomena for which (1;3)> symmetries pertain, use of QP-like radial coordinates can pertain.

3.2. Possibly, some uses of QE-like radial coordinates can pertain with regard to (1;3)< phenomena.

3.2.1. (See discussion pertaining to developing Table 2.13.33.)

3.2.2. (See discussion pertaining to developing Table 2.13.36.)

3.3. Possibly, some uses of QE-like radial coordinates can pertain regarding color charge.

3.3.1. People might say that (2.116) and (2.117) in Table 2.14.1 correlate with possible modeling involving the following.

3.3.1.1. #E + 1 QE-like radial coordinates.

3.3.1.2. Symmetries related to SU(#E + 1).

3.3.1.3. A relevant number of gluons or gluon-like analogs correlating with ...

3.3.1.3.1. 1, for #E = 0.

3.3.1.3.2. 8, for #E = 2.

3.3.1.3.3. 25, for #E = 4.

4. Possible uses with models correlating with masses for W-, H-, and O-family elementary particles.

4.1. People might say that models correlating with masses for WHO-family elementary particles could be based on more than one QE-like coordinate. (See Table 2.10.7, Section 3.7, and Section 3.8.)

4.1.1. Here, possibly, each of some uses of negative contributions toward the square of a mass correlates with a pair of QE-like coordinates.

5. Possible uses correlating with ranges of the weak interaction.

5.1. People might say that masses for W-family bosons correlate with the various R_0 for W-family bosons and, therefore, correlate with the range of the weak interaction. (See, for example, Table 2.10.8.)

Section 2.7 FERTRA and other models linking LADDER and DIFEQU models

Section 2.7 discusses correlations between LADDER models and DIFEQU models. We correlate LADDER solutions and DIFEQU solutions for the WHO-families of solutions. We discuss overlaps and gaps in the applicability of LADDER and DIFEQU models. We discuss the concept of composite particles. We show FERTRA LADDER solutions for the CN-families and for the QIRD-families.

~ ~ ~

This subsection discusses bridging between DIFEQU models and LADDER models.

Table 2.1.14 summarizes some applications for DIFEQU models and LADDER models. That table lists each of some applications more than once. This section discusses some overlaps between DIFEQU models and LADDER models.

~ ~ ~

This subsection starts to bridge between DIFEQU models and LADDER models for the WHO-families.

For $D^* = 3$ and $\nu = -1$, 2S equals an even integer (or, equivalently, S equals an integer).

Table 2.7.1 re-depicts, as LADDER ground-state solutions, DIFEQU ground states for $\nu = -1$ solutions that Table 2.5.16 shows. For each row in Table 2.7.1, each of the columns SUBF, S, Ω, and σ correlates with DIFEQU results. (See Table 2.5.16.) For each row in Table 2.7.1, each of the columns SUBF, S, and σ correlates with LADDER results. (See Table 2.3.1.) For each row in Table 2.7.1, the number of QP-like oscillators matches 2S + 1. Here, the symbol 'I denotes a 0. Uses of 'I correlate with instance-related harmonic oscillators. (See, for example, Table 2.13.4.) Here, the symbol 'G denotes a 0. Uses of 'G correlate with symmetries related to interactions with elementary fermions. A pair of 'G symbols correlates with an additional 3-generator (SU(2)-related) symmetry. [Physics:] People might say that the 3-generator symmetry related to the 2W-subfamily correlates with conservation of generation for interactions between leptons and 2W bosons. For example, in an interaction in which the emission of a W^- boson converts a muon into a neutrino, the neutrino is a muon-neutrino and not, for example, an electron-neutrino. People consider muons to be generation-2 leptons. Muon-neutrinos are generation-2 leptons. We associate an occurrence of four 'G symbols with the symmetry that correlates with #GEN(3/2). (See Table 2.6.4 and Section 2.8.)

Table 2.7.1 Ground states for D* = 3, ν = −1, and S ≤ 2

E	E	E	E	E	E	E	P	P	P	P	P	P	P	P	P				
6R	6L	4R	4L	2R	2L	0	0	2L	2R	4L	4R	6L	6R	8L	8R	SUBF	S	Ω	σ
				'G	'G	'I	0	0	0							2W	1	2	+1
						'I	0									0H	0	0	+1
						0	'I									00	0	0	−1
				0	0	0	'I	'G	'G							20	1	−2	−1
		0	0	0	0	0	'I	'G	'G	'G	'G					40	2	−6	−1

~ ~ ~

This subsection discusses some differences between fermions and bosons.

[Physics:] In physics, a boson state can be excited an indefinite number of times.

Ground states, such as those Table 2.7.1 shows, seem appropriate.

[Physics:] In physics, a fermion state can be either not excited or excited exactly once.

~ ~ ~

This subsection discusses concepts relevant to FERTRA LADDER solutions.

It might seem inappropriate to try to depict some aspects of each fermion solution via the type of LADDER representations Table 2.3.11 shows or Table 2.3.14 shows. (See 1C, 1N, and 3N in Table 2.3.11. See 1Q, 1R, 3Q, 3I, 3R, and 3D in Table 2.3.14.)

People might say that subtracting 1 from each N(..) might lead to LADDER solutions that represent some aspects of fermions. Each value of N(..) would be −1 or −2. A lone oscillator for which N(..) = −1 cannot be excited to a state with N(..) = 0. A lone oscillator for which N(..) = −2 can be excited to a state with N(..) = −1. A lone oscillator for which N(..) = −1 could be de-excited to a state with N(..) = −2. Such considerations seem at least somewhat appropriate for fermions.

To develop FERTRA LADDER solutions, we make such subtractions. For leptons, we maintain Œ = 0 by opening QE-like oscillators and assigning values of N(..) = −1. (See Table 2.7.3.)

Table 2.7.2 pertains.

Table 2.7.2 Concepts pertaining to FERTRA LADDER solutions

1. People might say that each FERTRA LADDER solution correlates with no or little more information than does the corresponding INTERN LADDER solution.

2. People might say that FERTRA LADDER solutions have some similarity to Y-family solutions.
   2.1. People might say that pairs of Y-family solutions correlate with nature. (See Section 3.6.)
      2.1.1. For example, a gluon correlates with a sum of terms, with each term correlating with a pair of Y-family solutions.
         2.1.1.1. One Y-family solution in the pair correlates with erasing a color charge from a quark.
         2.1.1.2. The other Y-family solution in the pair correlates with painting a color charge on to a quark.
   2.2. People might say that FERTRA LADDER solutions correlate with erasing some properties from fermions. (See Table 3.5.2, Table 3.5.3, and Table 3.5.5.)
   2.3. People might say that representations similar to FERTRA LADDER solutions correlate with painting some properties on to fermions. (See Table 3.5.4 and Table 3.5.5.)

~ ~ ~

This subsection discusses FERTRA LADDER solutions for the CN-families and the QIRD-families.

Table 2.7.3 shows some FERTRA LADDER solutions for the CN-families and the QIRD-families. We use the word some to allude to the need to develop similar results that include reversing, for each solution, N(P[#P – 1]) and N(P[#P]) and that include reversing, for each QIRD-family solution, N(E[#E]) and N(E[#E – 1]). Such reversals would double the number of C- and N-family rows and quadruple the number of QIRD-family rows.

Table 2.7.3 Some FERTRA LADDER solutions for the CN-families and the QIRD-families

	E	E	E	E	E	E	E	E	E	P	P	P	P	P	P	P	P	P	
σ	8R	8L	6R	6L	4R	4L	2R	2L	0	0	2L	2R	4L	4R	6L	6R	8L	8R	Subfamily
+1					−1	−1	−1	−1	−1	−1	−1	−2							1C
+1					−1	−1	−1	−1	−2	−2	−1	−2							1N
+1	-1	-1	-1	-1	-1	-1	-1	-1	-1	-2	-1	-1	-1	-2					3N
-1							-2	-1	-1	-1	-1	-2							1Q
-1							-2	-1	-2	-2	-1	-2							1R
-1					-2	-1	-1	-1	-1	-1	-1	-1	-1	-2					3Q
-1					-2	-1	-2	-2	-1	-1	-2	-2	-1	-2					3I
-1					-2	-1	-1	-1	-2	-2	-1	-1	-1	-2					3R
-1					-2	-1	-2	-2	-2	-2	-2	-2	-1	-2					3D

~ ~ ~

This subsection shows a parameter that correlates with whether, for a LADDER solution, this monograph discusses a matching DIFEQU solution.

Table 2.7.4 discusses correlations between rest mass (m') and N(P0).

Table 2.7.4 Correlations between m' and N(P0)

1. m' = 0, if ...
   1.1. For INTERN LADDER solutions, ...
      1.1.1. N(P0) ≤ −1.
   1.2. For FERTRA LADDER solutions, ...
      1.2.1. N(P0) = −2.
2. m' ≠ 0, if ...
   2.1. For INTERN LADDER solutions, ...
      2.1.1. N(P0) ≥ 0.
   2.2. For FERTRA LADDER solutions, ...
      2.2.1. N(P0) = −1.
3. [Physics:] m' denotes rest mass.
4. For elementary fermions, ...
   4.1. This monograph provides FERTRA LADDER solutions that correlate with the extent to which elementary fermions interact with elementary bosons of the 2W-, 2O-, and 4O-subfamilies. (See Section 3.5.)

Table 2.7.5 pertains.

Table 2.7.5 Correlations between solutions this monograph discusses and m', plus extrapolations from DIFEQU results for some solutions to similar results for other LADDER solutions

1. Regarding solutions for which m' ≠ 0, ...
   1.1. This monograph discusses LADDER solutions.
   1.2. This monograph discusses DIFEQU solutions.
2. Regarding solutions for which m' = 0, ...
   2.1. This monograph discusses LADDER solutions.
   2.2. We may infer characteristics (that people might correlate with DIFEQU solutions) by extrapolating from DIFEQU results that correlate with solutions for which m' ≠ 0.

~ ~ ~

This subsection notes that this monograph includes, in some tables, listings for composite particles.

[Physics:] This monograph contains tables that list, in columns with labels such as subfamily, at least two more types of concepts (other than subfamily). The concepts include mesons and baryons. (The concepts also encompass tetraquarks, pentaquarks, and other possible composite particles having more than one quark.) This monograph does not develop solutions correlating with these listings. The listings correlate with composite particles and not with elementary particles. For mesons, 2S is an even nonnegative integer. For baryons, 2S is an odd nonnegative integer.)

Section 2.8 Generation, color charge, and INTERN models for bosons

Section 2.8 discusses the extents to which elementary bosons impact, for elementary fermions, the properties of generation and color charge. We consider INTERN models correlating with G-family solutions. We discuss conservation of fermion generation for some interactions involving elementary fermions and G-family bosons. We consider INTERN models correlating with WHO-family solutions. We discuss conservation of fermion generation for some interactions involving elementary fermions and WHO-family bosons. We consider INTERN models correlating with Y-family solutions. We note a lack of conservation of fermion color charge for Y-family-related interactions involving fermions for which $\sigma = -1$.

~ ~ ~

This subsection shows ground-state INTERN LADDER solutions for the G-family.

Table 2.8.1 repeats (from Table 2.3.11) ground-state INTERN LADDER solutions for the G-family. Table 2.8.1 shows oscillators EG = E[16] through PG = P[16]. For each oscillator-related column other than columns E0 and P0, the column pertains to an oscillator pair. The symbol 00 correlates with (depending on the solution and the pair) values for the respective two N(..) of 0 and 0, 0 and @, @ and 0, or @ and @. Sometimes, the symbol @ denotes a 0 that does not change (from the ground-state value) for excited states correlating with the solution. Sometimes, the symbol @ denotes a 0 that does not change (from the ground-state value) for excited states correlating with a polarization mode that correlates with a solution. (See, for example, Table 3.2.2 and the explanation that precedes Table 3.2.2.) The case @-and-@ pertains for exactly (#P)/2 – |N(P0)| oscillator pairs.

Table 2.8.1 G-family ground-state INTERN LADDER solutions

E GF	E ED	E CB	E A9	E 87	E 65	E 43	E 21	E 0	P 0	P 12	P 34	P 56	P 78	P 9A	P BC	P DE	P FG	Solution or subfamily
								0	−1	00								022G2&
							00	0	−1	00	00							24..G

E	E	E	E	E	E	E	E	E	P	P	P	P	P	P	P	P	P	Solution or
GF	ED	CB	A9	87	65	43	21	0	0	12	34	56	78	9A	BC	DE	FG	subfamily
								0	−2	00	00							042G24&
						00	00	0	−1	00	00	00						46..G
							00	0	−2	00	00	00						26..G
								0	−3	00	00	00						064G246&
					00	00	00	0	−1	00	00	00	00					68..G
						00	00	0	−2	00	00	00	00					48..G
							00	0	−3	00	00	00	00					28..G
								0	−4	00	00	00	00					084G2468&

Table 2.8.2 restates Table 2.8.1. We convert each QE-like 00 to the symbol "G. We choose the symbol "G to correlate with the use of 'G for correlating with fermion generations. (This parallels use, regarding the 2W-subfamily, of 'G in Table 2.7.1.) Here, an instance of "G correlates with a pair of 'G. (Here, an occurrence of "G does not correlate with a notion of generations for bosons. The concept of generations does not pertain for bosons.) For a row in Table 2.8.2, the number of occurrences of "G correlates with the extent to which a boson (for which the row pertains) differentiates, when interacting with an elementary fermion, by generation of the fermion. For example, the electron, muon, and tauon correlate respectively with 3 generations of (in some sense) one fermion. 022G2& (or, a MM1MS1-photon) interacts equally strongly with the charge of an electron, muon, or tauon. For 022G2&, the table shows zero occurrences of "G. In contrast, 244G4& (or, a graviton) interacts more strongly with a tauon than with a muon and more strongly with a muon than with an electron. For 244G4&, the table shows one occurrence of "G. We denote the number of occurrences of "G by the symbol #"G.

Table 2.8.2 G-family ground-state INTERN LADDER solutions, showing QE-like possible interaction-related symmetries

E	E	E	E	E	E	E	E	E	P	P	P	P	P	P	P	P	P	Solution or
GF	ED	CB	A9	87	65	43	21	0	0	12	34	56	78	9A	BC	DE	FG	subfamily
								0	−1	00								022G2&
							"G	0	−1	00	00							24..G
								0	−2	00	00							042G24&
						"G	"G	0	−1	00	00	00						46..G
							"G	0	−2	00	00	00						26..G
								0	−3	00	00	00						064G246&
					"G	"G	"G	0	−1	00	00	00	00					68..G
						"G	"G	0	−2	00	00	00	00					48..G
							"G	0	−3	00	00	00	00					28..G
								0	−4	00	00	00	00					084G2468&

~ ~ ~

This subsection interprets aspects correlating with G-family INTERN LADDER solutions.

People might say that Table 2.8.3 correlates data with solutions. The leftmost five columns reflect work in this monograph. The S column shows spins. The SDI column shows spatial dependences of interactions. (See Table 2.13.23, Table 3.3.10, and Table 3.3.11.) Here, we count each "G symmetry independently from any other "G symmetry that pertains. We try to correlate known phenomena with G-family particles. Each of the electron, muon, and tauon has the same charge. (People might say that, for purposes of this table, we do not need to take into consideration anomalous magnetic dipole moments.) The rightmost column notes that, for electromagnetism, the strength of the interaction does not vary by generation of the lepton involved. The rightmost column notes that, for gravitation, the strength of the interaction varies by generation of the lepton involved. Each of the electron, muon, and tauon has a different mass. We think that, somewhat paralleling work regarding the weak interaction, the case in which exactly one "G pertains correlates with variation by generation for interaction vertices involving spin-1/2 fermions. (See discussion regarding Table 2.7.1.) We think that work, so far, covers electromagnetic interactions with stationary and moving charges but not necessarily electromagnetic interactions with elementary-particle magnetic dipole moments. Presumably, a boson interacting with an elementary-particle magnetic dipole moment carries information about the spin of the elementary particle. People might say that 2G2& cannot perform that function. We seek a particle for which % has at least 2 elements, spin-1 pertains, an interaction with a spatial dependence of r^{-4} pertains, and #"G is an even number. That spatial dependence matches characteristics of magnetic dipoles. So far, 2G24& and 2G68& might qualify. For 2G24&, $2 \in$ %. We think that aspects related to 2G24& can scale to describe magnetic dipole moments correlating with spins of systems of multiple spin-entangled charged particles. For 2G68&, $2 \notin$ %. We think that 2G68& couples with spin but not with charge. We correlate 2G24& with vertices in which electromagnetism interacts with the magnetic dipole moments of elementary particles.

Table 2.8.3 Interaction-related phenomena, G-family bosons that intermediate the interactions, and numbers of "G symmetries

G-family particle (%68both, per Table 3.3.3)	S	SDI	Subfamily or solution (per Table 2.8.2)	Number of "G (per Table 2.8.2)	Known phenomena	Known variation by fermion generation (for charged leptons)
2G2&	1	r^{-2}	022G2&	0	Electromagnetic interaction	No
4G4&	2	r^{-2}	24..G	1	Gravitational interaction	Yes

G-family particle (%68both, per Table 3.3.3)	S	SDI	Subfamily or solution (per Table 2.8.2)	Number of "G (per Table 2.8.2)	Known phenomena	Known variation by fermion generation (for charged leptons)
2G24&	1	r^{-4}	042G24&	0	Charged-lepton magnetic dipole moment	No
4G26&	2	r^{-4}	26..G	1		
2G46&	1	r^{-4}	26..G	1		
2G246&	1	r^{-6}	064G246&	0		
4G48&	2	r^{-4}	48..G	2		
2G68&	1	r^{-4}	48..G	2		
2G248&	1	r^{-6}	28..G	1		
4G268&	2	r^{-6}	28..G	1		
2G468&	1	r^{-6}	28..G	1		
4G2468&	2	r^{-8}	084G2468&	0		

Table 2.8.4 generalizes patterns from Table 2.8.3. We assume that the appearance in a Table 2.8.2 row of two "G symbols correlates with #GEN(3/2) (and, possibly with SU(4) symmetries) and elementary fermions for which 2S = 3. The symbol #"G denotes the relevant number of "G symbols.

Table 2.8.4 Assumptions about generation-based variation of elementary-fermion properties, regarding G-family interactions

#"G	Property (or, vertex-strength) variation by fermion generation (for spin-1/2 elementary fermions)	Property (or, vertex-strength) variation by fermion generation (for spin-3/2 elementary fermions)
0	No	No
1	Yes	No
2	No	Yes
3	Yes	Yes

~ ~ ~

This subsection discusses two distinct sets of models, each of which might pertain to G-family elementary particles.

This monograph notes possibilities for EACUNI models and for SOMMUL models for aspects of G-family elementary particles. (See, for example, Table 2.13.11 and Table 3.3.3.)

People might say that, for EACUNI models, results (in Table 2.8.3 and Table 2.8.4) related to #"G correlate with some known physics and are not incompatible with known physics. People might say that results (in Table 2.8.3 and Table 2.8.4) related to #"G do not correlate with the combination of known physics and SOMMUL models.

This monograph discusses some possibilities regarding SOMMUL models. This monograph does not try to model an analog (for SOMMUL models) to #"G. This monograph deemphasizes SOMMUL models.

~ ~ ~

This subsection points to aspects correlating with WHO-family INTERN LADDER solutions.

Table 2.7.1 shows information about ground-state WHO-family INTERN LADDER solutions. (People might say that, regarding that table, technically, the column labelled Ω might not pertain to LADDER solutions.) Table 2.8.5 pertains.

Table 2.8.5 Values of #"G for WHO-family elementary bosons

Subfamily	#"G
2W	1
0H	0
0O	0
2O	1
4O	2

~ ~ ~

This subsection interprets aspects correlating with WHO-family INTERN LADDER solutions.

Table 2.8.6 pertains.

Table 2.8.6 Aspects regarding generation-based variation of elementary-fermion properties, regarding interactions with WHO-family bosons

1. Regarding interactions in FRERAN environments, ...
   1.1. Each interaction vertex that includes a spin-1/2 elementary fermion and a 2W boson features ...
      1.1.1. Measuring the generation of the fermion.
      1.1.2. Conserving the generation of the fermion.
   1.2. Each interaction vertex comports with results from Table 2.8.4 and Table 2.8.5.
2. Regarding interactions in COMPAR environments, ...
   2.1. For interactions involving a pair of spin-1/2 elementary fermions and a pair of 2W bosons, ...
      2.1.1. At least one fermion can change generation.
   2.2. We are uncertain as to the extent that ...

2.2.1. Each interaction vertex that includes a spin-1/2 elementary fermion and a 20 boson features ...
    2.2.1.1. Measuring the generation of the fermion.
    2.2.1.2. Conserving the generation of the fermion.
2.2.2. Some interaction vertices might comport with results from Table 2.8.4 and Table 2.8.5.

~ ~ ~

This subsection interprets aspects correlating with Y-family INTERN LADDER solutions.

Table 2.8.7 summarizes relevant information from Table 2.3.14.

Table 2.8.7 Y-family ground-state INTERN LADDER solutions

E	E	E	E	E	E	E	P	P	P	P	P	P	P	P	P	(σ = −1)
6R	6L	4R	4L	2R	2L	0	0	2L	2R	4L	4R	6L	6R	8L	8R	Subfamily
				..	..	..	−1	−1	−1							2Y
		..	..	..	..	..	−1	−1	−1	−1	−1					4Y

Table 2.8.8 reinterprets Table 2.8.7. In Table 2.8.8, each occurrence of the symbol 'C correlates with a QP-like occurrence of −1 in Table 2.8.7. Regarding Table 2.8.8, we posit that #P + 1 correlates with the number of color charges for relevant elementary fermions. (See, for example, Table 2.14.1 and Section 3.6.)

Table 2.8.8 Reinterpretation of Y-family ground-state INTERN LADDER solutions

E	E	E	E	E	E	E	P	P	P	P	P	P	P	P	P	(σ = −1)
6R	6L	4R	4L	2R	2L	0	0	2L	2R	4L	4R	6L	6R	8L	8R	Subfamily
				..	..	..	'C	'C	'C							2Y
		..	..	..	..	..	'C	'C	'C	'C	'C					4Y

~ ~ ~

This subsection interprets aspects correlating with Y-family INTERN LADDER solutions.

Table 2.8.9 pertains.

Table 2.8.9 Lack of conservation of color charge

1. Y-family solutions differ from G-family and WHO-family solutions in the following regard.
    1.1. Each G-family solution does or might correlate with an elementary particle.

   1.2. Each WHO-family solution does or might correlate with an elementary particle.
   1.3. Each elementary particle that correlates with Y-family solutions correlates with a sum of terms, with each term being comprised of a product of two constructs, each correlating with a Y-family solution. One such solution correlates with erasing a color charge (or color-charge analog) from a spin-1/2 (or spin-3/2, respectively) fermion. The other such solution correlates with painting a color charge (or, color-charge analog).
2. Y-family solutions differ from G-family and WHO-family solutions in the following regard.
   2.1. For each G-family solution, the notion of 'G does not pertain to the E0 oscillator.
      2.1.1. Here, $\sigma = +1$.
   2.2. For each WHO-family solution, the notion of 'G does not pertain to the E0 oscillator (if $\sigma = +1$) and does not pertain to the P0 oscillator (if $\sigma = -1$).
   2.3. For each Y-family solution, the notion of 'C pertains regarding the P0 oscillator.
      2.3.1. Here, $\sigma = -1$.
3. People might say that such thinking does not correlate with a notion, for the Y-family, of conservation of color charge.
   3.1. People might say that the case of leptons provides an exception. (See Table 2.14.1.)

~ ~ ~

This subsection notes a possible lack of correlation between generation and color charge.

People might say that nature exhibits, for elementary fermions for which $\sigma = -1$, no correlation between generation and color charge.

Section 2.9 Schwarzschild radius, Planck length, and R_0

Section 2.9 discusses a series of lengths. The series features the Schwarzschild radius, the Planck length, and a length we call R_0. People might say that we show physics-relevance for R_0.

~ ~ ~

This subsection provides perspective about pattern recognition.

People turn perceived patterns into bases for models and theories. To some extent, the more elements that people think correlate with a possible pattern, the more confidence people might have that people can develop useful models.

Sometimes, even if the number of elements in a pattern is small, people can develop useful models.

In this section, we discuss patterns regarding small numbers of elements. We think the patterns point to useful results. Table 2.9.1 pertains.

Table 2.9.1 Types of patterns this section discusses

1. A series of formulas, with each formula based on a product of powers of factors common to all the formulas.
   1.1. Most of the factors are physics numbers..
2. A series of lengths with which the previous series correlates.
   2.1. Here, we look for physics phenomena that people might correlate with the lengths.

~ ~ ~

This subsection provides perspective about the Schwarzschild radius, the Planck length, and the rest mass of electrons.

In traditional classical physics pertaining to black holes, people ascribe significance to a distance people call the Schwarzschild radius. People say that (absent quantum effects) energy and matter that exist closer to the center of a (spherically symmetric, non-rotating) black hole than the Schwarzschild radius cannot escape from the black hole. (For this discussion, we ignore possible dissipation of the black hole. We discuss black-hole dissipation in Section 4.6.) A formula for that radius involves the so-called mass of the black hole. The formula does not involve $\hbar$.

In traditional quantum physics, people try to attribute significance to a distance people call the Planck length. The length is approximately 1.6×10^{-35} meters. The constant $\hbar$ appears in a formula for that length. That formula does not contain a factor correlating with the mass of anything. People correlate $\hbar/2$ with a minimum unit of spin.

In this monograph, the rest mass of an electron is the smallest rest mass for a known non-zero-mass elementary particle and may be the smallest rest mass for any non-zero-mass elementary particle. (We assume that MM1MS1-neutrinos have rest masses of zero.)

~ ~ ~

This subsection defines a series of lengths that includes the Schwarzschild radius and the Planck length.

Table 2.9.2 defines a series of lengths that includes the Schwarzschild radius and the Planck length.

Table 2.9.2 A series of lengths that includes the Planck length and a length derived by applying the formula for the Schwarzschild radius to the mass of an electron

1. The next equation notes a traditional physics length, the Planck length. This work provides a symbol, R_2, for that length. Here, G_N denotes the gravitational constant, ħ denotes Planck's constant (reduced), and c denotes the speed of light. This work adds to the traditional statement two factors, each of value 1. The first such factor is m_e^0. The symbol m_e denotes the rest mass of an electron. The second such factor is 2^0.
$$R_2 = G_N^{1/2}\, m_e^0\, \hbar^{1/2}\, c^{-3/2}\, 2^0 \tag{2.50}$$
2. The next equation applies a traditional formula to a property (rest mass) associated with electrons. The formula represents the Schwarzschild radius. Traditionally, people apply the Schwarzschild-radius formula to black holes. Traditionally, people do not apply the formula to objects people claim have not enough mass to form black holes. This work adds to the traditional statement one factor with value 1. That factor is $\hbar^0$.
$$R_4 = G_N^1\, m_e^1\, \hbar^0\, c^{-2}\, 2^1 \tag{2.51}$$
3. The next equation shows a ratio of the above two lengths.
$$Z = R_2/R_4 = G_N^{-1/2}\, m_e^{-1}\, \hbar^{1/2}\, c^{1/2}\, 2^{-1} \approx 1.1945\times10^{22} \tag{2.52}$$
4. The next equation defines a series of lengths.
$$R_j = R_2\, Z^{(2-j)/2}, \text{ with j being an even integer} \tag{2.53}$$
5. The next equation provides a formula for R_0.
$$R_0 = G_N^0\, m_e^{-1}\, \hbar^1\, c^{-1}\, 2^{-1} \tag{2.54}$$
6. The next equation restates the previous item.
$$R_0\, m_e c^2 = \hbar^1\, c^1\, 2^{-1} \tag{2.55}$$

Table 2.9.3 shows factors and values (for electrons and positrons) for a series of lengths. These approximate lengths are the products of the factors indicated by the five columns having labels k for l^k (for some l). Times are computed via time = length/c. The time-centric column shows the log-base-10 of times (in seconds). The time since the big bang is $\sim10^{17.6}$ seconds. The j column values indicate possibly interesting correlations between items and G-family solutions ejjGj&. (Elsewhere, this monograph does not use the notion of a 000G0& solution. Because each would correlate with S > 2, elsewhere this monograph deemphasizes ejjGj& solutions for which j > 4.) The j' column values indicate possibly interesting correlations between items and properties of objects. (See Table 3.13.2.) In effect, j' = j + 2.

Table 2.9.3 A series of lengths relevant to electrons and positrons

j	R_j	Length (m)	Log_{10} (time (sec))	Concept	k for G_N^k	k for m_e^k	k for $\hbar^k$	k for c^k	k for 2^k	j'
		3.3×10^{53}	+45		−1.5	−4	2.5	0.5	−4	
		2.7×10^{31}	+23		−1	−3	2	0	−3	
		2.3×10^{9}	+0.88		−0.5	−2	1.5	−0.5	−2	
0	R_0	1.9×10^{-13}	−21	R_0	0	−1	1	−1	−1	2
2	R_2	1.6×10^{-35}	−43	Planck length	0.5	0	0.5	−1.5	0	4
4	R_4	1.4×10^{-57}	−65	Schwarzschild radius	1	1	0	−2	1	6
6	R_6	1.1×10^{-79}	−87		1.5	2	−0.5	−2.5	2	8
8	R_8	9.5×10^{-102}	−109.5		2	3	−1	−3	3	
		7.9×10^{-124}	−132		2.5	4	−1.5	−3.5	4	

~ ~ ~

This subsection discusses possible significances of lengths R_0.

Possibly, the formula for R_0 pertains to other than electrons and positrons. For pions, R_0 may have significance. For Z and W bosons, R_0 may have significance. For the Higgs boson, R_0 may have significance. Table 2.9.4 pertains.

Table 2.9.4 Concepts indicating possible significance for lengths R_0

1. For a non-zero mass particle pp, this formula defines R_0(pp).

$$R_0(pp)\,(m'(pp))c^2 = \hbar c/2 \tag{2.56}$$

   1.1. m' denotes rest mass.
2. The charged-pion R_0 approximates the charged-pion charge radius.
   2.1. A charged-pion R_0 would be $\sim0.70\times10^{-15}$ meters.
      2.1.1. Here, we substitute, in the formula for R_0(electron), m'(charged pion) for m_e.
      2.1.2. The charged-pion R_0 is a factor ~139.6 / 0.511 or ~273.2 smaller than R_0 for electrons.
   2.2. An experimental charge radius for charged pions is $0.78^{+0.09}_{-0.10}\times10^{-15}$ meters. (See G. T. Adylov (1977).)

3. R_0 for weak-interaction bosons may correlate with a range of the weak interaction.
   - 3.1. A Z-boson R_0 is $\sim 2\times10^{-18}$ meters.
     - 3.1.1. A W-boson R_0 is $\sim 1.1\ R_0(\text{Z boson})$.
   - 3.2. People measure spatial dependence for interactions mediated by the weak interaction.
     - 3.2.1. For a separation of $\sim 10^{-18}$ meters between two interacting particles, the weak interaction and the electromagnetic interaction have similar magnitudes. (See Particle Data Group (2014).)
     - 3.2.2. At a separation of $\sim 3\times10^{-17}$ meters, the weak interaction is less by approximately a factor of 10^4. (See Particle Data Group (2014).)
4. These formulas pertain for the Higgs boson. (See remarks pertaining to Table 3.7.3 and see Table 3.7.4.)

$$(m'(\text{Higgs})c^2)^2 = (1/4)\ (\hbar c)^2\ /\ (R_0(m'(\text{Higgs})))^2 \quad (2.57)$$

$$\xi'/2 = (1/4)\ (\hbar c)^2\ /\ (R_0(m'(\text{Higgs})))^2 \quad (2.58)$$

~ ~ ~

This subsection shows values of R_0 for various particles.
Table 2.9.5 extends concepts in Table 2.9.3 and Table 2.9.4.

Table 2.9.5 Some values of R_0, for various particles

Particle	R_0(particle)
MM1MS1-photon	undefined or infinite
MM1MS1-neutrino	undefined or infinite
electron	$\sim 2\times10^{-13}$ m
pion	$\sim 7\times10^{-16}$ m
proton	$\sim 1\times10^{-16}$ m
Z boson	$\sim 2\times10^{-18}$ m
Higgs boson	$\sim 8\times10^{-19}$ m

~ ~ ~

This subsection posits possibilities for research regarding R_0 for various particles.

We do not much discuss further herein, some possible lines of inquiry that Table 2.9.6 notes. Regarding uncertainty, in traditional physics, uncertainty relationships often pertain to a superposition of amplitudes. Much of the work in this monograph focuses on base states.

Table 2.9.6 Possible lines of inquiry related to the concept of R_0 for elementary particles

1. To what extent does R_0 for a weak-interaction boson correlate with a size that characterizes a cloud of virtual particles that accompanies the boson? (See Table 2.10.1.)
2. To what extent does R_0 for an electron have significance?
    - 2.1. For example, to what extent does R_0 for electrons correlate with a size that characterizes the cloud of virtual particles that accompany an electron?
3. At least from a standpoint of theory, for zero-mass bosons, to what extent does an unbounded R_0 have significance?
    - 3.1. For example, to what extent might an unbounded R_0 correlate with some concept of (perhaps field-related) interconnectedness, for example, of particles correlating with gravity? Or, of photons? Or, of all G-family particles?
4. Regarding the following relationship, for non-zero-mass elementary bosons (See Section 2.10. See, for example, Table 2.10.9.), ...

$$<t^2> (m'c^2)^2 \sim \hbar^2/4 \tag{2.59}$$

    - 4.1. To what extent does the relationship ...
        - 4.1.1. Generalize?
        - 4.1.2. Correlate with a notion of uncertainty?
        - 4.1.3. Correlate with a notion of certainty?
    - 4.2. People might say that, here, $(<t^2>)^{1/2}$ correlates with a lifetime for the boson.
5. To what extent should people consider that, for non-zero-mass elementary particles and for composite particles, an uncertainty-like relationship for which lengths exceed the Planck length ($\sim 1.6 \times 10^{-35}$ m) pertains?
6. To the extent there is a meaningful such uncertainty-like relationship (regarding, at least, non-zero-mass elementary bosons), to what extent should people consider that this uncertainty-like relationship correlates with sums of some kind of base states that correlate with a concept (perhaps related to internal particle properties or to interaction-centric characteristics) that people might explore further?
7. To what extent should people consider that possible relevance of COMPAR models (See Section 2.13.) for phenomena that correlate with scale size R_0 augurs poorly (or well) for attempts to build theories based on possible phenomena for which the scale size R_2 (the Planck length) would be key?

~ ~ ~

Reference 2 G. T. Adylov (1977)
G. T. Adylov, et. al., A measurement of the electromagnetic size of the pion from direct elastic pion scattering data at 50 GeV/c, *Nuclear Physics B*, Volume 128, Issue 3, 3 October 1977, pages 461-505. (http://dx.doi.org/10.1016/0550-3213(77)90056-6)

Reference 3 Particle Data Group (2014)
Particle Data Group, Electroweak (web page), The Particle Adventure, Lawrence Berkeley National Laboratory, (2014), http://www.particleadventure.org/electroweak.html.

Section 2.10 Models related to vertices and to particle sizes, masses, and ranges

Section 2.10 discusses models related to sizes, masses, and ranges of elementary particles. We discuss clouds of virtual particles. We discuss MM1MS1-currents related to energy and momentum. We discuss interactions between gravity and MM1MS1-currents related to energy and momentum. We discuss aspects of models for interaction vertices (or, for vertices in Feynman diagrams). People might say that our work correlates with notions that, in some models, elementary particles have sizes of zero. We point to models and integers related to actual or approximate masses of elementary bosons. People might say that we show models that correlate with the spatial ranges of elementary bosons.

~ ~ ~

This subsection discusses aspects of people's thoughts about sizes of elementary particles.

People discuss possibilities that elementary particles have zero size. People discuss possibilities that elementary particle have non-zero sizes that correlate with the Planck length. People discuss the extent to which wave-like models may correlate with sizes relevant to particle-like models.

Perhaps, one interpretation of particle size correlates with models that address the topic of the extent (with respect to QE-like {or, temporal} coordinates and QP-like {or, spatial} coordinates) of interaction vertices.

In this section, we focus on some aspects of size related to interactions in which particles partake.

~ ~ ~

This subsection discusses aspects of interaction vertices.

In traditional quantum physics, people model interactions between (for example) elementary fermions and elementary bosons. A Feynman diagram may depict, for example, an interaction vertex in which an electron comes in and a neutrino and a W^- boson leave. For such vertices, the net number of entering fermions equals the net number of exiting fermions. Here, net denotes the number of matter fermion particles minus the number of antimatter fermion antiparticles.

~ ~ ~

This subsection provides perspective about this section.

This section discusses models for some aspects of how elementary particles interact. We suggest roles for clouds of virtual particles. Traditional physics provides concepts regarding such clouds.

This section provides results we later use for modeling approximate ratios of masses of non-zero-mass elementary bosons. (See Section 3.7 and Section 3.8.) As far as we know, traditional physics models do not estimate the ratios we approximate.

Some math this section shows pertains to models people use in traditional physics. Some of our applications of that math are non-traditional.

~ ~ ~

This subsection discusses the concept of clouds of virtual particles.

Table 2.10.1 discusses the concept of clouds of virtual particles and discusses some uses of the terms E and P.

Table 2.10.1 The concept of clouds of virtual particles, plus some uses of the terms E and P

1. [Physics:]
   - 1.1. For an elementary particle (or any object) for which $\sigma = +1$ and $m' \neq 0$ pertain, traditional quantum physics provides that ...
     - 1.1.1. The concept of a cloud of virtual particles that accompany any $m' \neq 0$ elementary particle or any object pertains.
     - 1.1.2. The following equation pertains.
       $$E^2 - c^2P^2 = (m')^2c^4 \tag{2.60}$$
     - 1.1.3. The value of E measured by one observer need not equal the value of E measured by another observer.
   - 1.2. We assume that E and P include effects of the traditional quantum-physics concept of such a cloud of virtual particles.

1.3. People might say that the following pertain.
 1.3.1. For $\sigma = +1$ and $m' = 0$, ...
  1.3.1.1. The term rest (as in not moving) in the phase rest mass does not apply.
  1.3.1.2. This monograph makes the following uses of the terms E and P. (See Table 2.1.1.)
   1.3.1.2.1. E = energy.
   1.3.1.2.2. P = the magnitude of the momentum.
  1.3.1.3. Here, the following equations pertain.

$$E = cP \tag{2.61}$$

$$E^2 - c^2P^2 = (m')^2c^4 \tag{2.62}$$

 1.3.2. The value of E measured by one observer need not equal the value of E measured by another observer.
 1.3.3. This monograph correlates E and P with the concept of a cloud of virtual particles.
1.4. For $\sigma = -1$, ...
 1.4.1. Questions such as those Table 6.3.2 mentions may pertain.
 1.4.2. People might say that distinguishing FRERAN SPATIM models from COMPAR SPATIM models may help regarding addressing such questions.
2. [Physics:]
 2.1. People might say that, regarding traditional quantum physics, models for aspects of clouds of virtual particles generally correlate with ...
  2.1.1. Invariances and symmetries compatible with special relativity.

~ ~ ~

This subsection discusses notions of currents.

In electromagnetism, people, when discussing motions of charged objects, may use the term current. People might use the term electric current.

More generally, physics applies the word current to a combination of the motion of an object and a property that object exhibits. For currents that people model via 4-vectors, a name for the current may correlate with the name for the property people associate with the QE-like component of the vector for an object that (relative to an observer) is not moving. The case of electric charge provides an example. People might use the term charge current.

Below, we consider other currents. For example, we consider a current correlating with the energy-momentum 4-vector pertaining (relative to an observer) to motion of energy related to an object.

~ ~ ~

This subsection discusses gravitational interactions with elementary particles and objects.

Table 2.10.2 extends discussion from Table 2.10.1.

Table 2.10.2 Energy-momentum MM1MS1-currents, rest mass m', and gravity

1. [Physics:]
   1.1. We posit that gravity interacts with energy-momentum MM1MS1-currents.
      1.1.1. Our interpretation correlates with gravity's interacting with light via the energy and momentum associated with clouds of virtual particles associated with photons.
      1.1.2. People might say that our interpretation does not correlate with the notion that gravity interacts with rest mass.
   1.2. The following statements pertain. (See, for example, Table 2.7.4.)
      1.2.1. Regarding some σ = +1 elementary particles for which m' = 0, ...
         1.2.1.1. Interactions between a MM1MS1-photon and gravity correlate with E and P (and the relationship E = cP) for, at least, the MM1MS1-photon.
            1.2.1.1.1. We use the term at least, because an interaction between a MM1MS1-photon and a graviton presumably correlates also with E and P for the graviton.
         1.2.1.2. Interactions between a MM1MS1-neutrino and gravity correlate with E and P (and the relationship E = cP) for, at least, the MM1MS1-neutrino.
      1.2.2. For all σ = +1 elementary particles for which m' = 0, ...
         1.2.2.1. Interactions between a particle and gravity correlate with E and P (and the relationship E = cP) for, at least, the particle.
      1.2.3. For σ = +1 elementary particles for which m' ≠ 0 specifically and for σ = +1 objects for which m' ≠ 0 generally, ...
         1.2.3.1. Strengths of interactions between an elementary particle or other object and gravity correlate with E and P for the particle or object.
      1.2.4. See, for example, Section 4.6.

~ ~ ~

This subsection discusses the possibility that some MM1MS1-currents correlate with tensors of rank greater than or equal to 2.

In traditional physics, gravity interacts with rest mass or RESENE. People might say that RESENE is a scalar quantity. People might say that this scalar quantity correlates with a rank-0 tensor.

In general relativity, effects of gravity correlate with a stress-energy tensor of rank-2. General relativity correlates that stress-energy tensor with curvature of space-time.

People might say that photons interact with a 4-vector charge current.

Above, we discuss the concept that gravity interacts with an energy-and-momentum 4-vector current. This MM1MS1-current correlates with a rank-1 tensor.

Later, we discuss matters related to curvature (and, thereby, possibly to rank-2 tensors). (See for example, Section 6.4. See, also, Table 6.1.5 and Table 6.1.6.)

~ ~ ~

This subsection discusses traditional quantum-physics notions of interaction vertices.

When working with quantum models for interactions between elementary particles, people may use techniques people correlate with the term Feynman diagrams. People model a specific vertex of an interaction as involving entering particles and exiting particles. The entering particles and their quantum states pertain before the specific vertex. The exiting particles and their quantum states pertain after the specific vertex. Away from and at vertices, the existence and motions of particles correlate with lines or other somewhat linear symbols.

An overall interaction can involve more than one vertex. For example, people envision an interaction in which an electron enters and a neutrino (more precisely, an electron-neutrino) exits as having 2 vertices. For example, at one vertex, the electron disappears, a neutrino appears, and a W^- both gets created and carries off a unit of negative charge. At another vertex, the W^- disappears and something else (for example, conversion of an up quark to a down quark) occurs.

For some such interactions, people diagram each vertex by using a point.

~ ~ ~

This subsection discusses notions regarding waves, particles, and sizes of elementary particles.

People might correlate the notion of an interaction vertex with notions of particles. Away from vertices, people might use terms such as particle, field, and/or wave.

We prefer to deemphasize some aspects of such discussion.

People might say that this monograph shows models for which notions of particle and zero-size correlate with interaction vertices.

~ ~ ~

This subsection discusses models that correlate with sizes that might correlate with interaction vertices.

Table 2.10.3 pertains.

Table 2.10.3 Some correlations between $\eta^2 \to 0$ and sizes associated with interaction vertices

1. For non-zero-mass elementary fermions, ...
   1.1. Work above correlates particles with DIFEQU solutions for which $\eta^2 \to 0$ pertains.
   1.2. People might say that models that require $\eta^2 \to 0$ correlate with the notion that, at least in the sense of the model, non-zero-mass elementary fermions have zero spatial size.
   1.3. People might say that Table 2.10.4 provides an alternative approach that leads to a model-based conclusion of zero-spatial size.
   1.4. People might say that a notion of zero size dovetails with concepts people traditionally correlate with models for interaction vertices.

Table 2.10.4 derives the result $\langle r^2 \rangle = 0$ for some solutions for which $\nu = -3/2$ and $D^* = 3$. Also, the table derives the result $\langle r^2 \rangle = 0$ for some solutions for which $\nu = -1$ and $D^* = 2$. (See Table 2.6.1.) Here, models feature $\eta^2 \to 0$.

Table 2.10.4 $\langle r^2 \rangle = 0$ for some $\nu = -3/2$, $D^* = 3$ solutions and $\langle r^2 \rangle = 0$ for some $\nu = -1$, $D^* = 2$ solutions

1. This result follows from Table 2.5.6.
$$\xi = (\xi'/2)\,(\,\eta^2\langle p_r^2\rangle + \eta^{-2}\langle r^2\rangle\,) \qquad (2.63)$$
   1.1. Here, $\langle j \rangle$ denotes the expected value of j.
   1.2. Here, p_r^2 denotes $r^{-(D-1)}(\partial/\partial r)(r^{D-1})(\partial/\partial r) - \Omega r^{-2}$.
2. We assume $\xi' \neq 0$.
3. This result follows from Table 2.5.8.
$$D + 2\nu = \eta^2\langle p_r^2\rangle + \eta^{-2}\langle r^2\rangle \qquad (2.64)$$
4. For 1C-subfamily particles, ...
   4.1. Models feature $D^* = 3$ and $\nu = -3/2$.
   4.2. $D = 3$. (See the $\nu = -3/2$ row in Table 2.6.5.)
   4.3. $D + 2\nu = 0$.
   4.4. We can consider each of η^2, $\langle p_r^2\rangle$, η^{-2}, and $\langle r^2\rangle$ to be non-negative.
   4.5. The next equation pertains.
$$\langle r^2\rangle = 0 \qquad (2.65)$$
5. For G-family particles, ...
   5.1. Models correlate with $D^* = 2$ and $\nu = -1$.

5.2. All uses (with respect to $D^* = 2$) of $D + 2\nu$ feature $D + 2\nu = 0$. (See Section 3.3.)

5.3. Somewhat similarly to results for 1C-subfamily particles, ...

$$\langle r_2^2 \rangle = 0 \tag{2.66}$$

5.3.1. Here, we use the symbol r_2 to call attention to the notion that this radial coordinate correlates with $D^* = 2$ (or, 2 dimensions).

People might say that, here, $(\langle r^2 \rangle)^{1/2}$ correlates with a size for an interaction vertex. People might say that, here, $(\langle r^2 \rangle)^{1/2}$ correlates with a size for an elementary particle.

Table 2.10.5 restates and extrapolates results from Table 2.10.4.

Table 2.10.5 Zero-length sizes for some aspects of some interaction vertices

1. $\langle r^2 \rangle = 0$ pertains for interaction vertices for leptons for which $m' \neq 0$.
2. $\langle r_2^2 \rangle = 0$ pertains for interaction vertices for the 2 dimensions that an observer would say are perpendicular to the direction of motion of a G-family boson.
   2.1. Here, we use the symbol r_2 to call attention to combination of notions that ...
      2.1.1. This radial coordinate correlates with $D^* = 2$ (or, 2 dimensions).
      2.1.2. For the vertex, $D^* = 3$.

~ ~ ~

This subsection discusses some types of interaction vertices, including vertices representing interactions between electrons and MM1MS1-photons.

We consider the example of an interaction vertex involving an electron and a G-family elementary boson. Table 2.10.6 pertains. Here, we extrapolate from Table 2.10.5. Here, QE-like variables come into play.

Table 2.10.6 Aspects of a model for a vertex correlating with an interaction between an electron and a G-family elementary boson

1. People might say that , for the electron, the next equation follows from $\langle r^2 \rangle = 0$.

$$\langle t^2 \rangle = 0 \tag{2.67}$$

2. People might say that, for the G-family particle, the next equations follows from $\langle r^2 \rangle = 0$ and $\langle t^2 \rangle = 0$ for the electron.

$$\langle r^2 \rangle = 0 \tag{2.68}$$

$$\langle t^2 \rangle = 0 \tag{2.69}$$

3. The next equation, for the G-family particle, follows from work above in this table.

$$<x^2> = 0 \quad (2.70)$$

3.1. Here, x aligns with the direction of motion (either before the vertex destroys the particle or after the vertex creates the particle) of the G-family particle. Compare with Table 2.10.5.)

4. The vertex correlates with a single point with respect to space-time coordinates.
5. People might say that, for an interaction vertex, models correlate with a collapse (or, disappearance) of the clouds of virtual particles that models otherwise would correlate with the vertex-entering particles that correlate with the vertex.

~ ~ ~

This subsection develops some aspects of quantum kinematics that dovetail with quantized masses for elementary bosons.

[Physics:] The operator $\partial^2/\partial^2 x$ correlates (within a factor) with P^2, in which P denotes an operator people associate with linear momentum. The operator $\partial^2/\partial^2 t$ correlates (within a factor) with E^2, in which E denotes an operator people associate with energy.

[Physics:] Each of those two factors includes a factor of $\hbar^2$.

For INTERN solutions that correlate with non-G-family elementary bosons, #E = #P.

[Physics:] People might say that Table 2.10.7 pertains. (See Table 2.6.2, Section 3.7, and Section 3.8.) Here, we, in effect, add (to work in Table 2.6.2 related to $D^* = 2$) concepts that people might correlate with a notion of a QE-like $D^* = 2$ and/or with a notion that ξ' can effectively be negative (at least, for some terms in a calculation for which other terms in the calculation correlate with contributions for which $\xi' > 0$). Here, c denotes the speed of light. For the G-family, Table 2.10.7 assumes $(D + 2\nu)_j = 0$ when j correlates with an oscillator pair EkR-and-EkL for which k is an even integer and $\#E < k \leq \#P$.

Table 2.10.7 Models for masses for elementary bosons for which either $\sigma = +1$ or the bosons are related to the Y-family of solutions

1. This equation pertains for the cases stated.

$$E^2 - c^2P^2 = (m')^2c^4 \quad (2.71)$$

   1.1. For $\sigma = +1$ elementary particles.
   1.2. For composite particles.
   1.3. For elementary particles correlating with Y-family solutions.

2. For elementary bosons for which $\sigma = +1$, we assume that the following equation pertains.

$$E^2 - c^2P^2 \approx (\xi'/2)\, \Sigma_j \pm_j (D + 2\nu)_j \quad (2.72)$$

- 2.1. Here, ...
  - 2.1.1. j ∈ { oscillator pairs E[#E]R-and-E[#E]L, ..., E2R-and-E2L, E0-and-P0, P2L-and-P2R, ..., P[#P]L-and-P[#P]R }.
  - 2.1.2. $\pm_j = \pm 1$.
  - 2.1.3. Each value of $(D + 2\nu)_j$ pertains to a $D^* = 2$ and $\nu = -1$ solution, as per Table 2.5.5, Table 2.5.17, and Table 2.6.2.
    - 2.1.3.1. We use the notation $\{D + 2\nu\}(2,S,\Omega)$ to denote the value of $D + 2\nu$ that correlates with $D^* = 2$, with a value of S, and with a value of Ω.
    - 2.1.3.2. People might express concern regarding applying this concept for the E0-and-P0 oscillator pair.
      - 2.1.3.2.1. Table 2.10.8 discusses this use.
- 2.2. For zero-mass elementary bosons (that is, particles correlating with the G- or Y-families), ...
  - 2.2.1. Each $(D + 2\nu)_j$ equals 0.
  - 2.2.2. That is, we use only $\{D + 2\nu\}(2,1,1)$.
- 2.3. For the known non-zero mass elementary bosons (that is, particles correlating with the H- or W-families), ...
  - 2.3.1. $\xi'/2$ correlates with experimental results via a formula for R_0. (See Table 2.9.4.)

People might say that Table 2.10.8 discusses concepts that correlate with some aspects of Table 2.10.7.

Table 2.10.8 $(D + 2\nu)_j$ correlating with the E0-and-P0 oscillator pair

1. People might say that this use of $\{D + 2\nu\}(2,S,\Omega)$ correlates with a selection of numbers that fits data about the masses of the Higgs, Z, and W bosons. (See Section 3.7.)
2. For oscillator pairs other than E0-and-P0, work correlating with Table 2.10.7 correlates with notions of (for coordinates for energy-momentum space) ...
   - 2.1. For QE-like aspects, ...
     - 2.1.1. $(e[\text{even}])^2 + (e[\text{even} - 1])^2$.
   - 2.2. For QP-like aspects, ...
     - 2.2.1. $c^2(p[\text{even} - 1])^2 + c^2(p[\text{even}])^2$.
3. People might say that, for oscillator pairs other than E0-and-P0, work correlating with Table 2.10.7 correlates with notions of (in extensions to space-time coordinates) ...
   - 3.1. $c^2(t[\text{even}])^2 + c^2(t[\text{even} - 1])^2$.

3.2. $(x[even - 1])^2 + (x[even])^2$.

4. People might say that, for each of the various oscillator pairs (other than the E0-and-P0 oscillator pair) this table mentions above, ...
   4.1. Math features a sum of two terms, with each term featuring the square of a coordinate.
   4.2. Math pertaining to $D^* = 2$ pertains.
5. For the oscillator pair E0-and-P0, traditional physics might correlate with notions of, ...
   5.1. In QE-like and QP-like coordinates for energy-momentum space, ...
      5.1.1. $(e0)^2 - c^2(p0)^2$.
   5.2. In space-time coordinates (relative to an interaction vertex), ...
      5.2.1. $c^2(t0)^2 - (x0)^2$.
6. For an interaction vertex involving a non-zero-mass boson, ...
   6.1. For r characterizing the spatial extent of the vertex, $\langle r^2 \rangle$ correlates with η for the fermion.
   6.2. The vertex correlates with $\eta^2 \rightarrow 0$.
   6.3. $\langle r^2 \rangle = 0$, for the vertex.
      6.3.1. This contrasts with $\langle r^2 \rangle \sim (R_0)^2$ (See Table 2.10.9.), for which ...
         6.3.1.1. $\langle r^2 \rangle$ pertains to the range of the non-zero-mass boson.
         6.3.1.2. R_0 pertains to the non-zero-mass boson.
   6.4. Therefore, ...

$$\langle (x0)^2 \rangle = 0 \tag{2.73}$$

   6.5. Similarly, for the variable t characterizing a temporal extent of the vertex, ...

$$c^2 \langle t^2 \rangle = 0 \tag{2.74}$$

   6.6. Therefore, ...

$$c^2 \langle (t0)^2 \rangle = 0 \tag{2.75}$$

   6.7. Therefore, ...

$$c^2 \langle (t0)^2 \rangle - \langle (x0)^2 \rangle = 0 \tag{2.76}$$

   6.8. We posit that ...
      6.8.1. $c^2 \langle (t0)^2 \rangle - \langle (x0)^2 \rangle = 0$ correlates with $E^2 - c^2 P^2 = (m')^2 c^4$.
      6.8.2. This difference (involving $c^2 \langle (t0)^2 \rangle$ and $\langle (x0)^2 \rangle$) contrasts with sums, pertaining to other relevant oscillator pairs, of terms, with each term involving two coordinates.
      6.8.3. $\{D + 2\nu\}(2,S',\Omega')$ numbers, for which S' need not match S, can pertain.

7. We note the following. (See Table 2.5.17, Section 3.7, and Section 3.8.)
   7.1. For the 2W family, the choice S' = 3 seems appropriate because ...
   $$\{D + 2\nu\}(2,3,\Omega') = 10 \quad (2.77)$$
   7.2. For the 0H family, the choice S' = 4 seems appropriate because ...
   $$\{D + 2\nu\}(2,4,\Omega') = 17 \quad (2.78)$$
8. We note that the rows of Table 2.7.1 show a sequence for which, respectively, ...
   $$S = 1, 0, 0, 1, 2 \ldots \quad (2.79)$$
   $$-\Omega = -2, 0, 0, 2, 6, \ldots \quad (2.80)$$
9. We note that Table 2.5.18 correlates with a concept for how to address a series of S for which ...
   $$|S| = 1, 0, 0, 1, 2, \ldots \quad (2.81)$$
   $$S = -1, 0, 0, 1, 2, \ldots \quad (2.82)$$
10. We posit that the following pertain.
   10.1. S' for the 2W-subfamily correlates with $\nu = -1$ and S = 3 in Table 2.5.17.
   10.2. S' for the 0H-subfamily correlates with $\nu = -1$ and S = 4 in Table 2.5.17.
   10.3. S' for the 0O-subfamily correlates with $\nu = -1$ and S = 5 in Table 2.5.17.
   10.4. S' for the 2O-subfamily correlates with $\nu = -1$ and S = 6 in Table 2.5.17.
   10.5. S' for the 4O-subfamily correlates with $\nu = -1$ and S = 7 in Table 2.5.17.

~ ~ ~

This subsection discusses models correlating with rest masses for elementary fermions.

For zero-mass elementary fermions (that is, the 1N-, 3N-, 1R-, 3R-, and 3D-subfamilies), as for zero-mass elementary bosons, LADDER solutions feature N(P0) ≤ −1. For zero-mass elementary fermions, in INTERN LADDER solutions, N(P0) = −1. Regarding rest mass, models similar to the models in Table 2.10.7 for zero-mass bosons may pertain. Here, each relevant $(D + 2\nu)_j$ equals 0.

For non-zero mass elementary fermions (that is, the 1C-, 1Q-, 3Q-, and 3I-subfamilies), relevant models differ from models for the WHO-family masses.

For each W-, H-, and O-family ground-state INTERN LADDER solution, each open oscillator pair has local-Œ = 1. For each 1C, 1Q, 3Q, and 3I INTERN LADDER solution, the P[#P − 1]-and-P[#P] oscillator pair is open and has local-Œ = 0.

For each of 1C, 1Q, 3Q, and 3I, the difference (compared to the WHO-families) regarding local-Œ for the P[#P − 1]-and-P[#P] oscillator pair is significant regarding the sizes of rest masses. For 1C and 1Q particles, this monograph shows (in Section 3.9) an approximate formula for rest masses. For generation-1 1C and 1Q particles, rest mass varies (at least approximately) as $m(0,0)\times e^{-k|Q'|}$, in which m(0,0) is a positive number with units of mass, k is a positive number, and Q' is the charge in units of $|q_e|/3$. Also, for other generations of 1C and 1Q particles, exponential functions involving functions of Q' pertain.

Perhaps, the rest mass for an elementary fermion particle for which m' ≠ 0 correlates conceptually with a notion of a degree of some sort of effort to maintain the P[#P − 1]-and-P[#P] oscillator pair as being open, compared to the many ways for this pair to be closed. Perhaps, such degree of effort correlates with an exponential function. Perhaps, such effort correlates with generation-1 m' ≠ 0 elementary fermions having less rest mass than m' ≠ 0 elementary bosons have. In Section 3.9, we possibly correlate generation-1 m' ≠ 0 elementary fermion particle masses with functions of the form $\exp(-k(D + 2\nu)_j)$, in which the $(D + 2\nu)_j$ correlate with $(D + 2\nu)_j$ for solutions for which D* = 2 and ν = −1/2, as per Table 2.5.17.

~ ~ ~

This subsection discusses spatial ranges for elementary bosons.

Work above correlates with the notion that elementary bosons have zero size with respect to an interaction vertex. Regarding elementary bosons for which the rest mass is non-zero, MM1MS1 models correlate with the notion that various $(D + 2\nu)_j$ are non-zero.

Table 2.10.9 pertains. The table repeats some results. (See Table 2.10.8.) Table 2.10.9 discusses spatial ranges for elementary bosons.

Table 2.10.9 Results regarding sizes pertaining to interaction vertices for interactions between an elementary-particle fermion and an elementary-particle boson and regarding distances between two related interaction vertices

1. For a non-zero-mass elementary fermion, vertex sizes feature $<r^2> = 0$.
   1.1. This matches the notion that the limit $\eta^2 \to 0$ pertains for fermion elementary particles.
2. For a non-zero-mass elementary fermion, vertex sizes feature $<t^2> = 0$.
   2.1. This matches a notion that $c^2<t^2> = <r^2> = 0$ pertains for non-zero-mass fermion elementary particles.
3. For a non-zero-mass elementary boson for which σ = +1, vertex sizes feature $<x^2> = 0$ and $<t^2> = 0$.
   3.1. This matches that notion that a projection, for the boson, to D* = 1 dimensions correlates with $<r^2> = 0$ for the fermion.
4. For a non-zero-mass elementary boson for which σ = +1, vertex sizes feature $<(r_2)^2> = 0$. (See Table 2.10.5.)

4.1. This matches that notion that a projection, for the boson, to $D^* = 2$ dimensions correlates with $<r^2> = 0$ for the fermion.

5. For a $\sigma = +1$ non-zero-mass elementary boson, at a vertex, ...
   5.1. Some $(D + 2\nu)_j$ are non-zero. (See Table 2.5.17.)
   5.2. Therefore, for the range associated with the boson, ...
      5.2.1. $<r^2> \neq 0$.
      5.2.2. We posit that that the following pertains for the field.

$$c^2<t^2> \sim <r^2> \sim R_0^2 \tag{2.83}$$

6. For a $\sigma = +1$ non-zero-mass elementary boson, near a vertex, ...
   6.1. We posit that ...
      6.1.1. The boson participates in another interaction vertex.
      6.1.2. People might say that MM1MS1 models correlate with the other vertex being close to the (original) vertex, with closeness correlating with (in a frame of reference that correlates with the pair of vertices) correlating with the following expressions.

$$c^2<t^2> \sim R_0^2 \tag{2.84}$$

$$<r^2> \sim R_0^2 \tag{2.85}$$

$$(<r^2>)^{1/2} \sim R_0 \tag{2.86}$$

   6.2. The leaves the possibility that, for the boson, $<r^2> \sim R_0^2$ correlates with a need for the other interaction (in which the boson participates) to be nearby.
   6.3. People might say that $c^2<t^2> \sim R_0^2$ for the boson correlates with the following approximation for a lifetime of the boson.

$$(<t^2>)^{1/2} \sim R_0/c \tag{2.87}$$

7. For a $\sigma = +1$ zero-mass elementary boson, ...
   7.1. Vertex sizes feature $<r^2> = 0$.
      7.1.1. This matches the notion that each $\{D + 2\nu\}_j$ that correlates with boson is 0.
      7.1.2. This also matches $<r^2> = 0$ for the fermion.
   7.2. Vertex sizes feature $<t^2> = 0$.
      7.2.1. This matches $<r^2> = 0$ for the boson and $<t^2> = 0$ for the fermion.
   7.3. People might say that $R_0 = \infty$.
      7.3.1. This correlates with the boson being able to travel an indefinitely large distance before interacting with another fermion.

Section 2.11 SPATIM symmetries

Section 2.11 discusses symmetries that correlate with the Poincare group and with special relativity. We provide perspective regarding possible uses of symmetries. We introduce SPATIM models correlating with symmetries that can correlate with solutions. We discuss rules for how, for a field that correlates with a combination of a fermion (elementary particle) field and a boson (elementary particle) field, to compute symmetries for the combined field based on the symmetries of the two component fields.

~ ~ ~

This subsection provides perspective about a possible use for symmetries.

People might say that LADDER solutions we show above tend to pertain to INTERN aspects of elementary particles and to INTERN aspects of the fields associated with the elementary particles. These solutions seem not to correlate with concepts such as a direction in which a particle might be traveling or such as a momentum that an observer might attribute to the particle.

We seek a technique for developing solutions that correlate with ENVIRO aspects related to LADDER solutions we develop above.

People might say that this technique correlates with a mathematical notion of mapping LADDER solutions from a quantum energy-momentum space into LADDER solutions in a space that correlates with applications of space-time coordinates.

The technique we develop correlates with the Poincare group, which is a mathematical construct that people correlate with symmetries that correlate with traditional physics models based on special relativity.

~ ~ ~

This subsection provides perspective about symmetries and about this section.

People are familiar with concepts of symmetries. People characterize some symmetries as exact. On a flat tabletop, rotate a circle in a way that the center of the circle stays at its original point. The circle seems the same. People characterize some symmetries as approximate. A mirror image of a person's face may seem somewhat the same as the original image.

Symmetries correlate with constraints. The more exact a symmetry, the smaller (in some sense) the set of symmetric entities.

Various aspects of physics theory feature symmetries. Some sets of symmetries correlate with models that use space-time coordinates that people correlate with concepts of time and space.

Models correlating with special relativity generally exhibit more constraints and more symmetries than do somewhat similar models correlating with Newtonian physics. People characterize symmetries correlating with special relativity by invoking a mathematical construct people call the Poincare group.

This section introduces concepts and notation this monograph correlates with symmetries related to special relativity.

~ ~ ~

This subsection defines SPATIM (or, space-time coordinate) symmetries.

[Physics:] People might say that SPATIM symmetries correlate with mappings from a quantum energy-momentum space into space-time coordinates.

Table 2.11.1 defines and discusses concepts related to some SPATIM symmetries. Work in subsequent sections focuses on (a`;b`)> symmetries and on (a`;b`)< symmetries.

Table 2.11.1 Definitions of and notation for SPATIM (or, space-time coordinate) symmetries for which a` ≤ 1 and b` ≤ 3

1. [Physics:] SPATIM abbreviates space-time coordinate (symmetries).
2. FRERAN SPATIM (or, free-ranging space-time coordinate) symmetries correlate with the Poincare group.
3. People correlate with the Poincare group the following symmetries.
    - 3.1. A 1-generator symmetry.
    - 3.2. Three 3-generator symmetries.
4. [Physics:] People might correlate the Poincare group symmetries with conservation laws and symmetries, as follows.
    - 4.1. The 1-generator symmetry correlates with conservation of energy.
    - 4.2. One 3-generator symmetry correlates with conservation of momentum.
    - 4.3. One 3-generator symmetry correlates with conservation of angular momentum.
        - 4.3.1. For this symmetry, we think that this monograph need not much address the distinction between each of the following two interpretations.
            - 4.3.1.1. Vector symmetry, which correlates with rotation symmetry and reflection symmetry.
            - 4.3.1.2. Pseudovector symmetry, which correlates with rotation symmetry but not reflection symmetry.
    - 4.4. One 3-generator symmetry correlates with the applicability of special relativity (as opposed to, say, Newtonian classical physics). (See Section 2.12.)
        - 4.4.1. For this symmetry, we use the term conservation of RESENE.
            - 4.4.1.1. We use the acronym RESENE to abbreviate the phrase rest energy.

5. For this monograph, we correlate with Poincare group symmetries (and solutions we discuss) the following.
   5.1. The symbol a` denotes the applicable integer number of 1-generator symmetries.
      5.1.1. 0 ≤ a` ≤ 1.
      5.1.2. This monograph does not attempt to correlate this 1-generator symmetry with a mathematical group.
   5.2. The symbol b` denotes the applicable integer number of 3-generator symmetries.
      5.2.1. 0 ≤ b` ≤ 3, for some SPATIM applications.
      5.2.2. Each 3-generator symmetry correlates with an occurrence of the group SU(2).
         5.2.2.1. The group SU(2) correlates with 3 generators.
      5.2.3. This monograph may correlate such a 3-generator symmetry with either of the following.
         5.2.3.1. A P[even – 1]-and-P[even] oscillator pair, for σ = +1.
         5.2.3.2. An E[even]-and-E[even – 1] oscillator pair, for σ = −1.
   5.3. For a solution for which σ = +1, the next item denotes relevant SPATIM symmetries.
      5.3.1. (a`;b`)>.
   5.4. For a solution for which σ = −1, ...
      5.4.1. Aspects of the next item can pertain for FRERAN SPATIM models.
         5.4.1.1. (a`;b`)>.
      5.4.2. The next item denotes relevant COMPAR SPATIM symmetries.
         5.4.2.1. (a`;b`)<.

Table 2.11.2 shows rules we assume pertain to determining ENVIRO symmetries for pairs of fields. (See, for example, the transitions from to Table 2.13.31 to Table 2.13.33 to Table 2.13.34.) In Table 2.11.2, j and k are integers.

Table 2.11.2 Rules for combining SPATIM symmetries

1. For identical or adequately similar fields for which (a`;b`)> pertain, the following combining of symmetries pertains.

   (a`;b`)> + (a`;b`)> = (a`;b`)> (2.88)

2. Otherwise, ...
   2.1. Regarding a`, ...

   (j;...)> + (k;...)> = (...;(j + k))>, for 0 ≤ j, 0 ≤ k, and j + k ≤ 1 (2.89)

   (j;...)< + (k;...)< = (...;(j + k))>, for 0 ≤ j, 0 ≤ k, and j + k ≤ 1 (2.90)

   2.2. Regarding b`, ...

$$(\ldots;j)> + (\ldots;k)> = (\ldots;(j + k))>, \text{ for } 0 \le j,\ 0 \le k, \text{ and } j + k \le 3 \quad (2.91)$$

$$(\ldots;j)< + (\ldots;k)< = (\ldots;(j + k))<, \text{ for } 0 \le j,\ 0 \le k, \text{ and } j + k \le 3 \quad (2.92)$$

2.3. Regarding COMPAR solutions and combining pre-composites (for which $\sigma = -1$ {See Table 2.13.33.}) to form composite particles (for which $\sigma = +1$), ...

$$(1;3)< + (1;3)< = (1;3)> \quad (2.93)$$

3. Regarding $\sigma = +1$, ...

$$(1;3)> + (1;3)> = (1;3)> \quad (2.94)$$

3.1. $(1;3)>$ correlates with the Poincare group.

Section 2.12 Invariances, symmetries, and conservation laws

Section 2.12 discusses the extent to which invariances, symmetries, and conservation laws pertain. We discuss relationships among concepts of invariances, symmetries, and conservation laws. We provide perspective regarding invariances, symmetries, and conservation laws that correlate with SPATIM-correlated phenomena. We show examples of MM1MS1-related symmetries that traditional physics models may not feature. One such symmetry correlates with conservation of fermion generation for some interactions. We suggest categories for conservation laws. We suggest a quantum-related basis for symmetries correlating with conservation of angular momentum. People might say that we provide new perspective regarding approximate conservation laws people associate with the terms C-symmetry, P-symmetry, and T-symmetry. We compare MM1MS1 models and traditional interpretations regarding conservation of charge.

~ ~ ~

This subsection discusses some aspects of conservation laws.

People entwine concepts of conservation laws with models for physics. People might say that, currently, it could be difficult to recast traditional physics in ways that do not include, for example, conservation of energy, conservation of momentum, and conservation of angular momentum. People entwine these three conservation laws with classical-physics theories and with quantum-physics theories.

~ ~ ~

This subsection discusses some aspects regarding invariances, symmetries, and conservation laws.

People entwine conservation laws with mathematics that correlates with models for physics. For example, people correlate conservation of energy with mathematics related to adding, regarding all measurements, an arbitrary constant to a coordinate for time. Or, people correlate conservation of momentum with mathematics related to adding an arbitrary three-dimensional displacement to coordinates for position in space. To the extent such displacements or other changes yield no change in the relevant physics, people say that a system is invariant under the changes.

People correlate notions of symmetry with such invariances.

People might also want to consider invariances regarding choice of observer.

~ ~ ~

This subsection points to results that correlate with and summarize some work in this section.

People might say that, based on context that Table 6.1.5 and Table 6.1.6 set, Table 6.1.7 summarizes some aspects of invariances that correlate with $\sigma = +1$, symmetries that correlate with $\sigma = +1$, conservation laws, and internal properties of elementary particles for which $\sigma = +1$.

~ ~ ~

This subsection provides perspective on math-based models related to uniform displacements.

Perhaps, at least two trains of thought correlate with models related to uniform displacements.

One train of thought correlates with math for which people use the term calculus. This train of thought leads to the series the people might characterize as position, velocity, acceleration, and so forth. People might discuss this series in terms of a number of derivatives (with respect to a coordinate correlating with time) of the expected value of position. A zero number of derivatives correlates with position. One derivative correlates with velocity. And so forth. People find many uses for applications based on this train of thought.

One train of thought correlates with classical-physics measurements for collective phenomena. We consider a series based on expected values of powers of position. The series consists of expected value of 1 (or, position to the zeroth power), expected value of position, expected value of the square of position (perhaps, as measured away from an expected value of position), and so forth. The first item does not change. People can say that this item evolves linearly with time, with a zero rate of change. For an object that does not experience significant forces, the second term changes linearly with time. Under some circumstances, for a collection of objects that diffuse from a small region, the third term evolves linearly with time. Here, each of the first few members of the series has applications, though not necessarily to the same specific physical systems.

People might say that the first train of thought correlates with classical mechanics (and extensions to classical mechanics). People might say that the second train of thought correlates with statistical mechanics.

~ ~ ~

This subsection provides perspective on FRERAN SPATIM invariances, symmetries, and conservation laws.

People tend to correlate applications of invariance with the first train of thought the previous subsection describes. Here, we look at invariance from perspectives that feature measurements.

Consider an observer who makes measurements of an object. The observer might be an astronomer. The object might be a star or galaxy. We focus on observations and deductions the astronomer might make. We do thought experiments based changes in temporal and spatial distances between the astronomer and the object. We idealize that the astronomer has full use of subtle data and not just partial use of optically available data. We idealize, for purposes of discussion, that the universe obeys fully deterministic laws of (say, classical) physics.

For a baseline, assume the astronomer does thorough work based on observations made around some time and from some point. Here, thorough work means, for example, deducing the evolution of the object during a significant period of the object's existence.

Under what circumstances might the astronomer do essentially the same thorough work? At a somewhat different time? Yes. From a somewhat different location? Yes. From a somewhat different velocity of motion perpendicular to a radius from the object to the astronomer? Yes. From a somewhat different velocity along the radius from the object to the astronomer? Yes. From a somewhat different rate of change of the tangential velocity? Yes. From somewhat different rate of change of radial velocity? Yes.

People might say that, correlating with the first train of thought, the first three of the invariances respectively correlate with conservation of energy, conservation of momentum, and conservation of angular momentum. Different observers can observe different energies, momenta, and/or angular momenta. For any one observer, each of conservation of energy, conservation of momentum, and conservation of angular momentum pertains.

For models that do not take into account special-relativistic effects (or, generally, for Newtonian models), for the something that correlates with fourth invariance, we might suggest the term conservation of relative velocity. Any relative velocity can pertain. In particular, speeds faster than light speed might pertain. Different observers can observe different relative velocities. For any one observer, conservation of relative velocity pertains.

For Newtonian physics, each of conservation of momentum, conservation of angular momentum, and conservation of relative velocity would correlate with a 3-vector.

For models based on special relativity, any relative velocity for which the speed does not equal or exceed the speed of light can pertain.

Perhaps, limiting speeds correlates with truncating the series we are discussing.

Perhaps, we can focus on an invariance that involves a quantity all observers can agree on.

We suggest that conservation of RESENE pertains. To the extent special relativity pertains, all observers can deduce the same the same rest energy.

For elementary particles, for rest energy, people use c^2 times rest mass. People might say that conservation of rest energy correlates with special relativity.

Regarding this phenomenon, for the notions of invariance and symmetry, people sometimes use the term boost.

For special relativity, models correlating with conservation of energy and conservation of momentum correlate with a 4-vector. Models correlating with conservation of angular momentum correlate with a 3-vector. Models correlating with conservation of relative velocity (or with conservation of RESENE) can correlate with a 3-vector.

~ ~ ~

This subsection provides perspective regarding invariances or symmetries correlating with SPATIM-related invariances and with FRERAN phenomena.

Table 2.12.1 pertains.

Table 2.12.1 SPATIM-related invariances (or symmetries) for FRERAN phenomena

1. FRERAN SPATIM-related symmetries.
    - 1.1. The symmetries (1;3)> pertain.
        - 1.1.1. In this table, we use notation that decomposes the symbol (1;3)> in to four components. Respectively, the components are the following.

          (1×1; ..., ..., ...)> (2.95)

          (...; 1×3, ..., ...)> (2.96)

          (...; ..., 1×3, ...)> (2.97)

          (...; ..., ..., 1×3)> (2.98)
    - 1.2. People might say that these symmetries correlate with mathematical-modeling possibilities for coordination between energy-momentum representations of fields (and/or particles) and space-time representations of wave functions.
    - 1.3. Invariance with respect to temporal displacement correlates with (1×1; ..., ..., ...)>.
        - 1.3.1. Here, displacement involves a 1-vector construct.

1.4. Invariance with respect to spatial displacement correlates with (...; 1×3, ..., ...)>.
    1.4.1. Here, displacement involves a 3-vector construct.
1.5. Invariance with respect to rotation correlates with (...; ..., 1×3, ...)>.
    1.5.1. Here, displacement involves a 3-vector construct.
1.6. Invariance with respect to boost correlates with (...; ..., ..., 1×3)>.
    1.6.1. Here, displacement involves a 3-vector construct.
    1.6.2. Here, relative velocity features a 3-vector value of velocity.
        1.6.2.1. People might think of the 3-vector as correlating with velocity/c. (See Table 2.12.8.)
    1.6.3. People might say that this invariance correlates with at least one (and perhaps all) of the following.
        1.6.3.1. The classical-physics special-relativistic notion of Lorentz invariance (or, of boost-related symmetry).
        1.6.3.2. That the speed of light, c, is a constant.
        1.6.3.3. Limitations that speeds cannot exceed c.
        1.6.3.4. Concepts of zero-curvature, with respect to notions of curvature inherent in uses of space-time coordinates.

Table 2.12.2 provides symbols we define to correlate with some point and global symmetries.

Table 2.12.2 The symbols #b' and #b" and correlations with point and global symmetries

1. The symbol #b' correlates with point symmetries.
    1.1. For point symmetries (a`;b`)>, the next equation pertains.

$$\#b' = b` \qquad (2.99)$$

        1.1.1. We assume that #b' ≥ 2.
            1.1.1.1. People might say that a value of less than 2 correlates with a violation of at least one of conservation of momentum and conservation of angular momentum.
2. The symbol #b" correlates with global symmetries.
    2.1. For global symmetries (a`;b`)>, the next equation pertains.

$$\#b'' = b` \qquad (2.100)$$

        2.1.1. We assume that #b' ≥ 2.

2.1.1.1. People might say that a value of less than 2 correlates with a violation of at least one of conservation of momentum and conservation of angular momentum.

Table 2.12.3 pertains.

Table 2.12.3 Assumptions regarding #b' and #b"

1. In Section 2.10, we show that elementary-particle interaction vertices correlate with points, with respect to space-time coordinates.
2. In Table 2.12.5, we correlate #b' = 3 with invariance (with respect to observers) of RESENE (or, rest energy). (See, for example, Table 2.11.1.)
3. For much of this monograph, ...
   3.1. We assume the following equation, for all interactions involving (at least) elementary particles for which $\sigma = +1$,

$$\#b' = 3 \qquad (2.101)$$

4. For some of this monograph (See Table 2.13.43.), ...
   4.1. We discuss possibilities correlating with the following equation.

$$\#b' = 2 \qquad (2.102)$$

5. We take up the topic of #b" in Section 6.1 and Section 6.4.

~ ~ ~

This subsection provides perspective regarding the extent to which some conservation laws pertain.

This monograph correlates (1;3)> local symmetries with a set of conservation laws. The set includes at least conservation of energy, conservation of momentum, and conservation of angular momentum.

For ENS48 models, ENS6 models, and ENS1 models, the set of conservation laws pertains for all $\sigma = +1$ phenomena.

Conservation of energy, momentum, and angular momentum correlate with symmetries people might symbolize by (1;2)>. People might raise the question of the extent to which the applicability of (1;3)> symmetries correlates with an additional QP-like conservation law (in addition to conservation of momentum and conservation of angular momentum). People might say that, to the extent $\sigma = +1$ and #b' = 3 pertain, that conservation law pertains to interactions between elementary particles.

MM1MS1 models point to other possibly useful possible new symmetries and conservation laws. Table 2.12.4 shows examples of possibly new symmetries. Elsewhere, we show a more comprehensive concept of conservation of fermion generation. (See Section 2.8.)

Table 2.12.4 Examples of symmetries that traditional ENVIRO models may not feature

1. For σ = −1 models, we symbolize various symmetries by (a`;b`)<.
   1.1. For example, sometimes (1;3)< pertains.
   1.2. These symmetries correlate with COMPAR solutions. (See Table 2.13.31.)
   1.3. People might say that various (a`;b`)< symmetries correlate with (possibly non-traditional) conservation laws or with other mathematical physics pertaining to QCD (or, quantum chromodynamics).
2. We might symbolize symmetries related to 2W-subfamily and to conservation of fermion generation by (1×3, ...; ..., ..., ...)>.
   2.1. Here, the notion of conversation of fermion generation pertains to an interaction between a boson and a spin-1/2 fermion such that the fermion does not change generation.
   2.2. Here, compared to Table 2.12.1, we add a QE-like component. (See Table 2.13.4.)
   2.3. People might say that (1×3, ...; ..., ..., ...)> correlates with (possibly non-traditional) conservation laws or with other mathematical physics pertaining to (at least) interaction vertices in which a 2W particle either ...
      2.3.1. Interacts with a lepton.
      2.3.2. Creates or destroys a matter-lepton-and-antimatter-lepton pair of particles.
3. We might symbolize some possible symmetries related to particles and fields in atoms and molecules by (...; ..., ..., ..., ...)>.
   3.1. To denote these symmetries, we use the symbol (a`;4)>.
   3.2. These symmetries correlate with ATOMOL solutions.
   3.3. Here, compared to Table 2.12.1, we add a QP-like component. (See Section 6.2.)
   3.4. People might say that those (a`; ..., ..., ..., ...)> symmetries correlate with the group SO(4).
   3.5. People might say that that these symmetries ...
      3.5.1. Do not correlate directly with the MM1 meta-model.

~ ~ ~

This subsection discusses categories of conservation laws and provides examples of possible non-traditional conservation laws.

We think that work above points to various categories of conservation laws. We think that work above points to possibilities for non-traditional conservation laws. Table 2.12.5 pertains.

Table 2.12.5 Categories of conservation laws, plus possible non-traditional conservation laws or symmetries

1. FRERAN SPATIM-related $\sigma = +1$ conservation laws.
   1.1. The symmetries (1;3)> pertain.
      1.1.1. In this table, we use notation that decomposes the symbol (1;3)> in to four components. (See Table 2.12.1.) Respectively, the components are ...
         1.1.1.1. (1×1; ..., ..., ...)>.
         1.1.1.2. (...; 1×3, ..., ...)>.
         1.1.1.3. (...; ..., 1×3, ...)>.
         1.1.1.4. (...; ..., ..., 1×3)>.
   1.2. People might say that these conservation laws correlate with mathematical-modeling possibilities for coordination between energy-momentum representations of fields (and/or particles) and space-time representations of wave functions.
   1.3. Conservation of energy correlates with (1×1; ..., ..., ...)>.
      1.3.1. Here, energy involves a 1-vector construct.
   1.4. Conservation of momentum correlates with (...; 1×3, ..., ...)>.
      1.4.1. Here, momentum involves a 3-vector construct.
   1.5. Conservation of angular momentum correlates with (...; ..., 1×3, ...)>.
      1.5.1. Here angular momentum involves a 3-vector construct.
   1.6. Conservation of relative velocity correlates with (...; ..., ..., 1×3)>.
      1.6.1. Here, relative velocity features a 3-vector construct.
   1.7. Conservation of RESENE correlates with (...; ..., ..., 1×3)>.
      1.7.1. People might say that conservation of RESENE correlates with a construct (rest energy) that is invariant with respect to observer.
2. Quantum-property-transfer-related conservation laws.
   2.1. People might say that these conservation laws correlate with aspects regarding interaction vertices.
   2.2. Conservation of charge (regarding transfers of charges between elementary particles and not regarding transfers of charges between, say, atoms) ...
      2.2.1. Correlates with transfers of charge via interactions intermediated by non-zero-charge 2W and 2O bosons.
         2.2.1.1. 2W and 2O bosons transfer charge in integer multiples of $|q_e|/3$.
            2.2.1.1.1. For example, a W^-, in effect, takes -3 units of $|q_e|/3$ from an electron, leaving a MM1MS1-neutrino with zero charge.
      2.2.2. Pertains to all interaction vertices.
         2.2.2.1. For a vertex that does not involve a 2W or 2O boson, no transfer of charge takes place.

3. Extensions to SPATIM-related symmetries.
    - 3.1. Table 2.12.4 shows examples of symmetries that might correlate with new conservation laws.
        - 3.1.1. For example, conservation of generation ...
            - 3.1.1.1. Correlates with interaction vertices in which both ...
                - 3.1.1.1.1. An elementary fermion absorbs or emits an elementary boson.
                - 3.1.1.1.2. The interaction cannot change the generation of the fermion.
            - 3.1.1.2. Pertains to some vertices and not to other vertices. (See Table 2.13.4 and Table 2.13.12.)
        - 3.1.2. For example, people might gain useful perspective regarding the topic of quantum chromodynamics (or, QCD) by considering the extent to which various (a`;b`)< symmetries correlate with conservation laws. (See Section 2.13.)

Table 2.12.6 pertains.

Table 2.12.6 Possible correlation between conservation of generation and variation by generation regarding interaction strength

1. We assume the following.
    - 1.1. For any type of FRERAN interaction between an elementary boson and elementary fermions that differ (essentially only) by generation, ...
        - 1.1.1. The type of interaction exhibits conservation of fermion generation if and only if the strength of such interaction varies with the generation of the fermion.
2. For example, for an MM1MS1-photon, ...
    - 2.1. Interaction strengths are the same (based on charge) for electrons, muons, and tauons.
    - 2.2. Conservation of generation need not pertain.
3. For example, for a graviton, ...
    - 3.1. Interaction strengths vary (with mass) for electrons, muons, and tauons.
    - 3.2. Conservation of generation pertains.
4. For example, for COMPAR interactions, ...
    - 4.1. Conservation of fermion generation need not pertain regarding some interactions involving pairs of quarks and pairs of 2W-subfamily bosons.

People might say that Table 2.12.7 provides an opportunity for research. Regarding 022G2&, conservation of charge pertains. Regarding 244G4&, conservation of generation pertains.

Table 2.12.7 Possible correlation between measurement of a property of an elementary fermion by an elementary boson and conservation of that property

1. To what extent does the following (or a variant of the following) correlation pertain?
   1.1. An interaction vertex in which an elementary boson essentially measures an internal property of an elementary fermion exhibits conservation of that property.

Table 2.12.8 discusses a thought experiment.

Table 2.12.8 A relationship possibly useful for thinking about boost-related symmetry or conservation of relative velocity

1. Consider a set of at least two observers. Consider a non-zero-mass object. Consider a frame of reference.
2. For each of the observers in the set, ...
   2.1. A 3-vector v characterizes the velocity of the observer relative to the object.
      2.1.1. Each component of the vector has units of velocity.
3. Compute a sum of the squares of the various v. Divide by the number of observers.
   3.1. The result is a number with units of square of velocity.
   3.2. Denote that number by $<v^2>$.
4. Compute an average of the velocities.
   4.1. Denote that average by a 3-vector $<v>$.
5. Denote the dot product of that 3-vector with itself by $<v>\cdot<v>$.
   5.1. $<v>\cdot<v>$ is a number with units of square of velocity.
6. The next result pertains.
   6.1. $<v^2> - <v>\cdot<v> \leq c^2$.
7. To the extent at least one observer has non-zero rest energy (or, non-zero rest mass), the next result pertains.

$$<v^2> - <v>\cdot<v> < c^2 \quad (2.103)$$

~ ~ ~

This subsection shows traditional and possible correlations between generalized notions of symmetries and FRERAN SPATIM-related $\sigma = +1$ conservation laws.

Table 2.12.5 shows four aspects of FRERAN SPATIM-related $\sigma = +1$ conservation laws. (See the decomposition, in Table 2.12.1, of (1;3)> symmetries into four components.) Table 2.12.9 pertains. People might say that people can gain insight about overlaps and differences between classical-physics models and quantum-physics models by comparing the first item in the table with subsequent items in the table.

Table 2.12.9 Concepts regarding symmetries and FRERAN SPATIM-related $\sigma = +1$ conservation laws

1. People might say that, regarding classical-physics models, ...
   - 1.1. People correlate conservation of energy with invariance (with respect to adding a constant to a coordinate for time) of models.
   - 1.2. People correlate conservation of momentum with invariance (with respect to adding constants to three coordinates for position) of models.
   - 1.3. People correlate conservation of angular momentum with invariance (with respect to adding constants to coordinates for rotational orientation) of models.
   - 1.4. People might correlate relative-velocity (or, boost-related) invariances (with respect to adding constants to vectors for relative velocities of various observers) with aspects of models. (See Table 2.12.8.)
2. People might say that, regarding quantum-physics models, people might envision possibilities for a set of four invariances.
   - 2.1. Each invariance would correlate, respectively, with quantum aspects of a FRERAN SPATIM-related $\sigma = +1$ conservation law.
   - 2.2. The first two invariances correlate with complementary-variable aspects of uncertainty.
3. The first member of such a set (of four invariances) would be ...
   - 3.1. Invariance under time translation.
     - 3.1.1. The complementary variables are time and energy.
4. The second member of such a set (of four invariances) would be ...
   - 4.1. Invariance under space translation.
     - 4.1.1. The complementary variables are position and momentum.
5. Possibly, the third member of such a set (of four invariances) would ...
   - 5.1. Correlate with the result that $\langle (2J)^2 \rangle - (\langle 2J \rangle)^2$ is, for a traditional quantum base state, a nonnegative integer.
     - 5.1.1. Here, J denotes the quantum operator people correlate with a term like total angular momentum divided by $\hbar$. Here, $(\langle 2J \rangle)^2$ is a vector dot product.

5.1.1.1. People may use the formula J = L + S, in which ...

5.1.1.1.1. L correlates with orbital angular momentum divided by ħ.

5.1.1.1.2. S correlates with MM1MS1 uses of the symbol S.

5.1.2. People might say that this relationship correlates with a concept somewhat related to uncertainty.

5.1.2.1. People might use the term self-uncertainty.

5.1.2.2. People might say that, from a quantum mechanical point of view, total angular momentum does not have a complementary variable.

6. Possibly, the fourth member of such a set (of four invariances) would ...

6.1. Correlate with conservation of RESENE.

6.2. Correlate with concepts Table 2.12.8 presents.

6.2.1. People might say that the relationship $<v^2> - <v>\cdot<v> < c^2$ correlates with a concept somewhat related to uncertainty.

6.2.1.1. People might use the term mutual certainty.

6.2.1.2. People might say that relative velocity does not have complementary variable.

7. For an elementary particle, ...

7.1. People might say that the following statements correlate with the respective four invariances.

7.1.1. Conservation of energy correlates with energy, which is an observer-specific quantity.

7.1.2. Conservation of momentum correlates with momentum, which is an observer-specific quantity.

7.1.3. Conservation of angular momentum correlates with the next expression, which (for S being a quantum operator and $(<2S>)^2$ being a vector dot product) is an observer-independent quantity.

$$<(2S)^2> - (<2S>)^2 \qquad (2.104)$$

7.1.4. Conservation of RESENE correlates with RESENE, which is an observer-independent quantity.

~ ~ ~

This subsection contrasts discussion above with some traditional interpretations of (1;3)> symmetries.

Some traditional discussion of what we call (1;3)> symmetries features a different interpretation of the (...; ..., ..., 1×3)> component. A traditional interpretation may include the terms CPT-symmetry and Lorentz invariance. (This contrasts with the term Lorentz invariance symmetry, which people might correlate with (1;3)> symmetries.) A traditional interpretation of Lorentz invariance may correlate each of the 3 generators with a different term from a list of three terms - charge inversion symmetry, position inversion symmetry, and time inversion symmetry. People may extend such discussion to include three approximate conservation laws. People use the terms invariance under charge conjugation, conservation of parity, and invariance under time reversal. Alternative names for these approximate conservation laws are, respectively, are C-symmetry (with C for charge), P-symmetry (with P for parity), and T-symmetry (with T for time).

~ ~ ~

This subsection provides a non-traditional interpretation of some aspects of C-symmetry, P-symmetry, and T-symmetry.

We use the term CPT-related symmetries to denote symmetries that include C-symmetry, P-symmetry, CP-symmetry, T-symmetry, and so forth.

Table 2.12.10 defines and interprets oscillator swap symmetries. People might say that (at least as shown in the table) these possibly physics-relevant symmetries correlate with swaps of values of various N(..) and not with swaps of oscillators. People might say that some aspects (such as portions of item 8) of this table are speculative.

Table 2.12.10 Oscillator swap symmetries

1. For an oscillator pair E[even]-and-E[even − 1] or an oscillator pair P[even − 1]-and-P[even], an oscillator swap involves the following.
   - 1.1. Swap the values of N(..) for the two oscillators.
     - 1.1.1. For example, for a case of pair E2-and-E1, ...
       - 1.1.1.1. Make the swap N(E2) $\leftrightarrow$ N(E1).
2. For elementary particles for which #E $\leq$ 2 and #P $\leq$ 2, the following provide notations for each of the only two possibly relevant oscillator swaps.
   - 2.1. Each of CQE-swap and T-swap denotes the swap correlating with the E2-and-E1 oscillator pair.
   - 2.2. Each of CQP-swap and P-swap denotes the swap correlating with the P1-and-P2 oscillator pair.
3. The following concepts pertain.
   - 3.1. For quarks specifically (and, thus, all #P $\leq$ 2 particles generally), ...
     - 3.1.1. C-symmetry (in traditional physics) correlates with the combination of CQE-swap and CQP-swap.
   - 3.2. P-symmetry correlates with P-swap.

   3.3. CPT-symmetry correlates with the combination of all four of CQE-swap, CQP-swap, P-swap, and T-swap.
   3.4. T-symmetry correlates with T-swap.
   3.5. CPT-symmetry pertains.
4. People might say that, based on the previous item, ...
   4.1. The term time reversal is misleading.
   4.2. The term time inversion might be misleading.
5. To extend the discussion to include all elementary particles, ...
   5.1. To maintain the relevance of G-family solutions, ...
      5.1.1. The notion of P-swap extends to (also) include oscillator pairs P3-and-P4, P5-and-P6, and P7-and-P8.
         5.1.1.1. For example, without P3-and-P4 swap, solutions correlating with the 2G24& particle become irrelevant.
   5.2. To maintain the relevance of Y-family solutions, ...
      5.2.1. The notion of T-swap extends to (also) include oscillator pair E4-and-E3.
         5.2.1.1. For example, without E4-and-E3 swap, differences between 4Y-subfamily matter-correlated solutions and 4Y-subfamily antimatter-correlated solutions break down.
   5.3. To maintain CPT-symmetry, ...
      5.3.1. CQE-swap extends to equal T-swap.
      5.3.2. CQP-swap extends to equal P-swap.
6. We can define C-swap to equal CQE-swap plus CQP-swap.
   6.1. C-swap equals T-swap plus P-swap.
   6.2. C-symmetry correlates with C-swap.
7. The G-family splits into sets. (See Section 3.2 and Section 3.3. In this table, we feature %68even solutions and deemphasize %68odd solutions.)
   7.1. The 1-swap G-family set ...
      7.1.1. Features particles for which the sub-list % has one of element.
      7.1.2. Includes 2G2& and 4G4&.
         7.1.2.1. Thereby, related symmetries include symmetries related to electromagnetism (perhaps, other than as correlates with traditional elementary-particle magnetic dipole moment) and gravity.
   7.2. The 2-swap G-family set ...
      7.2.1. Features particles for which the sub-list % has two elements.
      7.2.2. Includes 2G24& and 2G68&.
         7.2.2.1. Thereby, related symmetries may include symmetries related to traditional elementary-particle magnetic dipole moment.

- 7.3. The 3-swap G-family set ...
  - 7.3.1. Features particles for which the sub-list % has three elements.
  - 7.3.2. Includes 4G268& and 2G468&.
- 7.4. The 4-swap G-family set ...
  - 7.4.1. Features particles for which the sub-list % has four elements.
  - 7.4.2. Includes 4G2468&.

8. Perhaps, ...
   - 8.1. Physics does not correlate electromagnetism with the breaking of CPT-related symmetries.
   - 8.2. Physics does not correlate gravitation with the breaking of CPT-related symmetries.
   - 8.3. For ENS48 models and for ENS6 models, ...
     - 8.3.1. Within an ensemble, interactions mediated by G-family bosons adhere to all CPT-related symmetries.
     - 8.3.2. Across ensembles, interactions mediated by 2-swap, 3-swap, and 4-swap G-family bosons adhere to CPT-symmetry.
     - 8.3.3. Across ensembles, interactions mediated by 2-swap, 3-swap, and 4-swap G-family bosons might appear (with respect to each of the two relevant ensembles) not adhere to some CPT-related symmetries.
       - 8.3.3.1. Such apparent violations might correlate with apparent violations of at least one of C-swap symmetry, P-swap symmetry, or T-swap symmetry.
     - 8.3.4. Existence of apparent violations of at least one of C-swap, P-swap, or T-swap symmetry might suffice (assuming symmetry under C-swap plus P-swap plus T-swap) to imply ...
       - 8.3.4.1. Existence of apparent violations of each of C-swap, P-swap, and T-swap symmetry.
     - 8.3.5. Existence of apparent violations of at least one of C-swap, P-swap, or T-swap symmetry might correlate with ...
       - 8.3.5.1. Existence of apparent violations of each of P-symmetry, C-symmetry, T-symmetry, and CP-symmetry.
9. People might say that work in this monograph correlates with the term approximate (in the traditional physics notions that each of the following is an approximate symmetry) ...
   - 9.1. Conservation of parity.
   - 9.2. Invariance under charge conjugation.
   - 9.3. Invariance under time reversal. (Here, we use a traditional term.)
   - 9.4. CP-symmetry.

~ ~ ~

This subsection contrasts discussion above with traditional interpretations of conservation laws pertaining to charge.

In traditional physics, people correlate the conservation-of-charge conservation law with a symmetry people associate with the term Gauge invariance.

This monograph correlates conservation of charge with a lack of interactions that can change overall net charge.

Section 2.13 FRERAN and COMPAR applications of LADDER models

Section 2.13 discusses LADDER models that correlate with internal properties of elementary particles, motions of particles, and interactions between elementary particles. We define INSSYM7, an instance-related symmetry. We provide perspective about transforming INTERN LADDER solutions into SPATIM LADDER solutions. We transform INTERN LADDER solutions so as to include SPATIM symmetries and INSSYM7 symmetry. People might say that, for some interpretations, INSSYM7 symmetry correlates with the possibility the universe includes multiple ensembles, with each ensemble including a set of elementary particles that includes at least the Standard Model set of elementary particles. People might say that one ensemble correlates with ordinary matter, five ensembles correlate with dark matter, some ensembles possibly correlate with dark-energy stuff, and some elementary particles do not correlate with an ensemble. People might say that we correlate G-family solutions with various symmetries, with various force laws, and with the concept of the span on an instance of a force. People might say that COMPAR-related symmetries and solutions correlate with phenomena correlating with QCD (or, quantum chromodynamics) and with aspects of composite particles. We discuss possibilities for states of matter (other than as found in hadrons) that involve quarks.

~ ~ ~

This subsection provides perspective about this section.

This section discusses symmetries that pertain to solutions that correlate with fields that correlate with elementary particles. One such symmetry is an instance-related symmetry we call INSSYM7.

Table 2.13.1 lists two concepts that people might correlate with INSSYM7.

Table 2.13.1 Two concepts possibly correlating with INSSYM7

1. INSSYM7 pertains to mathematics underlying models this monograph shows.
2. INSSYM7 pertains to symmetry correlating with #ENS, which is the number of ensembles. (See Table 2.1.2.)

~ ~ ~

This subsection provides perspective about elementary particles and composite particles.

[Physics:] Some elementary particles, such as electrons and photons, can move over long distances. People apply the term free-ranging to such particles and to some other particles, such as the Z and W bosons, that, under normal circumstances, have limited ranges.

[Physics:] Some elementary particles have been found only in close association with other particles. Quarks and gluons provide examples. People apply the term non-free-ranging to these elementary particles.

For elementary particles, free-ranging correlates with $\sigma = +1$ and non-free-ranging correlates with $\sigma = -1$. Given that, under normal circumstances, the ranges of W- and H-family particles have similarities to the ranges of elementary bosons for which $\sigma = -1$, people might prefer, in some circumstances, to use terminology based on σ rather than use terminology related to free-ranging.

[Physics:] People apply the term free-ranging to some particles, such as composite particles, that contain non-free-ranging particles. Protons and pions are examples of composite particles.

For convenience, in tables, we list composite particles as correlating with $\sigma = +1$.

~ ~ ~

This subsection provides perspective about constructing FRERAN SPATIM LADDER solutions from INTERN LADDER solutions.

Table 2.13.2 pertains. For cases in which $\sigma = +1$, we assume that QE-like considerations correlate with instance symmetry and QP-like considerations correlate with b`. For cases in which $\sigma = -1$, we assume that QP-like considerations correlate with instance symmetry and QE-like considerations correlate with b`. (Regarding instance symmetry, see, for example, Table 2.7.1 and Table 2.13.1.)

Table 2.13.2 Steps for constructing a FRERAN SPATIM LADDER solution from one INTERN LADDER solution

1. We anticipate adding (to the relevant INTERN LADDER solution) ...
   1.1. 3 QE-like oscillator pairs.
   1.2. 3 QP-like oscillator pairs.

2. We anticipate that each 3-generator SPATIM symmetry correlates with an SU(2) group that correlates with an instance of an added oscillator pair. (See Table 2.11.1 and Table 2.12.1.)
   2.1. For σ = +1, ...
      2.1.1. We assume such an oscillator pair is QP-like.
   2.2. For σ = −1, ...
      2.2.1. We assume such an oscillator pair is QE-like.
3. We anticipate ...
   3.1. Setting each of the 12 new N(..) to the same value.
   3.2. For solutions not correlating with the QIRD-families, ...
      3.2.1. The 12 values of N(..) equal the value of ...
         3.2.1.1. N(E0), for cases for which σ = +1.
         3.2.1.2. N(P0), for cases for which σ = −1.
   3.3. For solutions correlating with the QIRD-families, ...
      3.3.1. Table 2.13.9 shows results.
4. We add the 6 oscillator pairs.
   4.1. Doing so effects the transition from an INTERN LADDER solution to a FRERAN SPATIM LADDER solution.
   4.2. Work above in this table results in Œ = 0 for the FRERAN SPATIM LADDER solution.
5. Nominally, we assume #b' = 3.

~ ~ ~

This subsection shows, for all families except the G-family, FRERAN SPATIM ground-state LADDER solutions and instance-related symmetries.

For σ = +1 non-G-family ground-state solutions, we follow steps Table 2.13.2 shows.

Table 2.13.3 shows ground-state FRERAN SPATIM LADDER solutions for non-G-family solutions for which σ = +1. (Regarding the symbol ~, see Table 2.3.11.)

Table 2.13.3 σ = +1 non-G-family ground-state FRERAN SPATIM LADDER solutions

E	E	E	E	E	E	E	E	E	P	P	P	P	P	P	P	P	P	P	P	(σ = +1)
8R	8L	6R	6L	4R	4L	2R	2L	0	0	2L	2R	4L	4R	6L	6R	8L	8R	AL	AR	Sub-family
		0	0	0	0	0	0	0	0	0	0	0	0	0	0					0H0
		0	0	0	0	0	0	0	0	~	~	0	0	0	0	0	0			1C
		−1	−1	−1	−1	−1	−1	−1	−1	~	~	−1	−1	−1	−1	−1	−1			1N
0	0	0	0	0	0	0	0	0	0	0	0	0	0	0	0	0	0			2W
		0	0	0	0	0	0	0	−1	0	0	~	~	0	0	0	0	0	0	3N

For each row in Table 2.13.3, the rightmost 3 open oscillator pairs correlate with 3 occurrences of SU(2) symmetries. These six oscillators are QP-like. We interpret each pair as correlating with a contribution to SPATIM symmetries. Thus, for each row in the table, #b' = 3. For each row in the table, the 7 QE-like oscillators E6R through E0 have identical values. INSSYM7 pertains. The term INSSYM abbreviates the term instance-related symmetry. For each row, a` = 1. For each of the rows in the table, (1;3)> symmetries pertain.

Table 2.13.4 restates Table 2.13.3. Table 2.13.4 uses the symbol 'I to point out aspects leading to occurrences of INSSYM7. For each row, the symbol 'I appears exactly 7 times. This table uses the symbol 'S to point out aspects correlating with the b` portion of (1;3)> symmetries. Here, for each row, b` = 3 pertains and the symbol 'S appears 6 times. (Here, use of the letter S correlates with the acronym SPATIM.) For each row, each of the oscillators for which N(..) = 'S does not pertain for the INTERN solution. For the 2W-subfamily, for each of the oscillators E8R and E8L, we assign the symbol 'G. This pair of 'G symbols correlates with an additional 3-generator (SU(2)-related) symmetry. [Physics:] People might say that this 3-generator symmetry correlates with conservation of generation for interactions between leptons and 2W bosons. For example, in an interaction in which the emission of a W^- boson converts a muon into a neutrino, the neutrino is a muon-neutrino and not, for example, an electron-neutrino. People consider muons to be generation-2 leptons. Muon-neutrinos are generation-2 leptons. (See, for example, Section 2.8.)

Table 2.13.4 A second depiction of σ = +1 non-G-family ground-state FRERAN SPATIM LADDER solutions

E 8R	E 8L	E 6R	E 6L	E 4R	E 4L	E 2R	E 2L	E 0	P 0	P 2L	P 2R	P 4L	P 4R	P 6L	P 6R	P 8L	P 8R	P AL	P AR	(σ = +1) Sub-family
		'I	'I	'I	'I	'I	'I	'I	0	'S	'S	'S	'S	'S	'S					0H0
		'I	'I	'I	'I	'I	'I	'I	0	~	~	'S	'S	'S	'S	'S	'S			1C
		'I	'I	'I	'I	'I	'I	'I	−1	~	~	'S	'S	'S	'S	'S	'S			1N
'G	'G	'I	'I	'I	'I	'I	'I	'I	0	0	0	'S	'S	'S	'S	'S	'S			2W
		'I	'I	'I	'I	'I	'I	'I	−1	0	0	~	~	'S	'S	'S	'S	'S	'S	3N

Table 2.13.5 restates Table 2.13.4. In Table 2.13.5, for each row, each of the three occurrences of "S correlates with the type of use we make of a pair of 'S in Table 2.13.4. For each row, b` = 3. Each "I correlates with the type of use we make of a pair of 'I in tables above. Each ~~ correlates with the type of use we make of a pair of ~ in Table 2.13.4. (See, for example, Table 2.3.11.) Each 00 correlates with the type of use we make of a pair of 0 in tables above. For the 2W row, the occurrence of "G correlates with the type of use we make of a pair of 'G in Table 2.13.4.

Table 2.13.5 A third depiction of σ = +1 non-G-family ground-state FRERAN SPATIM LADDER solutions

E A9	E 87	E 65	E 43	E 21	E 0	P 0	P 12	P 34	P 56	P 78	P 9A	(σ = +1) Subfamily
		"I	"I	"I	'I	0	"S	"S	"S			0H0
		"I	"I	"I	'I	0	~~	"S	"S	"S		1C
		"I	"I	"I	'I	−1	~~	"S	"S	"S		1N
	"G	"I	"I	"I	'I	0	00	"S	"S	"S		2W
		"I	"I	"I	'I	−1	00	~~	"S	"S	"S	3N

Table 2.13.6 shows instance-related symmetry and FRERAN SPATIM symmetries for solutions for which σ = +1. [Physics:] These solutions correlate with fields for all the free-ranging elementary particles, except for G-family particles, this monograph discusses.

Table 2.13.6 Instance-related symmetry and FRERAN SPATIM symmetries for σ = +1 non-G-family ground-state ENVIRO LADDER solutions

(σ = +1)	Instance-related symmetry	FRERAN SPATIM	Solution subfamily
	INSSYM7	(1;3)>	0H0
	INSSYM7	(1;3)>	1C
	INSSYM7	(1;3)>	1N
	INSSYM7	(1;3)>	2W
	INSSYM7	(1;3)>	3N

For elementary particles for which σ = −1, Table 2.13.7 pertains.

Table 2.13.7 Correlation of aspects of FRERAN modeling with aspects of phenomena involving elementary particles for which σ = −1

1. People might say that people have yet to observe FRERAN phenomena for aspects of nature that we correlate with σ = −1.
2. People might say that people use, for aspects of nature that we correlate with σ = −1, traditional models that correlate with symmetries that correlate with special relativity. (See Table 6.3.1.)
   2.1. Examples include models that correlate with Feynman diagrams.
3. People might say that those models correlate with modeling we correlate with σ = +1.
4. We think FRERAN models pertain to some aspects of modeling elementary particles that we correlate with σ = −1.
   4.1. COMPAR models pertain to some other aspects of modeling elementary particles that we correlate with σ = −1.

For elementary particles for which σ = −1, we follow steps Table 2.13.2 shows.

Table 2.13.8 shows results (paralleling results Table 2.13.5 shows) for σ = −1 boson ground-state FRERAN SPATIM LADDER solutions. Here, we use the symbol ..,.. to indicate two occurrences of the .. symbol. [Physics:] People might say that one occurrence of "G correlates with conservation of fermion generation for interactions with non-lepton spin-1/2 elementary fermions. People might say that two occurrences of "G correlates with conservation of fermion generation for interactions with non-lepton spin-3/2 elementary fermions. (See, for example, Table 2.13.12.)

Table 2.13.8 σ = −1 boson ground-state FRERAN SPATIM LADDER solutions

E A9	E 87	E 65	E 43	E 21	E 0	P 0	P 12	P 34	P 56	P 78	P 9A	(σ = −1) Subfamily
		"S	"S	"S	0	'I	"I	"I	"I			00
	"S	"S	"S	..,..	..	'I	"I	"I	"I	"G		2Y
	"S	"S	"S	00	0	'I	"I	"I	"I	"G		20
"S	"S	"S	..,..	..,..	..	'I	"I	"I	"I	"G	"G	4Y
"S	"S	"S	00	00	0	'I	"I	"I	"I	"G	"G	40

Table 2.13.9 shows results for σ = −1 fermion ground-state FRERAN SPATIM LADDER solutions. (See Table 2.13.2.) Here, we use the symbol −1,−1 to indicate two occurrences of the symbol −1. Here, regarding the symbol ~,'I, the symbol ~ pertains to the one oscillator (in the relevant oscillator pair) for which N(..) ≠ 'I. (For each row in the table, there are 4 cases. Here, 4 = 2 × 2, with one factor of 2 correlating with the symbol ~~. For each value of ~~, 2 solutions pertain. For example, for the 1Q-subfamily, for one solution, ~ correlates with oscillator P1 and 'I correlates with oscillator P2. More specifically, 'I = N(P2). For the other solution, ~ correlates with oscillator P2 and 'I correlates with oscillator P1. More specifically, 'I = N(P1).)

Table 2.13.9 σ = −1 fermion ground-state FRERAN SPATIM LADDER solutions

E A9	E 87	E 65	E 43	E 21	E 0	P 0	P 12	P 34	P 56	P 78	P 9A	(σ = −1) Subfamily
	"S	"S	"S	~~	0	0	~,'I	"I	"I	"I		1Q
	"S	"S	"S	~~	−1	−1	~,'I	"I	"I	"I		1R
"S	"S	"S	~~	00	0	0	00	~,'I	"I	"I	"I	3Q
"S	"S	"S	~~	−1,−1	0	0	−1,−1	~,'I	"I	"I	"I	3I
"S	"S	"S	~~	00	−1	−1	00	~,'I	"I	"I	"I	3R
"S	"S	"S	~~	−1,−1	−1	−1	−1,−1	~,'I	"I	"I	"I	3D

Table 2.13.10 shows instance-related symmetry and FRERAN SPATIM symmetries for solutions for which σ = −1. [Physics:] These solutions correlate with fields for all the non-free-ranging elementary particles this monograph discusses. (Starting with discussion pertaining to Table 2.13.26, we discuss, for σ = −1 solutions, COMPAR SPATIM symmetries.)

Table 2.13.10 Instance-related symmetry and FRERAN SPATIM symmetries for σ = −1 ground-state FRERAN SPATIM LADDER solutions

(σ = −1)	Instance-related symmetry	FRERAN SPATIM	Solution subfamily
	INSSYM7	(1;3)>	0O
	INSSYM7	(1;3)>	2YO
	INSSYM7	(1;3)>	4YO
	INSSYM7	(1;3)>	1QR
	INSSYM7	(1;3)>	3QIRD

~ ~ ~

This subsection provides perspective about the next few subsections.

Above, for each family (of elementary particles that work in this monograph could correlate with) except the G-family, we show models correlating with INSSYM7 symmetry and (1;3)> FRERAN SPATIM symmetries.

The next few subsections explore, for the G-family, various aspects and possible symmetries.

~ ~ ~

This subsection introduces the concept of channels.

[Physics:] Define a number β via the formula $(4/3)(\beta^6)^2$ = (electromagnetic repulsion between two electrons) / (gravitational attraction between the same two electrons). (See Table 3.9.2.) Numerically, β × (the rest mass of an electron) may provide a more accurate estimate (and a smaller standard deviation) for the rest mass of a tauon than do experiments (as of 2014). (See Table 3.9.5.) This monograph interprets the exponent 2 as correlating with the two interaction vertices in which a G-family boson participates. This monograph interprets β^6 as providing a per-channel-per-vertex ratio of vertex strengths - in terms of the strength of electromagnetism (in general) and the strength of gravitation (in general). (See, also, Table 3.9.3, Table 3.9.20, Table 3.9.21, and Table 3.13.4.) This thinking correlates with concepts related to channels. Perhaps, people will think of channels as parallel paths connecting the two vertices in an interaction between two fermions. MM1MS1-photon (or, 2G2&) vertices (and interactions) have four channels. Gravitational (or, 4G4&) vertices (and interactions) have three channels. #P = 6 G-family vertices would correlate with two channels. #P = 8 G-family vertices would correlate with one channel. #P = A (or, #P = [10]) vertices would have zero strength. (Table 3.3.3 shows definitions and uses of the symbols 2G2& and 4G4&.)

This monograph suggests two possibilities for modelling the number of channels. We correlate one possibility with the acronym EACUNI. We correlate one possibility with the acronym SOMMUL. Table 2.13.11 pertains. This monograph does not explore much the possibility that notions of channels pertain to other than G-family elementary particles.

Table 2.13.11 Two ways to model channels related to G-family elementary particles

1. Aspects common to EACUNI models and SOMMUL models.
   1.1. For each G-family elementary particle, ...
      1.1.1. A symbol of the form nG%& denotes the particle. (See, for example, Table 3.3.3.)
      1.1.2. One of more epnG%& solutions exist (with e = #E ≤ 6 and with p = #P ≤ 8).
      1.1.3. Either (EACUNI or SOMMUL) set of models correlates with the number of channels.
   1.2. Table 3.2.1 provides an example, based on 2G2&.
   1.3. Table 3.2.8 provides an example, based on 4G4&.
2. Aspects specific to EACUNI models.
   2.1. The acronym EACUNI abbreviates the words each G-family elementary particle correlates with a unique G-family solution.
   2.2. For a G-family particle nG%&, ...
      2.2.1. The solution with the least value of #P correlates with the particle.
         2.2.1.1. That solution also has the least value of #E.
      2.2.2. Any other solutions do not correlate with the particle.
   2.3. Table 2.13.12 shows ...
      2.3.1. How to represent channels.
      2.3.2. How to count the number of channels.
3. Aspects specific to SOMMUL models.
   3.1. The acronym SOMMUL abbreviates the words some G-family particles correlate with multiple G-family solutions (or, more than one G-family solution).
   3.2. For a G-family particle nG%&, ...
      3.2.1. Each epnG%& solution (with e = #E ≤ 6 and with p = #P ≤ 8) correlates with the particle.
      3.2.2. The number of such solutions equals the number of channels.

Work below (in this section specifically and in this monograph generally) may use EACUNI models and tends to deemphasize SOMMUL models.

~ ~ ~

This subsection starts development of G-family ground-state ENVIRO LADDER solutions.

We use the limits #E ≤ 6 and #P ≤ 8. We use the possible relevance of SU(7) and the possible relevance of SU(17). (See Table 8.2.1.)

In Table 2.13.12, we show 17 QE-like oscillators and 17 QP-like oscillators. We introduce the symbols #E" and #P". We let #E" = #P" = [G] = 16 denote limits regarding oscillators.

We assume that #b' = 3 for each G-family elementary particle. (See Table 2.12.3.)

Table 2.13.12 shows ground-state FRERAN SPATIM LADDER solutions for G-family subfamilies. (Compare with Table 2.8.2.) Each x denotes a closed oscillator pair. For each row in the table, the number of such pairs of closed oscillators equals (1/2) (#P – #E). For each row in the table, based on the rightmost four QE-like columns (that is, column E6-and-E5 through column E0), one might think that, for each solution, INSSYM7 pertains. Per the rightmost three QP-like columns (that is, column PB-and-PC through column PF-and-PG) of Table 2.13.12, #b' = 3 pertains for each row. The symbol 00 symbolizes a 0 for each of the two oscillators to which the column pertains. The table uses EACUNI models. QP-like use of the symbol * correlates with the concept of channels. The number of QP-like uses of the symbol * equals the number of channels. (For each row in Table 2.13.12, the number of QE-like instances of the symbol * equals the number of QP-like instances of the symbol *.) People might say that each * correlates with a closed oscillator pair. (See Table 2.2.8.) The symbol "G denotes QE-like instances of 00 for oscillators for which "I does not pertain in ground-state FRERAN SPATIM LADDER solutions. (See Table 2.8.2.)

Table 2.13.12 Channels and (bases for) FRERAN SPATIM symmetries for G-family ground-state LADDER solutions (assuming EACUNI models and that #b' = 3 pertain for each solution or subfamily)

E GF	E ED	E CB	E A9	E 87	E 65	E 43	E 21	E 0	P 0	P 12	P 34	P 56	P 78	P 9A	P BC	P DE	P FG	Solution or subfamily
x	*	*	*	*	"I	"I	"I	'I	−1	00	*	*	*	*	@'	@'	@'	022G2&
x	*	*	*	"G	"I	"I	"I	'I	−1	00	00	*	*	*	@'	@'	@'	24..G
x	x	*	*	*	"I	"I	"I	'I	−2	00	00	*	*	*	@'	@'	@'	042G24&
x	*	*	"G	"G	"I	"I	"I	'I	−1	00	00	00	*	*	@'	@'	@'	46..G
x	x	*	*	"G	"I	"I	"I	'I	−2	00	00	00	*	*	@'	@'	@'	26..G
x	x	x	*	*	"I	"I	"I	'I	−3	00	00	00	*	*	@'	@'	@'	064G246&
x	*	"G	"G	"G	"I	"I	"I	'I	−1	00	00	00	00	*	@'	@'	@'	68..G
x	x	*	"G	"G	"I	"I	"I	'I	−2	00	00	00	00	*	@'	@'	@'	48..G
x	x	x	*	"G	"I	"I	"I	'I	−3	00	00	00	00	*	@'	@'	@'	28..G
x	x	x	x	*	"I	"I	"I	'I	−4	00	00	00	00	*	@'	@'	@'	084G2468&

~ ~ ~

This subsection discusses models for interpreting aspects of FRERAN SPATIM symmetries for G-family ground-state LADDER solutions.

Table 2.13.13 discusses some math possibly related to interpretations of INSSYM7.

Table 2.13.13 Math related to possible interpretations of INSSYM7

1. INSSYM7 correlates with SU(7).
2. The following expressions correlate with subgroups of SU(7).

$$SU(7) \supset SU(5) \times SU(2) \times U(1) \quad (2.105)$$

$$SU(7) \supset SU(3) \times SU(4) \times U(1) \quad (2.106)$$

3. The following pertain.
   - 3.1. For SU(7), ...
     - 3.1.1. 48 generators pertain.
     - 3.1.2. 48 / 48 = 1.
   - 3.2. For SU(5), ...
     - 3.2.1. 24 generators pertain.
     - 3.2.2. 48 / 24 = 2.
   - 3.3. For SU(3), ...
     - 3.3.1. 8 generators pertain.
     - 3.3.2. 48 / 8 = 6.
   - 3.4. For no symmetry, ...
     - 3.4.1. 48 / 1 = 48.

Table 2.13.14 discusses possible physics-relevance of ensembles. Motivation for this table includes physics data that the first item in the table discusses.

Table 2.13.14 Possible correlation between observations and instance-related symmetries

1. [Physics:] People say that gravity intermediates interactions between ordinary matter and dark matter. Regarding the recent universe, inferred ratios of dark-matter density of the universe to ordinary-matter density of the universe exceed 5, but not by much. Possibly, inferred ratios (pertaining to earlier times) have not varied much over most of the history of the universe. (See Section 4.3.)
2. Gravitational interactions occur between ordinary matter and dark matter.
3. Perhaps, a QE-like SU(3) symmetry correlates with gravitons (and gravity). (See the 24..G row in Table 2.8.2. There, #E = 2.)
   - 3.1. 244G4& correlates with gravitons. (See Section 3.2 and Section 3.3.)
4. Perhaps, the 8 generators correlating with that symmetry correlate with the 8 generators correlating with an item in Table 2.13.13.
5. Perhaps the number 6 (correlating with that item in Table 2.13.13) correlates with 6 similar units of stuff.
6. Perhaps, dark matter correlates (in essence) with 5 copies of ordinary matter.
7. Perhaps, the universe includes at least 6 ensembles. (See Table 2.1.2.)
   - 7.1. Numbers of physics-relevant ensembles might be (See Table 2.13.13.) either of ...

7.1.1. 48.

7.1.2. 6.

7.2. Here, we ignore the following (because each is less than 6). (Elsewhere, we discuss ENS1 models, which feature just 1 physics-relevant ensemble. Generally, we deemphasize the case that would feature exactly 2 physics-relevant ensembles.)

7.2.1. 2.

7.2.2. 1.

8. Perhaps, one of ENS48 models and ENS6 models pertains.
9. For each row in Table 2.8.2 for which #E = 2, 4, or 6, ...

9.1. Let #E` = #E.

9.2. Perhaps, the number of generators of SU(#E` + 1) correlates with a number of physics-relevant instances of each solution that correlates with the row.

9.2.1. Above, this table uses #E` = 2 for 244G4&.

10. The notions that 8 instances of 244G4& are physics-relevant and that each instance correlates with 6 ensembles correlate with ...

10.1. 48 instances of 022G2& are physics-relevant.

10.2. For 022G2&, ...

10.2.1. #E = 0.

10.2.2. We can assume ...

10.2.2.1. #E` = 6.

10.2.2.2. ENS48 models pertain.

11. The notions that 1 instance of 244G4& is physics-relevant and that the one instance correlates with 6 ensembles correlate with ...

11.1. For gravitons, ...

11.1.1. #E` = 2.

11.1.2. Only 1 of the 8 instances of 244G4& is physics-relevant.

11.2. 6 instances of 022G2& are physics-relevant.

11.3. For 022G2&, ...

11.3.1. #E = 0.

11.3.2. We can assume ...

11.3.2.1. #E` = 6.

11.3.2.2. Only 6 of the possible 48 instances of 022G2& are physics-relevant.

11.3.2.3. ENS6 models pertain.

~ ~ ~

[Physics:] This subsection shows a method for cataloging, relative to ordinary matter, instances (correlating with ENS48 models) of elementary particles and composite particles.

Traditional physics discusses at least three possible types of stuff - ordinary matter (or, baryonic matter), dark matter, and (possibly) dark-energy stuff. (See Section 4.3.)

So far, we interpret, for ENS48 models, INSSYM7 as correlating with the existence of 48 instances of each elementary particle (other than G-family particles) and of each composite particle. We discuss the existence of 48 ensembles. Each ensemble correlates with an instance of 2G2&, of each elementary particle other than non-2G2& G-family particles, and of each composite particle.

One ensemble correlates with ordinary matter. Table 2.13.15 shows a means to catalog ensembles and to hypothetically include (in such a catalog) various instances of G-family particles other than 2G2&. Table 2.13.16 discusses possible concerns regarding these definitions and adds some definitions. Later, we use the term G#XY to symbolize individual symbols (and all such symbols) items 4, 5, and 6 in Table 2.13.15 show.

Table 2.13.15 Definitions of OME, DME, DEE, and hypothetical definitions of sets of instances of non-2G2& G-family particles (ENS48 models)

1. The acronym OME denotes ordinary-matter ensemble.
   1.1. OME abbreviates ordinary-matter ensemble.
   1.2. 1 OME exists.
      1.2.1. People sometimes use the term baryonic matter instead of the term ordinary matter.
2. The acronym DME denotes dark-matter ensemble or dark-matter ensembles.
   2.1. DME abbreviates dark-matter ensemble(s).
   2.2. 5 DME exist.
   2.3. One instance of 4G4& intermediates interactions within and between each of the one OME and five DME.
3. The acronym DEE denotes dark-energy ensemble or dark-energy ensembles.
   3.1. DEE abbreviates dark-energy ensemble(s).
   3.2. 42 DEE exist.
   3.3. The instance of 4G4& that intermediates interactions within and between each of the one OME and five DME does not interact with DEE.
4. Regarding G-family particles with spans of 2, ...
   4.1. G2OMOM correlates with interactions within the OME.
   4.2. G2OMDM correlates with interactions between the OME and a DME.
   4.3. G2DMDM correlates with interactions between a DME and a DME.
      4.3.1. The two DME may be different ensembles.
      4.3.2. The two DME may be the same ensemble.
   4.4. G2DEDE correlates with interactions between a DEE and a DEE.
      4.4.1. The two DEE may be different ensembles.
      4.4.2. The two DEE may be the same ensemble.
5. Regarding G-family particles with spans of 6, ...
   5.1. G6OMOM correlates with interactions within the OME.

- 5.2. G6OMDM correlates with interactions between the OME and a DME.
- 5.3. G6DMDM correlates with interactions between a DME and a DME.
  - 5.3.1. The two DME may differ.
  - 5.3.2. The two DME may be the same ensemble.
- 5.4. G6DEDE correlates with interactions between a DEE and a DEE.
  - 5.4.1. The two DEE may differ.
  - 5.4.2. The two DEE may be the same ensemble.

6. Regarding G-family particles with spans of 48, ...
   - 6.1. G48OMOM correlates with interactions within the OME.
   - 6.2. G48OMDM correlates with interactions between the OME and a DME.
   - 6.3. G48OMDE correlates with interactions between the OME and a DEE.
   - 6.4. G48DMDM correlates with interactions between a DME and a DME.
     - 6.4.1. The two DME may differ.
     - 6.4.2. The two DME may be the same ensemble.
   - 6.5. G48DMDE correlates with interactions between a DME and a DEE.
   - 6.6. G48DEDE correlates with interactions between a DEE and a DEE.
     - 6.6.1. The two DEE may differ.
     - 6.6.2. The two DEE may be the same ensemble.

Table 2.13.16 extends Table 2.13.15 and notes possible concerns about the catalog that Table 2.13.15 suggests.

Table 2.13.16 Discussion regarding a possible cataloging method for ensembles and other particles, plus definitions of DME-1, DME-4, and OMDME (ENS48 models)

1. The definition of DME correlates with notions that ...
   - 1.1. The OME is one of six ensembles spanned by one instance of 4G4&.
   - 1.2. The term DME correlates with the other five such ensembles.
2. One possible concern could be that, for at least one of 4G268& and 2G468&, the instance (of that particle) that correlates with the OME does not correlate exactly with the five ensembles we designate as DME (and, thus, does correlate with a different set of five non-OME ensembles, of which some least some ensembles would be DEE).
   - 2.1. We acknowledge this possibility.
   - 2.2. This monograph deemphasizes this possibility.
3. Similar concerns could arise regarding alignment of spans between 4G4& and G-family bosons with spans of 2.
   - 3.1. This monograph assumes that each instance of a G-family boson with a span of 2 intermediates interactions only within a set of two ensembles for which one instance of 4G4& intermediates interactions between the same two ensembles.

4. Similar concerns could arise regarding possible non-alignment of spans between two types of G-family bosons, each having span 2. (These concerns would parallel those regarding alignment of spans regarding two types of G-family bosons for which each type has a span of 6.)
   4.1. Again, we acknowledge and deemphasize the possibility.
5. Given discussion above in this table, we define the following.
   5.1. The acronym DME-4 denotes the four DME for which no span-2 G-family boson intermediates interactions between those ensembles and the OME.
   5.2. The acronym DME-1 denotes the one DME with which those instances of G-family bosons with span 2 for which the instances interact with the OME interact with this one DME.
6. The acronym OMDME denotes ordinary-matter plus dark-matter ensemble(s).
   6.1. 6 OMDME exist.
7. Another type of possible concern involves G#XY cataloguing of non-2G2& G-family bosons. For non-2G2& G-family bosons that have been emitted and absorbed, (at least conceptually) the G#XY cataloguing scheme might be useful. But what about, for example, a 4G4& particle that the OME has emitted and nothing has yet to absorb? Presumably, eventually, the particle would correlate with one of G6OMOM and G6OMDM. But, until absorption occurs, (conceptually) one cannot determine which one.
   7.1. For this monograph, this issue pertains only to the topic of densities of the universe of various types of stuff.
      7.1.1. Possibly, regarding known and anticipated data, the issue has little significance.

The second, third, and fourth items in Table 2.13.16 reflect an assumption that people might characterize by the term telescoping or by the term nesting. Table 2.13.17 explains.

Table 2.13.17 Principles and characteristics of assumptions, regarding interactions mediated by G-family bosons, of telescoping or nesting (ENS48 models)

1. Instances of G-family particles with spans of 2 align.
   1.1. We assume that, if one instance of a 2-ensemble-span G-family boson intermediates interactions between two ensembles, any instance of another 2-ensemble-span G-family boson that interacts with one of those ensembles interacts with the other one of those two ensembles (and not with a third ensemble).
   1.2. This assumption groups ensembles into 24 sets such that each instance of a 2-ensemble-span G-family boson correlates with exactly one set of two ensembles.

2. Instances of G-family particles with spans of 6 align.
    2.1. We assume that, if one instance of a 6-ensemble-span G-family boson intermediates interactions between two ensembles, any instance of another 6-ensemble-span G-family boson that interacts with one of those ensembles interacts with the other one of those two ensembles.
    2.2. This assumption groups ensembles into 8 sets such that each instance of a 6-ensemble-span G-family boson correlates with exactly one set of the six ensembles.
3. Instances of G-family particles with spans of 2 align with instances of G-family particles with spans of 6.
    3.1. We assume that each instance of a 2-ensemble-span G-family boson intermediates interactions with two ensembles within the span of a relevant instance of each 6-ensemble-span G-family boson.
    3.2. The assumption correlates with each of the 24 sets of two ensembles being a subset of just one of the 8 sets of six ensembles.

~ ~ ~

This subsection discusses the possibility that non-G-family elementary bosons mediate interactions that span ensembles.

People might say that, for example, if 244G4& can mediate interactions between fermions in two different ensembles, perhaps, 2W can mediate interactions between fermions in two different ensembles. A reason might be that similar #"G = 1 considerations pertain for each of 244G4& and 2W.

Table 2.13.18 pertains.

Table 2.13.18 Possible aspects correlating with deemphasizing the notion that non-G-family elementary bosons mediate interactions that span ensembles

1. Practicality.
    1.1. Regarding conditions for which people make or infer measurements, ...
        1.1.1. The amount of stuff (from ensembles other than the OME) within the range of a WHO-family boson or a Y-family boson (emitted or absorbed by an OME fermion) could be minimal. (Regarding ranges, see, for example, elementary bosons that Table 2.9.5 lists.)
        1.1.2. People might say that cross-ensemble interactions would ...
            1.1.2.1. Be hard to detect.
            1.1.2.2. Have essentially no effect, either locally or on (anything other than the very early) evolution of universe.
2. Theory.

   - 2.1. The G-family correlates with concepts of polarization modes.
   - 2.2. Each of the WHO-families and the Y-family does not similarly correlate with concepts of polarization modes.
   - 2.3. Perhaps, this distinction leads to a relevant difference regarding relevant models.
3. Theory.
   - 3.1. The G-family correlates with concepts of channels.
   - 3.2. Each of the WHO-families and the Y-family may not similarly correlate with concepts of channels.
   - 3.3. Perhaps, this distinction leads to a relevant difference regarding relevant models.

~ ~ ~

This subsection discusses possible numbers of physics-relevant ensembles and possible values of #E`.

Table 2.13.19 pertains.

Table 2.13.19 Possible values of #E` for G-family bosons for which #E = 0 (ENS1 models, ENS6 models, and ENS48 models)

1. For ENS1 models, ...
   - 1.1. #ENS = 1.
   - 1.2. Only 1 ensemble (the OME) is physics-relevant.
     - 1.2.1. No elementary bosons intermediate interactions between two different ensembles.
   - 1.3. In effect, we can assume ...
     - 1.3.1. #E` = 6 for each G-family boson for which #E = 0.
2. For ENS6 models, ...
   - 2.1. #ENS = 6.
   - 2.2. 6 ensembles (the OME and the 5 DME) are physics-relevant.
     - 2.2.1. No elementary bosons intermediate interactions between an OMDME and a DEE.
   - 2.3. In effect, we can assume ...
     - 2.3.1. #E` = 6 for 222G2&.
     - 2.3.2. #E` = 6, 4, or 2 for each other G-family boson for which #E = 0.
3. For ENS48 models, ...
   - 3.1. #ENS = 48.
   - 3.2. All 48 ensembles (the OME, the 5 DME, and the 42 DEE) are physics-relevant.
     - 3.2.1. At least one G-family elementary boson intermediates interactions between the OME and a DEE.

3.2.1.1. Thus, at least one G-family boson correlates with #E` = 0
3.2.2. None of the G-family bosons for which #E ≠ 0 exhibits #E` = 0.
3.3. We require ...
3.3.1. #E` = 6 for 222G2&.
3.3.2. #E` = 0 for at least one G-family boson for which #E = 0.
3.4. In effect, we can assume ...
3.4.1. #E` = 6, 4, or 2 for each other G-family boson for which #E = 0.

~ ~ ~

This subsection discusses ENS48 models this monograph features.

For ENS48 models, this monograph features one set of choices for #E` for G-family bosons for which #E = 0. We call this choice ENS48". This monograph deemphasizes other possibilities. (See Table 2.13.19.)

Table 2.13.20 shows assumptions regarding #E` for G-family solutions for which #E = 0. Here, the solution 084G2468& correlates with an elementary particle (4G2468&) that can intermediate interactions between fermions in any 2 of 48 ensembles. In effect, we interpolate between 022G2& and 084G2468&. We correlate this interpolation with the term ENS48" models. With this interpretation, the solution 084G2468& provides the only example of a solution for which the span is 48 ensembles.

Table 2.13.20 Values of #E` for G-family solutions for which #E = 0 (ENS48" models)

1. For 022G2&, #E` = 6.
2. For 042G24&, #E` = 4.
3. For 064G246&, #E` = 2.
4. For 084G2468&, #E` = 0.

~ ~ ~

This subsection discusses ENS6 models this monograph features.

For ENS6 models, this monograph features one set of choices for #E` for G-family bosons for which #E = 0. We call this choice ENS6'. This monograph deemphasizes other possibilities. (See Table 2.13.19.)

Table 2.13.21 shows assumptions regarding #E` for G-family solutions for which #E = 0. Here, we set (the other) three values to match the value pertaining to 022G2&. We correlate this extrapolation with the term ENS6' models.

Table 2.13.21 Values of #E` for G-family solutions for which #E = 0 (ENS6' models)

1. For 022G2&, #E` = 6.

2. For 042G24&, #E` = 6.
3. For 064G246&, #E` = 6.
4. For 084G2468&, #E` = 6.

~ ~ ~

This subsection discusses some aspects of ENS48 models.

Table 2.13.22 discusses aspects regarding solutions correlating with various values of #E`. For each value of #E`, the table shows a group that correlates with a relevant symmetry, the number of generators that correlates with the group, the number instances of each G-family solution for which #E` pertains, and the span of such a solution. Here, span denotes the number of ensembles that an instance of a G-family solution connects.

Table 2.13.22 Instances and spans, by #E`, for G-family solutions (ENS48 models)

1. A #E` = 6 G-family solution correlates with an SU(7) symmetry.
    1.1. SU(7) correlates with 48 generators.
    1.2. 48 / 48 = 1.
    1.3. An instance of a #E` = 6 G-family boson can intermediate interactions between fermions in 1 of 48 ensembles (assuming the specific fermions interact via the boson).
    1.4. A #E` = 6 G-family boson cannot intermediate interactions between the 1 ensemble and any other ensembles.
    1.5. 48 instances of such a boson exist, with each instance pertaining to 1 ensemble.
2. A #E` = 4 G-family solution correlates with an SU(5) symmetry.
    2.1. SU(5) correlates with 24 generators.
    2.2. 48 / 24 = 2.
    2.3. An instance of a #E` = 4 G-family boson can intermediate interactions between fermions in any 2 of 2 ensembles (assuming the specific fermions interact via the boson).
    2.4. A #E` = 4 G-family boson cannot intermediate interactions between those 2 ensembles and any other ensembles.
    2.5. 24 instances of such a boson exist, with each instance pertaining to 1 set of 2 ensembles.
3. A #E` = 2 G-family solution correlates with an SU(3) symmetry.
    3.1. SU(3) correlates with 8 generators.
    3.2. 48 / 8 = 6.
    3.3. An instance of a #E` = 2 G-family boson can intermediate interactions between fermions in any 2 of 6 ensembles (assuming the specific fermions interact via the boson).

3.4. A #E` = 2 G-family boson cannot intermediate interactions between those 6 ensembles and any other ensembles.
3.5. 8 instances of such a boson exist, with each instance pertaining to 1 set of 6 ensembles.

4. The #E` = 0 G-family solution correlates with 1 instance.
4.1. 48 / 1 = 48.
4.2. The #E` = 0 G-family boson can intermediate interactions between fermions in any 2 of 48 ensembles (assuming the specific fermions interact via the boson).
4.3. 1 instance of the boson exists, with that instance pertaining to the 1 set of 48 ensembles.

For ENS48" models (and not for other ENS48 models), Table 2.13.23 summarizes results from above and brings forward elementary-particle names from Table 3.3.4. (See, for example, Table 2.8.3, Table 2.13.12, Table 2.13.20, and Table 2.13.22.) The SPAN column provides the number of ensembles an instance of the particle spans. The INST column shows the number of instances of the particle. The CHN column shows the number of channels. The #"G column shows that number of instances of the symbol "G.

Table 2.13.23 Spans and instances for G-family particles (ENS48" models)

G-family particle (%68both)	S	SDI	SPAN	INST	CHN	#"G
2G2&	1	r^{-2}	1	48	4	0
4G4&	2	r^{-2}	6	8	3	1
2G24&	1	r^{-4}	2	24	3	0
4G26&	2	r^{-4}	6	8	2	1
2G46&	1	r^{-4}	6	8	2	1
2G246&	1	r^{-6}	6	8	2	0
4G48&	2	r^{-4}	2	24	1	2
2G68&	1	r^{-4}	2	24	1	2
2G248&	1	r^{-6}	6	8	1	1
4G268&	2	r^{-6}	6	8	1	1
2G468&	1	r^{-6}	6	8	1	1
4G2468&	2	r^{-8}	48	1	1	0

~ ~ ~

This subsection discusses some aspects of ENS6 models.

For ENS6' models (and not for other ENS6 models), Table 2.13.24 summarizes results from above and brings forward elementary-particle names from Table 3.3.4. (See, for example, Table 2.8.3, Table 2.13.12, and Table 2.13.21.) The SPAN column provides the number of ensembles an instance of the particle spans. The INST column shows the number of instances of the particle. The CHN column shows the number of channels. The #"G column shows that number of instances of the symbol "G.

Table 2.13.24 Spans and instances for G-family particles (ENS6' models)

G-family particle (%68both)	S	SDI	SPAN	INST	CHN	#"G
2G2&	1	r^{-2}	1	6	4	0
4G4&	2	r^{-2}	6	1	3	1
2G24&	1	r^{-4}	1	6	3	0
4G26&	2	r^{-4}	6	1	2	1
2G46&	1	r^{-4}	6	1	2	1
2G246&	1	r^{-6}	1	6	2	0
4G48&	2	r^{-4}	2	3	1	2
2G68&	1	r^{-4}	2	3	1	2
2G248&	1	r^{-6}	6	1	1	1
4G268&	2	r^{-6}	6	1	1	1
2G468&	1	r^{-6}	6	1	1	1
4G2468&	2	r^{-8}	1	6	1	0

~ ~ ~

This subsection discusses some aspects of ENS1 models.

For ENS1 models, Table 2.13.25 summarizes results from above and brings forward elementary-particle names from Table 3.3.4. (See, for example, Table 2.8.3 and Table 2.13.12.) The SPAN column provides the number of ensembles an instance of the particle spans. The INST column shows the number of instances of the particle. The CHN column shows the number of channels. The #"G column shows that number of instances of the symbol "G.

Table 2.13.25 Spans and instances for G-family particles (ENS1 models)

G-family particle (%68both)	S	SDI	SPAN	INST	CHN	#"G
2G2&	1	r^{-2}	1	1	4	0
4G4&	2	r^{-2}	1	1	3	1
2G24&	1	r^{-4}	1	1	3	0
4G26&	2	r^{-4}	1	1	2	1
2G46&	1	r^{-4}	1	1	2	1
2G246&	1	r^{-6}	1	1	2	0
4G48&	2	r^{-4}	1	1	1	2
2G68&	1	r^{-4}	1	1	1	2
2G248&	1	r^{-6}	1	1	1	1
4G268&	2	r^{-6}	1	1	1	1
2G468&	1	r^{-6}	1	1	1	1

G-family particle (%68both)	S	SDI	SPAN	INST	CHN	#"G
4G2468&	2	r^{-8}	1	1	1	0

~ ~ ~

This subsection discusses further uses of the terms ENS48 models, ENS48" models, ENS6 models, and ENS6' models and further uses of concepts related to those terms.

People might say that, generally, some details of subsequent work in this monograph correlate with assuming that ENS48 models are ENS48" models. We think that people can, if desired, modify such work to explore ENS48 models that are not ENS48" models. Modifications would involve changing one or more of the values of #E` that Table 2.13.20 shows.

People might say that, generally, some details of subsequent work in this monograph correlate with assuming that ENS6 models are ENS6' models. We think that people can, if desired, modify such work to explore ENS6 models that are not ENS6' models. Modifications would involve changing one or more of the values of #E` that Table 2.13.21 shows.

~ ~ ~

This subsection shows COMPAR LADDER solutions for elementary particles for which $\sigma = -1$.

For each elementary particle for which $\sigma = -1$, the term free-ranging does not pertain. People might say that physics of composite particles correlates with models based on special relativity and that such models correlate with (1;3)> symmetries. People might say that people can expect that (1;3)> symmetries pertain for fields correlating with the entirety of a composite particle. We explore the possibility that (a`;b`)< symmetries pertain for fields related to elementary particles (for which $\sigma = -1$) that combine to form composite particles.

People might say that, for bosons for which $\sigma = -1$, people should explore the notion that INSSYM7 pertains. (See Table 2.13.10.) Starting from ground-state INTERN LADDER solutions, we add enough QP-like harmonic oscillators to bring the total number of QP-like oscillators to 7. We assume a` = 1. We set N(..) for each added oscillator to match N(..) for each previously relevant QP-like oscillator. To maintain Œ = 0, we add the same (as we added QP-like oscillators) number of QE-like oscillators and use, for each of these added oscillators, the value N(..) that we use for added QP-like oscillators. Let b` denote the number of added QE-like pairs of oscillators. The number b` is an integer. We assume that (1;b`)< symmetries pertain.

Table 2.13.26 shows ground-state COMPAR ENVIRO LADDER solutions for elementary bosons for which $\sigma = -1$.

Table 2.13.26 σ = −1 boson ground-state COMPAR ENVIRO LADDER solutions

E	E	E	E	E	E	E	E	E	P	P	P	P	P	P	P	P	P	(σ = −1)
8R	8L	6R	6L	4R	4L	2R	2L	0	0	2L	2R	4L	4R	6L	6R	8L	8R	Subfamily
		0	0	0	0	0	0	0	0	0	0	0	0	0	0			00
		−1	−1	−1	−1	..	..	..	−1	−1	−1	−1	−1	−1	−1			2Y
		0	0	0	0	0	0	0	0	0	0	0	0	0	0			20
		−1	−1	..	..	..	..	..	−1	−1	−1	−1	−1	−1	−1			4Y
		0	0	0	0	0	0	0	0	0	0	0	0	0	0			40

Table 2.13.27 restates Table 2.13.26. Table 2.13.27 uses the symbol 'I to point out aspects leading to occurrences of INSSYM7. For each row, the symbol 'I appears exactly 7 times. For each row, a` = 1. The table uses the symbol 'S in ways that correlate with b`.

Table 2.13.27 A second depiction of σ = −1 boson ground-state COMPAR ENVIRO LADDER solutions

E	E	E	E	E	E	E	E	E	P	P	P	P	P	P	P	P	P	(σ = −1)
8R	8L	6R	6L	4R	4L	2R	2L	0	0	2L	2R	4L	4R	6L	6R	8L	8R	Subfamily
		'S	'S	'S	'S	'S	'S	0	'I	'I	'I	'I	'I	'I	'I			00
		'S	'S	'S	'S	..	..	..	'I	'I	'I	'I	'I	'I	'I			2Y
		'S	'S	'S	'S	0	0	0	'I	'I	'I	'I	'I	'I	'I			20
		'S	'S	..	..	..	..	..	'I	'I	'I	'I	'I	'I	'I			4Y
		'S	'S	0	0	0	0	0	'I	'I	'I	'I	'I	'I	'I			40

For elementary fermions for which σ = −1, starting from ground-state INTERN LADDER solutions, we add enough QP-like harmonic oscillators to match considerations related to generations. (That is, for 1QR particles, we add one pair of QP-like oscillators. For 3QIRD particles, we add two pairs of QP-like oscillators. See Table 2.6.4.) We add the same number of QE-like oscillators. Let b` denote the number of added QE-like pairs of oscillators. The number b` is an integer. We assume that (0;b`)< symmetries pertain. (Assuming that (1;b`)< symmetries pertain would not change results that we show starting with Table 2.13.33. But, assuming a` = 1 would change results we show in Table 2.13.42.) Here, we think that the value of each of the N(..) for added oscillators need not be relevant, as long as the values are the same. (For example, perhaps N(..) = 0 pertains for each added oscillator.) In tables, for each N(..) for such an added oscillator, we use the notation =.

Table 2.13.28 shows ground-state COMPAR ENVIRO LADDER solutions for fermions for which σ = −1.

Table 2.13.28 σ = −1 fermion ground-state COMPAR ENVIRO LADDER solutions

E 8R	E 8L	E 6R	E 6L	E 4R	E 4L	E 2R	E 2L	E 0	P 0	P 2L	P 2R	P 4L	P 4R	P 6L	P 6R	P 8L	P 8R	(σ = −1) Subfamily
				=	=	~	~	0	0	~	~	=	=					1Q
				=	=	~	~	−1	−1	~	~	=	=					1R
=	=	=	=	~	~	0	0	0	0	0	0	~	~	=	=	=	=	3Q
=	=	=	=	~	~	−1	−1	0	0	−1	−1	~	~	=	=	=	=	3I
=	=	=	=	~	~	0	0	−1	−1	0	0	~	~	=	=	=	=	3R
=	=	=	=	~	~	−1	−1	−1	−1	−1	−1	~	~	=	=	=	=	3D

Table 2.13.29 restates Table 2.13.28. In Table 2.13.29, the symbol 'I does not appear. The table uses the symbol 'G in ways that correlate with the number of generations. For a row in which 'G appears exactly 2 times, an SU(2) symmetry pertains to generations, the number of generators of SU(2) is 3, and there are 3 generations. Regarding a row in which 'G appears exactly 4 times, see Table 2.6.4. Table 2.13.29 uses the symbol 'S in ways that correlate with b`.

Table 2.13.29 A second depiction of σ = −1 fermion ground-state COMPAR ENVIRO LADDER solutions

E 8R	E 8L	E 6R	E 6L	E 4R	E 4L	E 2R	E 2L	E 0	P 0	P 2L	P 2R	P 4L	P 4R	P 6L	P 6R	P 8L	P 8R	(σ = −1) Subfamily
				'S	'S	~	~	0	0	~	~	'G	'G					1Q
				'S	'S	~	~	−1	−1	~	~	'G	'G					1R
'S	'S	'S	'S	~	~	0	0	0	0	0	0	~	~	'G	'G	'G	'G	3Q
'S	'S	'S	'S	~	~	−1	−1	0	0	−1	−1	~	~	'G	'G	'G	'G	3I
'S	'S	'S	'S	~	~	0	0	−1	−1	0	0	~	~	'G	'G	'G	'G	3R
'S	'S	'S	'S	~	~	−1	−1	−1	−1	−1	−1	~	~	'G	'G	'G	'G	3D

Table 2.13.30 restates Table 2.13.29.

Table 2.13.30 A third depiction of σ = −1 fermion ground-state COMPAR ENVIRO LADDER solutions

E A9	E 87	E 65	E 43	E 21	E 0	P 0	P 12	P 34	P 56	P 78	P 9A	(σ = −1) Subfamily
			"S	~~	0	0	~~	"G				1Q
			"S	~~	−1	−1	~~	"G				1R
	"S	"S	~~	00	0	0	00	~~	"G	"G		3Q
	"S	"S	~~	−1, −1	0	0	−1, −1	~~	"G	"G		3I
	"S	"S	~~	00	−1	−1	00	~~	"G	"G		3R
	"S	"S	~~	−1, −1	−1	−1	−1, −1	~~	"G	"G		3D

Table 2.13.31 shows instance-related symmetry and COMPAR SPATIM symmetries for solutions for which σ = −1. [Physics:] These solutions correlate with all the non-free-ranging elementary particles this monograph discusses.

Table 2.13.31 Instance-related symmetry and COMPAR SPATIM symmetries for σ = −1 fermion ground-state COMPAR ENVIRO LADDER solutions

(σ = −1)	Instance-related symmetry	COMPAR SPATIM	Solution subfamilies
	INSSYM7	(1;3)<	0O
	INSSYM7	(0;1)<	1QR
	INSSYM7	(1;2)<	2YO
	INSSYM7	(0;2)<	3QIRD
	INSSYM7	(1;1)<	4YO

People might say that, based on Table 2.11.2 rules for combining symmetries and on Table 2.13.31, Table 2.13.32 provides combinations correlating with (1;3)< symmetries. The table lists pairs. The table does not list triplets, quadruplets, and so forth. Assuming a` = 0 pertains for the relevant fermions, of the combinations the table shows, the only combinations that combine to (1;3)< without, in effect, encountering a redundant symmetry, are 1QR+2YO, 3QIRD+4YO, and 0O. [Physics:] People might say that 1Q+2Y correlates with a component of known composite particles.

Table 2.13.32 Pairs that might combine to correlate with (1;3)<

1. 0O+x, for which x can be any one of the following.
   - 1.1. 0O.
   - 1.2. 1QR.
   - 1.3. 2YO.
   - 1.4. 3QIRD.
   - 1.5. 4YO.
2. 1QR+2YO.
3. 2YO+x, for which x can be any one of the following.
   - 3.1. 3QIRD.
   - 3.2. 4YO.
4. 3QIRD+4YO.

Table 2.13.33 and Table 2.13.34 show and discuss results from combining COMPAR-related symmetries for σ = −1 ground-state ENVIRO LADDER solutions. ENVIRO LADDER solution-related COMPAR symmetries combine so as to produce items in Table 2.13.33. For example, for 1Q+2Y, (0;1)< (for 1Q) and (1;2)< (for 2Y) combine term-wise to yield (1;3)<.

Table 2.13.33 shows COMPAR SPATIM for combinations of solutions for which σ = −1 for each solution. We use the term pre-composite to correlate with such combinations of solutions. We use notation that Table 2.3.5 shows. People might say that 0O does not correlate with the term pre-composite.

Table 2.13.33 COMPAR SPATIM symmetries for $\sigma = -1$ combinations of solutions for which $\sigma = -1$

($\sigma = -1$)	Instance-related symmetry	COMPAR SPATIM	Pre-composite
	INSSYM7	(1;3)<	0O
	INSSYM7	(1;3)<	1QR+2Y
	INSSYM7	(1;3)<	1QR+2O
	INSSYM7	(1;3)<	3QIRD+4Y
	INSSYM7	(1;3)<	3QIRD+4O

Table 2.13.34 shows SPATIM symmetries for combinations of solutions for which $\sigma = +1$ for each combination. Here, the term SPATIM symmetries denotes both of COMPAR SPATIM symmetries and FRERAN SPATIM symmetries. Here, n denotes any integer ≥ 2. ([Physics:] For the first row in the table, 2 × 0O would correlate with two occurrences of a particle that is its own antiparticle. For the second row in the table, n = 2 correlates with mesons, n = 3 correlates with baryons, n = 4 correlates with tetraquarks, and n = 5 correlates with pentaquarks.) For n ≥ 2, n instances of (1;3)< combine to yield (1;3)>. (See Table 2.11.2.) Here, we use the term composite to correlate with and extend traditional use of the term composite. ([Physics:] People use the term composite particles to describe mesons and baryons.) People might say that n × 0O does not correlate with the term composite.

Table 2.13.34 Instance-related symmetry and SPATIM symmetries for $\sigma = +1$ combinations of solutions for which $\sigma = -1$

($\sigma = +1$)	Instance-related symmetry	SPATIM	Composite (n ≥ 2)
	INSSYM7	(1;3)>	n × 0O
	INSSYM7	(1;3)>	n × 1QR+2Y
	INSSYM7	(1;3)>	n × 1QR+2O
	INSSYM7	(1;3)>	n × 3QIRD+4Y
	INSSYM7	(1;3)>	n × 3QIRD+4O

Perhaps, Table 2.13.35 provides a possible basis for research.

Table 2.13.35 Speculation regarding relationships among $\sigma = -1$, $\sigma = +1$, and abilities of QIRD-family elementary particles to interact with YO-family elementary particles

1. To what extent might the following notions correlate with each other?
    - 1.1. The following set of concepts.
        - 1.1.1. The relationship (1;3)< + (1;3)< = (1;3)>.
        - 1.1.2. That (1;3)< correlates with $\sigma = -1$ (or, non-free-ranging).

1.1.3. That (1;3)> correlates with σ = +1 (or, free-ranging).

1.2. The possibility that a QIRD-family elementary particle cannot absorb a YO-family particle that it emits.

1.3. The possibility that a sufficiently isolated QIRD-family particle cannot interact with YO-family particles.

~ ~ ~

This subsection shows COMPAR ENVIRO solutions for non-G-family elementary particles for which σ = +1.

[Physics:] People might say that violations of CP-symmetry correlate with weak interactions occurring under circumstances in which the physics of composite particles pertains. Under some such circumstances, people observe effects correlating with generation changes for fermions. Some examples correlate with an asymmetry between zero-charge kaons and their antiparticles. Some examples correlate with decays of neutral B mesons.

Regarding such observations pertaining to kaons, people show Feynman diagrams that depict interactions mediated by two bosons, each from the 2W-subfamily.

We posit correlating these interactions with COMPAR SPATIM solutions and symmetries.

For FRERAN SPATIM LADDER solutions, results Table 2.13.5 states correlate with conservation of fermion generation.

For interactions (such as those people correlate with CP-symmetry violation and we correlate with COMPAR environments) mediated by more than one W-family boson, we think COMPAR SPATIM solutions and symmetries need not correlate completely with FRERAN SPATIM solutions and symmetries. People might say that the behavior of the two (or presumably possibly more than two) W-family bosons are correlated. (See, also, Table 3.10.2.)

Perhaps, for COMPAR solutions, Table 2.13.36 pertains. Compared to Table 2.13.5, Table 2.13.36 shows changes only for the 2W-subfamily. In Table 2.13.36, the QE-like occurrence of "G and the P7-and-P8 occurrence of "S are not present. People might say that the P1-and-P2 occurrence of the symbol "S` correlates with a denaturing of symmetry, namely a lack of reflection symmetry. This lack of reflection symmetry correlates with the topic of parity. The E2-and-E1 occurrence of the symbol "I` correlates with another lack of a reflection symmetry. People might say that the QE-like lack of reflection symmetry correlates with the topic of charge conjugation. People might say that, together, the two lacks correlate with possibilities for violations of CP-symmetry. The lack of a "G correlates with lack of conservation of fermion generation.

Table 2.13.36 σ = +1 non-G-family ground-state COMPAR LADDER solutions

E 87	E 65	E 43	E 21	E 0	P 0	P 12	P 34	P 56	P 78	P 9A	(σ = +1) Subfamily
	"I	"I	"I	'I	0	"S	"S	"S			0H0

E 87	E 65	E 43	E 21	E 0	P 0	P 12	P 34	P 56	P 78	P 9A	(σ = +1) Subfamily
	"I	"I	"I	'I	0	~~	"S	"S	"S		1C
	"I	"I	"I	'I	-1	~~	"S	"S	"S		1N
	"I	"I	"I`	'I	0	00("S`)	"S	"S			2W
	"I	"I	"I	'I	-1	00	~~	"S	"S	"S	3N

Perhaps, Table 2.13.37 provides an alternative possibility regarding the W-family and interactions involving two W-family members. Here, a lack of conservation of generation (regarding spin-1/2 elementary fermions) correlates with #"G = 2. We do not discuss this possibility further.

Table 2.13.37 Speculative alternative solution for 2W bosons in a COMPAR environment

E A9	E 87	E 65	E 43	E 21	E 0	P 0	P 12	P 34	P 56	P 78	P 9A	(σ = +1) Subfamily
"G	"G	"I	"I	"I`	'I	0	00("S`)	"S	"S	"S	"S	2W

~ ~ ~

This subsection summarizes COMPAR ENVIRO solutions for elementary particles for which σ = +1.

Table 2.13.38 pertains.

Table 2.13.38 Assumptions regarding COMPAR LADDER solutions for elementary particles for which σ = +1

1. For elementary particles for which σ = +1, ...
   - 1.1. We assume that behavior in COMPAR environments correlates with behavior that correlates with solutions for which FRERAN SPATIM symmetries pertain, except ...
     - 1.1.1. Interactions involving more than one W-family boson ...
       - 1.1.1.1. Can correlate with results Table 2.13.36 shows.
       - 1.1.1.2. Need not correlate with results Table 2.13.5 shows.
2. Thus, elementary particles for which σ = +1, we, in effect, assume ...
   - 2.1. Except for some interactions mediated by W-family bosons, ...
     - 2.1.1. COMPAR SPATIM equals FRERAN SPATIM.
   - 2.2. For some interactions mediated by W-family bosons, each of the following pertains.
     - 2.2.1. COMPAR SPATIM does not equal FRERAN SPATIM.
     - 2.2.2. Violations of CP-symmetry can occur.

~ ~ ~

This subsection summarizes instance-related symmetry and SPATIM symmetries for elementary particles and for composite particles.

Table 2.13.39 shows instance-related symmetry and SPATIM symmetries for elementary particles and for composite particles.

Table 2.13.39 Instance-related symmetry, FRERAN SPATIM symmetries, and COMPAR SPATIM symmetries for elementary particles and for composite particles

1. Each $\sigma = +1$ ground-state FRERAN LADDER solution or COMPAR LADDER solution correlates with ...
   - 1.1. An instance-related symmetry of INSSYM7.
     - 1.1.1. For the 2W-subfamily in a COMPAR environment (at least, regarding some interactions intermediated by more than one W-family particle), ...
       - 1.1.1.1. The instance-related symmetry lacks reflection symmetry.
   - 1.2. SPATIM symmetries of (1;3)>.
     - 1.2.1. For the 2W-subfamily in a COMPAR environment (at least, regarding some interactions intermediated by more than one W-family particle), ...
       - 1.2.1.1. One of the three SU(2)-related symmetries lacks reflection symmetry.
2. Each $\sigma = +1$ composite correlates with ...
   - 2.1. An instance-related symmetry of INSSYM7.
   - 2.2. SPATIM symmetries of (1;3)>.
3. Each $\sigma = -1$ ENVIRO LADDER solution correlates with ...
   - 3.1. An instance-related symmetry of INSSYM7.
   - 3.2. (Perhaps hypothetical) FRERAN symmetries of (1;3)>.
   - 3.3. COMPAR symmetries of (a`;b`)< with the following.

$$a` = 1, \text{ for bosons} \quad (2.107)$$

$$a` = 0 \text{ (or, possibly, 1), for fermions} \quad (2.108)$$

$$b` = 3 - S, \text{ for bosons} \quad (2.109)$$

$$b` = S + 1/2, \text{ for fermions} \quad (2.110)$$

~ ~ ~

[Physics:] This subsection discusses possibilities that single-quark composite particles might exist and that hybrid composites might exist.

Possibly, people should consider single-quark or hybrid composite particles. Table 2.13.40 pertains. (See, also, Table 4.11.1.)

Table 2.13.40 Possibilities regarding single-quark composite particles and other hybrid composite particles

1. Each row in Table 2.13.34 combines n items from a single row in Table 2.13.33. In Table 2.13.34, n must be greater than 1.
2. A question arises.
   - 2.1. To what extent does nature exhibit composites that combine items from differing rows in Table 2.13.33?
3. For example, ...
   - 3.1. Possibly, nature could include (1 × 0O) + (1 × 1Q+2Y) or could include (1 × 0O) + (1 × 1Q+2O).
     - 3.1.1. These could be free-ranging composites, some with a charge with absolute value of either $(1/3)|q_e|$ or $(2/3)|q_e|$.

Table 2.13.41 shows some possible examples for which (1;3)> symmetries might pertain. The last two examples in the table could correlate with free-ranging fractionally charged particles. These two examples might pertain to a plasma (or, sea) of quarks, before the formation nucleons. (See Section 4.5.)

Table 2.13.41 Possibilities for hybrid composite particles

1. (1 × 1Q+2Y) + (1 × 1Q+2O).
2. (1 × 0O) + (1 × 1Q+2Y).
3. (1 × 0O) + (1 × 1Q+2O).

~ ~ ~

This subsection discusses possible limitations regarding interactions among elementary particles for which σ = −1.

Work above correlates with the notion that, assuming a\` = 0 for the relevant fermions, 1QR+2YO, 3QIRD+4YO, and 0O are the only combinations of fields (that correlate with elementary particles for which σ = −1) that correlate with COMPAR (1;3)< symmetries without a redundancy in a\` and without a redundancy in b\`. (See, for example, discussion before Table 2.13.32.)

Possibly, Table 2.13.42 pertains.

Table 2.13.42 Possible limitations on interactions between elementary fermions for which σ = −1 and elementary bosons for which σ = −1

1. 1QR-subfamily particles ...
   - 1.1. Interact with 2YO-subfamily particles.
   - 1.2. Do not interact with 4YO-subfamily particles.
2. 3QIRD-subfamily particles ...
   - 2.1. Do not interact with 2YO-subfamily particles.
   - 2.2. Interact with 4YO-subfamily particles.

~ ~ ~

This subsection discusses possible implications of assuming #b' = 2.

People might say that steps Table 2.13.43 suggests might correlate with opportunities for research.

Table 2.13.43 Possible steps for research regarding possibilities that #b' =2

1. Assume #b' = 2 (and #b' ≠ 3) for all phenomena for which $\sigma = +1$.
2. Try to parallel work this monograph presents in Section 2.11, in Section 2.12, and above in Section 2.13.
3. Perhaps explore possibilities such as the following.
   - 3.1. Instance-related symmetry (or, INSSYM) correlates with concepts people might correlate with or denote by the following.

     SU(5) (2.111)

     INSSYM5

     - 3.1.1. Here, 24 correlates with the number of generators of SU(5).

       24 mathematically possible ensembles (2.112)

       #ENS = 24 (2.113)

       ENS24 models

       #ENS = 24 / 8 = 3 (2.114)

       ENS3 models

     - 3.1.2. Here, 8 correlates with the number of generators of SU(3).

       #ENS = 24 / 24 = 1 (2.115)

       ENS1 models

     - 3.1.3. Here, 24 correlates with the number of generators of SU(5).
   - 3.2. For each of the cases ENS24 models and ENS3 models, ...
     - 3.2.1. The number of DME is 2 (= 3 − 1, in which the 3 comes from above in this table and equals the number of OMDME).
     - 3.2.2. Each ensemble contains significant amounts of SEDMS. (See Section 4.3.)

Section 2.14 EXTINT LADDER models for fermion generations and color charge

Section 2.14 discusses EXTINT LADDER solutions that correlate with numbers of generations for elementary fermions and with numbers of color charges for elementary fermions existing in COMPAR environments. People might say that MM1MS1 models correlate with 5 color charges for elementary fermions for which $\sigma = -1$ and $S = 3/2$. People might say that MM1MS1 models provide, for spin-1/2 elementary fermions for which $\sigma = +1$, some correlation between numbers of generations and numbers of charges.

~ ~ ~

This subsection provides perspective about internal properties and INTERN LADDER solutions.

People might say that generation is an internal property of elementary fermions. People might say that color charge is an internal property of elementary fermions for which $\sigma = -1$.

INTERN LADDER solutions do not explicitly show information regarding generations.

In this section, we discuss possibilities for EXTINT LADDER solutions that correlate with number of generations and numbers of color charges.

~ ~ ~

This subsection provides perspective about work in this section.

People might say that work in this section correlates with possible theory. (See, for example, Table 2.14.5.) This monograph tends not to directly use EXTINT LADDER solutions in trying to correlate MM1MS1 models with experimental or observational endeavors.

~ ~ ~

This subsection discusses color charge and gluons.

People might say that, in the course of developing traditional physics theory regarding quarks bound in composite particles, people developed the concept of color charge to overcome otherwise evident theoretical problems regarding multiple quarks existing in the same fermion state. Traditional physics provides that no more than one fermion can occupy any one state. The notion of multiple color charges provides a construct for overcoming the problem. For example, two quarks that occupy an otherwise identical state can differ with respect to color charge.

For a quark, S = 1/2. People might say that people have observed quarks only in COMPAR environments. People might say that for quarks in COMPAR environments, the number of color charges (not counting antimatter color charges that pertain to antimatter quarks) is 3.

We know of no discussion regarding color-charge analogs that might pertain regarding possible spin-3/2 elementary fermions.

We use the term color charge (or, color-charge analog) regarding 3QIRD solutions. People might say that, here, MM1MS1 models correlate with 5 color charges.

People might say that gluons change the color charges of individual quarks. People might say that gluons provide interactions (the strong interaction) that bind quarks into hadrons.

People might say that quarks can exist in conditions other than COMPAR environments. For example, the early universe would have been too dense for the formation of hadrons. People might speculate regarding the extent to which quarks not governed by COMPAR considerations exist in neutron stars.

Perhaps, Table 2.14.1 points to opportunities for research. Here, #CC denotes the number of possible color charges. (See, also, Table 2.6.6.)

Table 2.14.1 Possible concepts regarding color charge

1. For non-lepton elementary fermions, ...

$$\#CC = \#E + 1 \tag{2.116}$$

   1.1. #CC denotes the number of possible color charges.

2. People might say that, for leptons, ...

$$\#CC = \#E + 1 = 1 \tag{2.117}$$

   2.1. No interaction changes that color charge.

   2.2. The lack of an interaction that could change that color charge correlates with ...

      2.2.1. A lack of physics-relevance of 0Y solutions.

~ ~ ~

This subsection discusses symmetries for EXTINT solutions for elementary fermions for which σ = −1.

Table 2.14.2 pertains. We start with Table 2.13.30. We replace each occurrence of the symbol "S with the symbol "C. For each QE-like occurrence of the symbol ~~, we replace the QE-like occurrence of ~~ with the symbol 'C,~. Here, we use the symbol −1,−1 to indicate two occurrences of the symbol −1. Here, regarding the symbol 'C,~, the symbol ~ pertains to the one oscillator (in the relevant oscillator pair) for which N(..) ≠ 'C. (For each row in the table, there are 4 cases. Here, 4 = 2 × 2, with one factor of 2 correlating with the symbol ~~. For each value of ~~, 2 solutions pertain. For example, for the 1Q-subfamily, for one solution, ~ correlates with oscillator E2 and 'C correlates with oscillator E1. Here, 'C = N(E1). For the other solution, ~ correlates with oscillator E1 and 'C correlates with oscillator E2. Here, 'C = N(E2).) In general, we use the symbol #'C to denote the number of occurrences of 'C. In general, we count each "C as correlating with two occurrences of 'C. For the 1QR-subfamilies, #'C = 3. People might say that, for the 1QR-subfamilies, #'C is the number of color charges. That is, #'C = #CC. For the 3QIRD-subfamilies, #'C = 5. People might say that, for the 3QIRD-subfamilies, #'C is the number of color-charge analogs. That is, #'C = #CC. Occurrences of "G correlate with numbers of generations.

Table 2.14.2 A depiction of σ = −1 fermion ground-state EXTINT LADDER solutions

E A9	E 87	E 65	E 43	E 21	E 0	P 0	P 12	P 34	P 56	P 78	P 9A	(σ = −1) Subfamily
			"C	'C,~	0	0	~~	"G				1Q
			"C	'C,~	−1	−1	~~	"G				1R
	"C	"C	'C,~	00	0	0	00	~~	"G	"G		3Q
	"C	"C	'C,~	−1, −1	0	0	−1, −1	~~	"G	"G		3I
	"C	"C	'C,~	00	−1	−1	00	~~	"G	"G		3R
	"C	"C	'C,~	−1, −1	−1	−1	−1, −1	~~	"G	"G		3D

Table 2.14.2 directly extends Table 2.3.14. Table 2.3.14 characterizes ground-state INTERN LADDER solutions for which σ = −1.

~ ~ ~

This subsection discusses symmetries for EXTINT solutions for elementary fermions for which σ = +1.

Table 2.14.3 recasts relevant rows from Table 2.3.11.

Table 2.14.3 Ground-state fermion INTERN LADDER solutions for σ = +1

E A9	E 87	E 65	E 43	E 21	E 0	P 0	P 12	P 34	P 56	P 78	P 9A	(σ = +1) Subfamily
					0	0	~~					1C
					−1	−1	~~					1N
					0	−1	00	~~				3N

Table 2.14.4 adds (to Table 2.14.3) QE-like occurrences of symbols of the form "G to correlate with results from Table 2.6.4. Each occurrence of "G correlates with two occurrences of N(E0). For each QE-like occurrence of "G, we add a QP-like occurrence of the symbol "X. Each occurrence of "X correlates with two occurrences of a symbol 'X. For each QP-like occurrence of the symbol ~~, we replace the QP-like occurrence of ~~ with the symbol ~,'X. Each occurrence of 'X correlates with an occurrence of N(E0). That is, 'X = N(E0). Somewhat paralleling considerations for 'C,~ in Table 2.14.2, in Table 2.14.4, each ~,'X correlates with two cases.

Table 2.14.4 Ground-state fermion EXTINT LADDER solutions for $\sigma = +1$, correlating with generations

E	E	E	E	E	E	P	P	P	P	P	P	($\sigma = +1$)
A9	87	65	43	21	0	0	12	34	56	78	9A	Subfamily
				"G	0	0	~,'X	"X				1C
				"G	−1	−1	~,'X	"X				1N
			"G	"G	0	−1	00	~,'X	"X	"X		3N

Table 2.14.4 directly extends Table 2.6.4. In Table 2.14.4, occurrences of "G correlate with numbers of generations.

In Table 2.14.4, 1C correlates with 3 occurrences of 'X. 1N correlates with 3 occurrences of 'X. 3N correlates with 5 occurrences of 'X.

People might say that, for 1C and 1N, the 3 occurrences of 'X correlate with the 3 charges that correlate with the 2W family. 1C correlates with the two charges that people consider to be non-zero. One of these charges is the charge of the W^+ boson. One of these charges is the charge of the W^- boson. 1N correlates with the one charge that people consider to be zero. This is the charge of the Z boson.

People might say that, for 3N, the 5 occurrences of 'X correlate with a property that might pertain, if 4W solutions correlated with nature. For MM1MS1 models, 4W solutions do not correlate with nature.

~ ~ ~

This subsection suggests possible opportunities for research regarding, for elementary fermions, various models for and relationships among spin, generations, color charge, charge, and SPATIM symmetries.

Table 2.14.5 pertains.

Table 2.14.5 Possible opportunities for research regarding, for elementary fermions, relationships among spin, generations, color charge, charge, and SPATIM symmetries

1. For modeling regarding elementary fermions, determine correlations (and lacks of correlations) among ...
   - 1.1. Any or all of the following properties.
     - 1.1.1. Spin.
     - 1.1.2. Numbers of generations.
     - 1.1.3. Numbers of color charges.
     - 1.1.4. Numbers of charges.
     - 1.1.5. Other.
   - 1.2. Any or all of the following modeling techniques.
     - 1.2.1. DIFEQU.
     - 1.2.2. INTERN LADDER.
     - 1.2.3. EXTINT LADDER.

1.2.4. FRERAN SPATIM LADDER.
1.2.5. COMPAR SPATIM LADDER (or, COMPAR ENVIRO LADDER).
1.2.6. FERTRA LADDER.
1.2.7. Other.
1.3. Any or all of the properties and any or all of the modelling techniques.

Section 2.15 The MM1 meta-model and various MM1MS1 models

Section 2.15 discusses using the MM1 meta-model. We define the MM1 meta-model by featuring a list of topics that correlate with choices people can make when using the MM1 meta-model to define models. We provide a process for using the MM1 meta-model to select and work with models. We list some key parameters (such as #ENS) that people can set in order to specify a model. We list some characteristics that models share. We discuss some aspects of ENS48 models, ENS6 models, and ENS1 models. We describe models we discuss in subsequent chapters.

~ ~ ~

This subsection discusses aspects of the MM1 meta-model.

People might say that work above provides the essence of a meta-model. (See Section 2.1 through Section 2.14.)

People might say that Table 2.15.1 notes topics that correlate with limitations this monograph places on the MM1 meta-model.

Table 2.15.1 Topics that correlate with limitations this monograph places on the MM1 meta-model

1. The extent to which models should be based on math that has discrete solutions.
2. The extent to which models should be based on math regarding isotropic pairs of isotropic quantum harmonic oscillators.
3. Œ = 0.

Table 2.4.1 discusses relationships between the MM1 meta-model and MM1MS1 models.

Below, we discuss aspects (of the MM1 meta-model) correlating with choices people can make when selecting or developing MM1MS1 models. (See, for example, Table 2.15.5, Table 2.15.6 , and Table 2.15.7.)

~ ~ ~

This subsection outlines steps for using the MM1 meta-model to select a model and for using the model.

Table 2.15.2 pertains.

Table 2.15.2 Steps for using the MM1 meta-model and using a MM1MS1 model

1. Choose each of the following.
    1.1. Information related to some relevant aspects of nature. The information might include one or more of ...
        1.1.1. Data.
        1.1.2. Past, ongoing, or planned experiments or observations.
        1.1.3. Various theories or models.
    1.2. A model (based on the MM1 meta-model) that might correlate with those aspects.
        1.2.1. Perhaps, consider limiting the set of topics (to explore) to topics that correlate with steps Table 2.15.5 and Table 2.15.6 show.
        1.2.2. Perhaps, consider making choices that correlate with examples this monograph shows.
        1.2.3. Perhaps, consider topics Table 2.15.7 shows.
2. Try to correlate aspects of the information with aspects of the model.

~ ~ ~

This subsection discusses topics for which people might have one choice or limited choices when using the MM1 meta-model.

Table 2.15.3 notes topics for which people might not be able to make choices when using the MM1 meta-model to develop MM1MS1 models. (Regarding the terms %68even, %68odd, and %68both, see discussion preceding Table 3.3.3 and see Table 3.3.3.)

Table 2.15.3 Some topics correlating with the MM1 meta-model and possibly with all MM1MS1 models

1. A most comprehensive list of candidate elementary particles.
    1.1. This list includes the G-family %68both set.
2. $D^* = 3$, for DIFEQU aspects correlating directly with cataloging non-zero-mass elementary particles.
    2.1. In this monograph, uses of $D^* = 2$ and uses of $D^* = 1$ correlate with projections of functions (for example, functions for which $D^* = 3$).

Table 2.15.4 notes a topic for which people might not be able to make choices when using the MM1 meta-model to develop one or more types of MM1MS1 models.

Table 2.15.4 A topic for which people might not be able to make choices when using the MM1 meta-model to develop one or more types of MM1MS1 models

1. For LADDER models, ...
    1.1. For physics-relevant solutions other than G-family solutions, ...
        1.1.1. Relationships among spin, S, and #P.

~ ~ ~

This subsection illustrates steps that people might take to specify a model.

Table 2.15.5 shows a list of steps that, for some situations, people might want to take to specify MM1MS1 models. For example, people making models that correlate with the nature of dark-energy stuff and dark matter may want to do the second step in the table.

Table 2.15.5 Steps for specifying a model (an example)

1. Choose aspects of nature to try to model.
2. Choose #ENS from one of the following possibilities.
    2.1. #ENS = 48.
        2.1.1. This choice correlates with ENS48 models (including ENS48" models).
        2.1.2. People might say that this selection is not incompatible with much physics developed before the year 2016.
        2.1.3. People might say that this selection correlates with more data than does the choice of #ENS = 1.
            2.1.3.1. People might say that, for example, #ENS = 48 provides possible insight regarding ratios of densities of dark matter to densities of ordinary matter.
    2.2. #ENS = 6.
        2.2.1. This choice correlates with ENS6 models.
        2.2.2. People might say that this selection is not incompatible with much physics developed before the year 2016.
        2.2.3. People might say that this selection correlates with more data than does the choice of #ENS = 1.
            2.2.3.1. People might say that, for example, #ENS = 6 ...
                2.2.3.1.1. Provides possible insight regarding ratios of densities of dark matter to densities of ordinary matter.
                2.2.3.1.2. Requires dark-energy stuff to correlate with SEDES. (See, for example, Table 4.3.2 and Table 4.3.13.)
    2.3. #ENS = 1.
        2.3.1. This choice correlates with ENS1 models.

2.3.2. People might say that this selection correlates with much physics developed before the year 2016.
 2.3.2.1. For example, this choice correlates with exactly one instance of Standard Model particles.
2.3.3. People might say that this selection ...
 2.3.3.1. Requires dark-energy stuff to correlate with SEDES. (See, for example, Table 4.3.2 and Table 4.3.13.)
 2.3.3.2. Requires dark matter stuff to correlate with SEDMS. (See, for example, Table 4.3.2 and Table 4.3.14.)

3. Choose a set of solutions to consider including in the model.
 3.1. The following items show possibilities for limiting the choices.
 3.1.1. Assumptions regarding lack of physics-relevance for some solutions. (See Table 2.15.8.)
 3.1.2. The possibility that %68odd G-family solutions are not physics-relevant.

Table 2.15.6 shows some further steps that, for some situations, people might want to take to specify MM1MS1 models. For example, people making models that correlate with aspects of nature that people traditionally model via general relativity may want to do the first three steps in the table.

Table 2.15.6 Possible further steps for specifying a model (an example)

1. List solutions correlating with elementary particles (for which $\sigma = +1$) for which to choose the extent, for each SPATIM solution, to which to assume that ...
 1.1. #b" = 2 (and, therefore, #b" ≠ 3).
 1.1.1. Perhaps, only G-family solutions appear on the list.
 1.1.1.1. For example, to the extent phenomena people model via general relativity are relevant, people might include in the list 244G4&.
2. List solutions correlating with elementary particles (for which $\sigma = +1$) for which to choose the extent, for each solution, to which to assume that ...
 2.1. #b' = 3.
 2.1.1. People might say that this choice correlates with being able to use, on at least not too large a scale, models that correlate with special relativity.
 2.2. #b' = 2.
 2.2.1. People might say that this choice implies that #b" = 2 also pertains for the specific solution.
3. Choose, for each solution for which $\sigma = +1$ and at least one of #b' and #b" equals 2, ...

    3.1. To what extent and how to modify relevant results above (such as results Table 2.13.12 shows). (See Table 2.13.43.)
4. Choose, for each of the 3QIRD-subfamilies, a number of generations, #GEN(3/2). (See Table 2.6.4.)
5. Choose a value of #E` for each G-family solution for which #E = 0. (See, for example, Table 2.13.20.)

~ ~ ~

This subsection provides a list of topics about which people might make choices when using the MM1 meta-model to develop one or more models.

Table 2.15.7 notes topics about which people might make choices when using the MM1 meta-model to develop one or more models. People might say that we define and limit, via this monograph, the MM1 meta-model by choices such as choices Table 2.15.1 shows. People might say that, for many purposes, work including Table 2.15.5 and Table 2.15.6 provides a more tractable list than does Table 2.15.7.

Table 2.15.7 Some topics correlating with the MM1 meta-model and with choices available when developing a MM1MS1 model

1. Aspects of nature to try to model.
2. Limits regarding #E, #P.
3. The extents to which ground-state solutions correlate with aspects with which the model should correlate.
4. Types of models to explore and the extents to which results might correlate with nature. Examples of types of models include ...
    4.1. INTERN, EXTINT, ENVIRO, and FERTRA.
        4.1.1. (ENVIRO:) SPATIM, FRERAN, and COMPAR.
    4.2. LADDER and DIFEQU.
    4.3. CORMAT and CORPHY.
5. Relevance of and interpretations regarding various groupings of oscillators. For example, ...
    5.1. QE-like and/or QP-like.
    5.2. Various oscillator pairs.
    5.3. Other sets of oscillators.
6. Relationships between MM1MS1 models and other models.
7. Physics-relevance of work and/or solutions. (See, for example, Table 2.15.8).
8. Families and subfamilies of solutions to include (or exclude).
9. Representations specific to various families.
10. Appropriateness of use, in models, of even numbers of QE-like or QP-like harmonic oscillators.
11. Ranges and interpretations of N(P0) for ground states.
12. Applicability of traditional and/or non-traditional solutions.

13. Needs for normalization (regarding DIFEQU solutions).
14. Correlations (for numbers such as S) between solutions and physics quantities.
15. Roles and significances of σ (regarding FRERAN solutions, COMPAR solutions, and possibly other solutions).
16. Correlations between LADDER solutions and DIFEQU solutions.
17. Correlations between solutions and fields.
18. Correlations between solutions and particles.
19. Correlations between field solutions and particle solutions.
20. Possibilities that multiple elementary particles correlate (beyond correlations we discuss) with a solution.
21. Limits regarding parameters (such as S).
22. The extent to which to consider DIFEQU solutions (that might correlate directly with fields and/or particles) for which $D^* \neq 3$.
23. The extent of relevance for projections from solutions for which $D^* = 3$ to mathematical constructs for which $D^* = 2$ or $D^* = 1$.
24. Other uses for solutions which $D^* = 2$ or $D^* = 1$.
25. Correlations between relevant values of η and the notions of fields and particles.
26. Correlations between phenomena that might correlate with the E0-and-P0 oscillator pair and applications of DIFEQU math for which $D^* = 2$.
27. Significance of aspects of solutions for which $D^* + 2\nu = 0$.
28. The extent to which some integers related to $D^* + 2\nu = 0$ solutions correlate with approximate relative masses of at least the Z, W, and Higgs bosons.
29. The extent to which, for fermion elementary particles and their fields, min(D, 2(2S + 1)) provides a factor relevant to calculating a maximum number of candidate solutions.
30. The extent to which, for fermion elementary particles, generations provides a factor relevant to calculating a maximum number of candidate solutions.
31. Modeling that correlates with a value for #GEN(3/2).
32. Roles for models that involve more than one QE-like coordinate.
33. The extent to which $\sigma = -1$ correlates with phenomena and/or models.
34. The extent of correlations between FERTRA representations and fermion INTERN representations.
35. The extent of correlations between SPATIM symmetries, INSSYM7 symmetry, uses of symmetries in models, and nature.
36. Algebra for combining various (a`;b`)< and/or (a`;b`)>.
37. Correlations between various components of (1;3)> symmetries and conservation laws.
38. Possible new conservation laws, such as conservation of fermion generation for some vertices.

39. Relationships between conservation of angular momentum and quantization related to angular momentum.
40. Relationships between FRERAN boost-related symmetries and various concepts (such as, limitation on velocities and conservation of rest energy).
41. Possible significance of oscillator swap symmetries.
42. Interpretations of INSSYM7.
43. Physics-relevant values for #ENS.
44. Values of #E` for G-family solutions for which #E = 0.
45. Construction and interpretations of various sets of solutions (for example, FRERAN LADDER solutions).
46. Relationships between various sets of solutions.
47. Choices of G-family solutions, for example ...
    47.1. EACUNI set or SOMMUL set.
    47.2. %68even or %68both.
    47.3. For ENS48 models, definitions and roles of OME, DME, DEE, various G#XY. (See Table 2.13.15.)
        47.3.1. Here, # = 2, 6, or 48; X = OM, DM, or DE; and Y = OM, DM, or DE.
    47.4. For ENS48 models, relationships between OMS, DMS, DES, and other stuff and OME, DME, DEE, and various G#XY. (See Table 4.3.2 and Table 2.13.15.)
    47.5. For ENS6 models, considerations similar to those (just above) for ENS48 models.
48. Correlations between oscillators (in representations for non-G-family elementary particles) and various G-family forces.
    48.1. For example, between P0 and 4G4&.
49. For various particles and/or fields, choices of #b'.
50. For various particles and/or fields, choices of #b".
51. Metrics to consider as pertaining to energy-momentum space coordinates and/or to space-time coordinates.
52. The extents to which to consider that MM1MS1-currents should be treated as tensors (with rank ≥ 2) and not just as vectors ...
    52.1. Especially, for cases in which a relevant #b' = 2 and/or a relevant #b" = 2.
    52.2. Possibly, especially, regarding currents that interact with 4G4&.
53. Details regarding models pertaining to currents that correlate with the symbol 3v (for example, currents that correlate with 2G24&). (See Table 6.1.3.)
54. Details regarding interactions between elementary bosons and zero-mass elementary particles.
55. Details regarding modeling interactions between individual G-family bosons and an object (for example, an astrophysical object) that contains many elementary particles and/or composite particles.

56. Details regarding modeling interactions between (classical-physics-like) fields based on G-family bosons and an object (for example, an astrophysical object) that contains many elementary particles and/or composite particles.
57. Strengths for interactions mediated by G-family bosons other than 2G2& and 4G4&.
58. Strengths for interactions mediated by non-G-family bosons (perhaps, other than W-family bosons).
59. Possible significance of 17 oscillators and/or SU(17).
60. Roles and mechanisms regarding G-family channels.
61. Possibilities of other families having channels.
62. Aspects correlating #"G with conservation of fermion generation.
63. The extent to which INSSYM7 might be incompatible with COMPAR solutions, symmetries, and interpretations.
64. Limits on applicability, for elementary particles for which $\sigma = -1$, for FRERAN solutions and for COMPAR solutions.
65. Treatments regarding quark-and-boson states other than quark-and-gluon composites.
66. Aspects regarding color-charge analogs for spin-3/2 fermions and 4Y-related bosons.
67. Possibly, other.

~ ~ ~

This subsection notes some criteria people might use to decide to assume that solutions are not physics-relevant.

People might say that Table 2.15.8 pertains.

Table 2.15.8 Assumptions regarding lack of physics-relevance for some solutions

1. Particle (that would correlate with solution) shown not to exist in nature.
2. Lack of an appropriate D.
    2.1. For a specific solution, the number of dimensions, D, correlating with work similar to work that Table 2.5.16 summarizes would be undefined or negative.
    2.2. Examples include ...
        2.2.1. The 3C-subfamily. (See Table 2.3.1.)
        2.2.2. The 4W-subfamily. (See Table 2.3.1.)
3. Inappropriate redundancy.
    3.1. For a specific solution, this solution and at least one other similar solution would correlate with one known elementary particle or one candidate elementary particle.
    3.2. Examples include ...

3.2.1. The 3C-subfamily. (See Table 2.3.1.)
3.2.2. G-family solutions for which #P ≥ A = [10]. (See Table 2.3.13.)

4. Inappropriate square of rest mass.
   4.1. For a subfamily of solutions, at least one solution within the subfamily would correlate with an elementary particle for which applicable models would calculate an inappropriate square of the mass.
      4.1.1. For example, models for rest mass calculate, for a supposedly non-zero mass particle, a rest mass of 0.
      4.1.2. For example, models for rest mass calculate, for a particle, a negative number for the square of the rest mass.
   4.2. Examples may include ...
      4.2.1. The 4W-subfamily. (See Table 2.3.1 and Table 3.8.2.)
5. Non-excitable.
   5.1. For a specific solution, math correlates with a lack of excitations of the ground state.
   5.2. Examples include ...
      5.2.1. The 0Y-subfamily. (See Table 2.3.3.)
6. Too large a value of S.
   6.1. S would be > 2.
      6.1.1. [Physics:] The limit of S ≤ 2 for elementary particles comes from field theory. (See Table 2.3.10.)
7. Redundancy specific to a relevant model.
   7.1. This pertains to G-family solutions and some models.
      7.1.1. This correlates with two choices regarding models. (See, for example, Table 3.2.1.)
         7.1.1.1. We denote one choice by the term EACUNI set (of solutions), as in ...
            7.1.1.1.1. For each G-family elementary particle, exactly one solution correlates with the particle.
         7.1.1.2. We denote one choice by the term SOMMUL set (of solutions), as in ...
            7.1.1.2.1. For each of some G-family elementary particles, more than one solution correlates with the particle.
               7.1.1.2.1.1. That number of solutions equals the number of channels that correlate with the particle. (See Section 2.13.)
      7.1.2. People might say that the set of G-family elementary particles does not depend on this choice regarding models.
   7.2. For example, ...
      7.2.1. People might say that people can assume that ...

7.2.1.1. Out of an entire set of epnG%& solutions for which all the values of n are identical and all the values of % are identical (and for which p = #P is not greater than 8), ...

7.2.1.1.1. The solution with the least value of p (and, it turns out, the least value of e) is physics-relevant.

7.2.1.1.2. Each solution with a less-than-minimal value of p (and, it turns out, a less-than-minimal value of e) is not physics-relevant.

7.2.2. People might say that this approach correlates with a definition of channel that correlates with Table 2.13.12.

7.2.3. For such approaches, we use the term EACUNI models (or EACUNI-set models).

7.3. However, ...

7.3.1. People might say that people might assume that ...

7.3.1.1. Out of an entire set of epnG%& solutions for which all the values of n are identical and all the values of % are identical (and for which p = #P is not greater than 8), ...

7.3.1.1.1. Each solution is physics-relevant.

7.3.2. People might say that each solution correlates with a channel.

7.3.3. For such approaches, we use the term SOMMUL models (or, SOMMUL-set models).

People might say that Table 2.15.9 pertains. (Regarding the terms %68even, %68odd, and %68both, see discussion preceding Table 3.3.3 and see Table 3.3.3. See, also, Table 3.13.2.)

Table 2.15.9 Possible lack of physics-relevance for some solutions, plus definitions of the terms %68even, %68both, and %68odd

1. Inappropriate appearance of 6 ∈ % or 8 ∈ %.

1.1. This pertains to G-family solutions. (See Table 3.13.2.)

1.2. Each G-family solution correlates with a symbol of the form epnG%&.

1.2.1. % denotes a sub-list that can include up to four even integers.

1.2.2. The even integers are 2, 4, 6, and 8.

1.2.3. The spin correlating with a G-family solution correlates with 2S equaling (the absolute value of) an arithmetic combination that features, for each element of %, plus or minus that element.

1.3. The criterion S ≤ 2 disqualifies from being physics-relevant ...

1.3.1. % = 6.

1.3.2. % = 8.

- 1.4. People might say that, for a % for which 6 ∈ % or for which 8 ∈ % to be physics-relevant, ...
  - 1.4.1. Both 6 ∈ % and 8 ∈ % must pertain.
  - 1.4.2. The arithmetic contribution to 2S must include one of ...
    - 1.4.2.1. $\pm 2 = \pm|-6 + 8|$.
    - 1.4.2.2. $\pm 2 = \pm|+6 - 8|$.
- 1.5. For the choice that, for each physics-relevant G-family solution, either none or both of 6 and 8 must appear in %, ...
  - 1.5.1. We use the term %68even.
    - 1.5.1.1. Here the word even denotes that the total number of appearances and 6 and/or 8 in each solution is 0 or 2.
- 1.6. For those solutions for which exactly one of 6 and 8 appears in %, ...
  - 1.6.1. We use the term %68odd.
- 1.7. Possibly, ...
  - 1.7.1. All %68odd solutions are not physics-relevant.
- 1.8. We use the term %68both to denote the union of the %68even set and the %68odd set.

~ ~ ~

This subsection suggests considerations and results that may be relevant for many models correlating with the MM1 meta-model.

Table 2.15.10 pertains. Here and elsewhere, we call attention to a gravitational somewhat analog to electromagnetism's magnetic field. We do so, in part, because we are uncertain as to the extent traditional physics models take into account effects that would correlate with such an analog. (See, for example, Section 4.7, Section 4.8, and Section 4.9.)

Table 2.15.10 Characteristics shared by many models that correlate with the MM1 meta-model

1. For all families except the G- and Y-families, ...
   - 1.1. Aspects related to energy and momentum correlate with the E0-and-P0 oscillator pair.
   - 1.2. For (at least) spin-1/2 and spin-1 elementary particles, ...
     - 1.2.1. Aspects related to charge correlate with correlate with the E2-and-E1 oscillator pair and with the P1-and-P2 oscillator pair.
   - 1.3. Aspects related to spin correlate with #P.
2. For interactions mediated by G-family bosons, ...
   - 2.1. Aspects related to charge (and charge-related MM1MS1-currents) of the interacting (generally, non-G-family) particles correlate with (regarding the G-family) the P1-and-P2 oscillator pair.

2.2. Aspects related to energy-and-momentum MM1MS1-currents of the interacting (generally, non-G-family) particles correlate with (regarding the G-family) the P3-and-P4 oscillator pair.
   2.2.1. Regarding a gravitational (or 4G4&-mediated) interaction between two objects that move parallel (not antiparallel) to each other (and, therefore produce energy-and-momentum currents that parallel {not antiparallel} each other), ...
      2.2.1.1. People might say that a gravitational somewhat analog to electromagnetism's magnetic field provides for attraction between the two currents. (See, for example, Section 4.7.)

2.3. Aspects related to magnetic dipole moments of the interacting (generally, non-G-family) elementary particles correlate with (regarding the G-family) the P1-and-P2 and P3-and-P4 oscillator pairs.

2.4. Aspects related to spins of the interacting (generally, non-G-family) particles correlate with (regarding the G-family) the P5-and-P6 and P7-and-P8 oscillator pairs.

~ ~ ~

This subsection provides perspective about some models we discuss in this monograph.

Table 2.15.11 pertains. Because the OME might contain DMS (or, SEDMS or dark matter stuff) and/or DES (or, SEDES or dark-energy stuff) and because each DME might contain DES, the table uses the phrase (that) most significantly correlates with. (For definitions of DMS and DES, see Table 4.3.2. For possible examples of SEDMS, see Table 4.3.14. For possible examples of SEDES, see Table 4.3.13. Compare with definitions of OME, DME, and DEE, per Table 2.13.15.)

Table 2.15.11 #ENS-related models we discuss in this monograph

1. We discuss ENS48 models (including ENS48" models).
   1.1. We think these models can correlate with known data and might correlate with future physics.
   1.2. Here, each ensemble that most significantly correlates with dark-energy stuff has similarities to each ensemble that most significantly correlates with dark matter and to the ensemble that most significantly correlates with ordinary matter.
2. We discuss ENS6 models (including ENS6' models).
   2.1. We think these models can correlate with known data and might correlate with future physics.

    2.2. Here, dark-energy stuff correlates with aspects of each of the 6 ensembles that most significantly correlate with dark matter (5 ensembles) or most significantly correlate with ordinary matter (1 ensemble).

3. We discuss ENS1 models.
    3.1. We think these models can correlate with known data and might correlate with future physics.
    3.2. Here, dark-energy stuff correlates with aspects of the one physics-relevant ensemble.
    3.3. Here, dark matter correlates with aspects of the one physics-relevant ensemble.

Chapter 3 From the MM1 meta-model to particles and properties

Chapter 3 (From the MM1 meta-model to particles and properties) starts from solutions we identify in the previous chapter, shows possibilities for listing known and possible elementary particles, and shows possibilities for modeling properties of particles and for modeling interactions in which particles partake.

~ ~ ~

People might say that results in Chapter 3 (From the MM1 meta-model to particles and properties) might correlate with or provide insight about the following aspects of physics. A list of known elementary particles. A list of possible elementary particles. Some interactions in which the particles partake. Spins for known and possible elementary particles. Possible charges, for some possible elementary bosons. The weak mixing angle. The ratio of the mass of the Z boson to the mass of the Higgs boson. Possible approximate masses, for possible boson elementary particles. A possible mass, for the tauon, that could be more accurate than a mass determined from experimental results. Correlations between measurements of the tauon mass and the gravitational constant. Mechanisms correlating with neutrino oscillations. Possible new appearances of the fine-structure constant.

Section 3.1 Introduction

Section 3.1 discusses some results that span the previous chapter and this chapter. We list families of known and possible elementary particles. We discuss notation for solutions and notation for elementary particles. We note that the catalog of known and candidate elementary particles correlates with the MM1 meta-model and, therefore, with all MM1MS1 models.

~ ~ ~

This subsection discusses families of known and possible particles, elementary-particle subfamilies, and types of known and possible composite particles.

This subsection summarizes work in Chapter 2 and work that runs from Section 3.2 through Section 3.6.

T.J. Buckholtz, *Models for Physics of the Very Small and Very Large*,
Atlantis Studies in Mathematics for Engineering and Science 14,
DOI 10.2991/978-94-6239-166-6_3

Table 3.1.1 shows families of particles. S denotes particle spin/ħ. Odd and even denote types of nonnegative integers. The m' column shows whether particle rest masses are non-zero or zero. In the examples column, non-parenthesized wording reflects traditional terminology for known particles. Parentheses denote traditional terminology for a particle that some people predict and that has not yet been found. Correlations between Y-family solutions and elementary particles differ from correlations between solutions in other families and elementary particles. (See Table 3.1.3.) In the family column, symbols of the form X- match symbols for families of solutions. (See Table 2.3.4.) Composite particles are not elementary particles. Each of the first two G-family rows features one particle. The last G-family row provides information about more than one subfamily of the G-family. For a row that does not pertain to composites or to the G-family, the number of subfamilies equals the number of numbers the S column shows. The generations column shows numbers of generations for elementary fermions. Regarding neutrinos, the use of quotation marks recognizes that people sometimes use the term flavor (and not the term generation) when discussing neutrinos. For generations, if a row shows a value of S = 1/2, 3 generations pertains for S = 1/2. For generations, if a row shows a value of S = 3/2, see Table 2.6.4 regarding the value of #GEN(3/2).

Table 3.1.1 Families of elementary particles, plus some composite particles

S	σ	m'	Examples	Family	Generations
1/2	+1	≠ 0	Charged leptons	C-	3
1/2; 3/2	+1	0	Neutrinos	N-	"3"; #GEN(3/2)
0	+1	≠ 0	Higgs boson	H-	
Odd/2	+1	≠ 0	Baryons	Composite	
Even/2	+1	≠ 0	Mesons	Composite	
1	+1	≠ 0	W and Z bosons	W-	
1	+1	0	Photon	G-	
2	+1	0	(Graviton)	"	
1, 2	+1	0		"	
1/2; 3/2	−1	≠ 0	Quarks	Q-	3; #GEN(3/2)
3/2	−1	≠ 0		I-	#GEN(3/2)
1/2; 3/2	−1	0		R-	3; #GEN(3/2)
3/2	−1	0		D-	#GEN(3/2)
1, 2	−1	0	Gluons	Y-	
0, 1, 2	−1	≠ 0		O-	

The families Table 3.1.1 lists include all (as of 2015) known elementary particles.

Work in this monograph points to possibilities for yet-to-be-found elementary particles correlating with N-, G-, QIRD-, YO-family solutions. Table 3.1.2 pertains.

Table 3.1.2 Notes about the possible existence of suggested yet-to-be-found elementary particles

1. This monograph may attempt to correlate, with some solutions, particles that nature does not exhibit.
2. In some cases, non-existence of such candidate particles would not significantly contradict other aspects of this monograph.
3. Table 7.4.2 discusses sets of elementary particles needed to match data about nature.
4. In places, when referring to possible elementary particles, this monograph may omit mentioning the notions of possible or candidate.

Table 3.1.3 pertains.

Table 3.1.3 Notes about correlations between and notation for solutions and elementary particles

1. For each family other than the G-family and the Y-family, the notation for an elementary particle equals the notation for the solution correlating with the particle.
   1.1. For cases in which the antiparticle differs from a particle, such notation distinguishes the solution for the antiparticle from the solution for the particle.
2. For the G-family, ...
   2.1. Notation of the form epnG%& correlates with solutions.
      2.1.1. Here, each of e, p, and n is an even integer.
      2.1.2. Here, % stands for an ordered set of even integers.
         2.1.2.1. We correlate the symbol % with the term sub-list.
   2.2. Notation of the form nG%& correlates with elementary particles.
   2.3. For each of epnG%& and nG%&, the next equation pertains.

$$n = 2S \tag{3.1}$$

   2.4. For each solution or particle, ...
      2.4.1. Two modes pertain.
         2.4.1.1. One mode is left-circular polarization.
         2.4.1.2. One mode is right-circular polarization.
      2.4.2. People might say that each mode is an anti-mode for the other mode.
         2.4.2.1. Here, we use the term anti-mode instead of antiparticle.
3. For the Y-family, ...
   3.1. For the 2Y-subfamily, ...
      3.1.1. The corresponding elementary particles are gluons.

- 3.1.2. Each gluon correlates with a sum of terms.
- 3.1.3. Each term in a sum combines ...
  - 3.1.3.1. An element that erases a color charge from a fermion.
  - 3.1.3.2. An element that paints a color charge onto a fermion.
- 3.1.4. The number of color charges is 3.
- 3.1.5. Each element correlates with a Y-family solution.

3.2. For the 4Y-subfamily, ...

- 3.2.1. The number of color-charge analogs is 5.
- 3.2.2. Considerations similar to considerations for the 2Y-subfamily pertain.
  - 3.2.2.1. People might choose not to use the terminology color charge.
  - 3.2.2.2. People might choose not to use the terminology gluon.

~ ~ ~

This subsection provides perspective regarding discussion in this chapter, the MM1 meta-model, and MM1MS1 models.

Table 3.1.4 pertains.

Table 3.1.4 Correlations between #ENS and aspects of elementary-particle physics

1. People might say that some aspects (especially some aspects of Section 3.2 through Section 3.6) regarding the various families of elementary particles couple closely enough with INTERN LADDER solutions that people might consider those aspects to be part of an extended notion of the MM1 meta-model.
   - 1.1. For example, a catalog of candidate solutions correlates with the meta-model. People may, for specific MM1MS1 models, focus on just subsets of that catalog. (See Table 2.4.1.)
2. People might say that many aspects (of Section 3.2 through Section 3.13) regarding properties of elementary particles and regarding interactions in which elementary particles partake do not depend on a value of #ENS.

Section 3.2 Aspects of models for MM1MS1-photons and gravitons

Section 3.2 discusses some aspects of models for MM1MS1-photons and gravitons. We discuss two concepts (the EACUNI set and the SOMMUL set) for sets of solutions that might correlate with MM1MS1-photons and gravitons. We show INTERN LADDER solutions that correlate with MM1MS1-photons and INTERN LADDER solutions that correlate with gravitons. We discuss aspects of mappings from an energy-momentum space that correlates with INTERN LADDER solutions to space-time coordinates.

~ ~ ~

This subsection provides perspective about this section.

Traditional physics includes four fundamental interactions. One of those interactions is electromagnetism. The phenomenon of light correlates with electromagnetism. Another of those interactions is gravitation.

Traditional physics provides detailed models for various aspects of light and photonics. People consider that there is much understanding of quantum photonics.

Traditional physics provides detailed models for various aspects of gravity. People use general relativity to describe some of those aspects.

People might say that the topic of quantum gravity is not settled. Some people mention the notion that there may not be much practical use for a quantum theory of gravity. Some people try to develop theories of quantum gravity.

Generally, work on the topic of quantum gravity seems to consider photonics and gravity as being quite different. Yet, traditional classical physics exhibits similarities between electromagnetism and gravity. For example, each concept features an interaction strength that people characterize as one divided by the square of the distance between the centers of relevant property of two interacting objects. For gravity, the property is often characterized as mass. For electromagnetism the property is often characterized as charge.

This section finds similarities between aspects of solutions that we correlate with MM1MS1-photons and aspects of solutions that we correlate with gravitons (or, with quantum gravity).

~ ~ ~

This subsection discusses a non-traditional way to represent some aspects of photons.

One traditional representation for photons features two harmonic oscillators. People think of each oscillator as correlating with an axis perpendicular to the direction of motion of the photon and perpendicular to the direction of the axis correlating with the other oscillator. No correlation exists between oscillators or the respective quantum numbers that represent excitations. People might say that people think of this representation in a context of space-time coordinates.

In this monograph, solution 022G2& correlates with MM1MS1-photons. Table 3.2.1 pertains. People might say that people can think of this representation in a context of a quantum energy-momentum space. (See Table 2.15.8.)

Table 3.2.1 EACUNI-set models, SOMMUL-set models, and MM1MS1-photon representations

1. For EACUNI-set models, ...
   - 1.1. The following solution correlates with a MM1MS1-photon.
     - 1.1.1. 022G2&.
   - 1.2. The following solutions are not used.
     - 1.2.1. 242G2&.
     - 1.2.2. 462G2&.
     - 1.2.3. 682G2&.
   - 1.3. Section 2.13 discusses channels.
2. For SOMMUL-set models, ...
   - 2.1. The following solutions combine to correlate with a MM1MS1-photon.
     - 2.1.1. 022G2&.
     - 2.1.2. 242G2&.
     - 2.1.3. 462G2&.
     - 2.1.4. 682G2&.
   - 2.2. These four solutions correlate with the concept of four channels.
3. For of EACUNI-set models and SOMMUL-set models, ...
   - 3.1. The symbol 2G2& represents a MM1MS1-photon.

People can envision solution 022G2& as follows. A quantum-centric concept of energy-momentum space provides context. N(P2L) (or, N(P1)) counts the number of left-circular polarized excitations. N(P2R) (or, N(P2)) counts the number of right-circular polarized excitations. 022G2& cannot be both left-circularly polarized and right-circularly polarized at the same time. At least one of N(P2L) and N(P2R) must be 0. For example, if N(P2L) = 2 and N(P2R) = 0, an addition of a unit of right-circular polarization results in N(P2L) = 1 and N(P2R) = 0. Each mode (the left-circular polarization mode and the right-circular polarization mode) functions somewhat as an antiparticle for the other mode. (In Table 3.1.3, we suggest the term anti-mode.)

The antiparticle-like aspect pertains for each individual 022G2&.

The possibility for adding a unit to an N(P2L) for a 022G2& solution correlating with a specific momentum (magnitude and direction, relative to an observer) does not correlate with the N(P2R) for a solution correlating 022G2& with a different momentum (relative to the same observer).

A complete description of a MM1MS1-photon includes a momentum and values of N(P2L) and N(P2R). The momentum is relative to an observer. Different observers could measure different momenta. Polarization values are invariant with respect to choice of observer. In part because observers cannot travel (relative to light or to each other) at speeds exceeding c (or, the speed of light), all observers can agree on the values of N(P2L) and N(P2R). (Technically, this discussion relies on some idealizations. For example, the discussion assumes that a second (in time) observer can observe the same MM1MS1-photon without a first observer having disturbed the MM1MS1-photon.)

Table 3.2.2 shows INTERN LADDER solutions for ground states for the two 022G2& modes. The first mode has left-circular polarization. The second mode has right-circular polarization. The symbol @ denotes a 0 that does not change for the mode. N(E0) + 1/2 correlates with energy.

Table 3.2.2 022G2& modes and their ground-state INTERN LADDER solutions

E	E	E	E	E	E	E	P	P	P	P	P	P	P	P	P	
6R	6L	4R	4L	2R	2L	0	0	2L	2R	4L	4R	6L	6R	8L	8R	Mode
						0	−1	0	@							022G2L
						0	−1	@	0							022G2R

Table 3.2.3 shows the Nth excited state for each mode. N(E0) + 1/2 correlates with energy. Here, N is a nonnegative integer.

Table 3.2.3 022G2& modes and their Nth excited states

E	E	E	E	E	E	E	P	P	P	P	P	P	P	P	P	
6R	6L	4R	4L	2R	2L	0	0	2L	2R	4L	4R	6L	6R	8L	8R	Mode
						N	−1	N	@							022G2L
						N	−1	@	N							022G2R

Table 3.2.4 shows the ground state and the symbol for 022G2&. (Regarding the symbol 'I, see, for example, Table 2.7.1.)

Table 3.2.4 022G2& ground-state INTERN LADDER solution

E	E	E	E	E	E	E	P	P	P	P	P	P	P	P	P	
6R	6L	4R	4L	2R	2L	0	0	2L	2R	4L	4R	6L	6R	8L	8R	Solution
						'I	−1	0	0							022G2&

~ ~ ~

This subsection discusses how to connect the above representation for aspects of MM1MS1-photons with energy-momentum-space coordinates and with space-time coordinates.

Work above regarding 022G2& features four harmonic oscillators. People can associate four energy-momentum-space coordinates - e0, p0, p1, and p2 - with the 022G2& representation. People might associate four space-time coordinates - t0, x0, x1, and x2 - with the four energy-momentum-space coordinates. Here, the space-time coordinate-x0 axis parallels the motion of the 022G2&. Table 3.2.5 pertains. The first statement regarding each of p1 and p2 correlates with developing linear-polarization constructs from circular-polarization constructs. Such steps parallel work in traditional quantum physics.

Table 3.2.5 Coordinates relevant for 022G2&

1. The coordinate e0 associates with oscillator E0.
2. The coordinate p0 associates with oscillator P0.
3. The coordinate p1 associates with oscillators P2L and P2R via (022G2L + 022G2R)/$2^{1/2}$.
4. The coordinate p2 associates with oscillators P2L and P2R via (022G2L − 022G2R)/({−i}$2^{1/2}$).
5. The coordinate t0 associates with e0.
6. The coordinate x0 associates with p0.
7. The coordinate x1 associates with p1.
8. The coordinate x2 associates with p2.

Table 3.2.6 correlates coordinates with the terms QE-like and QP-like.

Table 3.2.6 QE-like coordinates and QP-like coordinates, for a specific 022G2&

Term	Energy-momentum-space coordinates	022G2&-specific space-time coordinates
QE-like	e0	t0
QP-like	p0, p1, p2	x0, x1, x2

More generally, people use space-time coordinates that are not specific to a specific 022G2&. Table 3.2.7 shows notation for general space-time coordinates.

Table 3.2.7 QE-like coordinates and QP-like coordinates, for a specific 022G2& and a specific observer

Term	022G2&-specific space-time coordinates	Observer-specific space-time coordinates
QE-like	t0	t
QP-like	x0, x1, x2	x, y, z

Symmetries that link 022G2&-specific coordinates and observer-specific coordinates can include time translation, spatial translation, rotation, and boost. These symmetries combine to form (1;3)> symmetries, the Poincare-group symmetries. (See Table 2.11.2.) These symmetries correlate with FRERAN SPATIM symmetries.

~ ~ ~

This subsection discusses a representation for 244G4&.
In this monograph, 244G4& correlates with gravitons. Table 3.2.8 pertains.

Table 3.2.8 EACUNI-set models, SOMMUL-set models, and graviton representations

1. For EACUNI-set models, ...
    1.1. The following solution correlates with a graviton.
        1.1.1. 244G4&.
    1.2. The following solutions are not used.
        1.2.1. 464G4&.
        1.2.2. 684G4&.
    1.3. Section 2.13 discusses channels.
2. For SOMMUL-set models, ...
    2.1. The following solutions combine to correlate with a graviton.
        2.1.1. 244G4&.
        2.1.2. 464G4&.
        2.1.3. 684G4&.
    2.2. These three solutions correlate with the concept of three channels.
3. For both sets of models, ...
    3.1. The symbol 4G4& represents a graviton.

Table 3.2.9 shows ground states for INTERN LADDER solutions for 244G4& modes. People expect that, if gravitons exist, gravitons are spin-2 particles. Above, this monograph associates S = 2 with the P4L and P4R oscillators.

Table 3.2.9 244G4& modes and their ground-state INTERN LADDER solutions

E	E	E	E	E	E	E	P	P	P	P	P	P	P	P	P	
6R	6L	4R	4L	2R	2L	0	0	2L	2R	4L	4R	6L	6R	8L	8R	Mode
				'I	'I	'I	−1	@	@	0	@					244G4L
				'I	'I	'I	−1	@	@	@	0					244G4R

Section 3.3 Elementary particles correlating with the G-family

Section 3.3 describes the G-family of elementary particles. People might say that G-family particles include all zero-mass elementary bosons except for particles (including gluons) correlating with Y-family solutions. We list two possible sets of G-family elementary particles. For %68both models, we list 12 particles. For %68even models, we list 7 particles. We discuss models for excitations of G-family bosons. We note that G-family particles provide for interactions with spatial dependences of r^{-2}, r^{-4}, r^{-6}, and r^{-8}. People might say that we point to a G-family member, 2G24&, that interacts with elementary-particle magnetic dipole moments. People might say that we anticipate possible correlations between Standard Model photons and a pair consisting of MM1MS1-photons and 2G24&. People might say that we anticipate interactions, intermediated by 2G24& between ordinary matter and dark matter. People might say that we anticipate the interactions intermediated by 2G68& couple to spin and do not couple to charge or to energy. We discuss a possibility for mechanisms related to G-family boson excitations and channels. We correlate zero-mass for each G-family elementary particle with work above in this monograph.

~ ~ ~

This subsection provides perspective about this section.

In this section, we develop concepts for a family of solutions that could correlate with (not only MM1MS1-photons and gravitons but ...) a more-than-two-particle family of zero-mass, free-ranging elementary particles.

We discuss possible new interactions. These interactions would, under most present circumstances, have significant impact mostly on very small scales and/or on very large scales. Black holes provide a present circumstance for which other-scale significant impact is possible.

~ ~ ~

This subsection shows G-family INTERN LADDER solutions that could correlate with elementary particles.

Table 3.3.1 shows candidates for ground-state INTERN LADDER solutions for G-family bosons for which $2 \leq \#P \leq 8$. Of the candidates this table shows, only 466G6&, 686G6&, 688G8&, and 486G28& would have $S > 2$. The table includes these solutions so as to improve its visual symmetries and overall utility. The table uses strikethrough text to denote these four solutions. We think these solutions are not physics-relevant. We deemphasize these solutions. (See Table 2.15.8.)

Table 3.3.1 Ground-state INTERN LADDER solutions for the G-family

E 6R	E 6L	E 4R	E 4L	E 2R	E 2L	E 0	P 0	P 2L	P 2R	P 4L	P 4R	P 6L	P 6R	P 8L	P 8R	Solution
						'I	−1	0	0							022G2&
				'I	'I	'I	−1	0	0	@	@					242G2&
				'I	'I	'I	−1	@	@	0	0					244G4&
						'I	−2	0	0	0	0					042G24&
		'I	'I	'I	'I	'I	−1	0	0	@	@	@	@			462G2&
		'I	'I	'I	'I	'I	−1	@	@	0	0	@	@			464G4&
		~~'I~~	~~'I~~	~~'I~~	~~'I~~	~~'I~~	~~−1~~	~~@~~	~~@~~	~~@~~	~~@~~	~~0~~	~~0~~			~~466G6&~~
				'I	'I	'I	−2	0	0	0	0	@	@			262G24&
				'I	'I	'I	−2	0	0	@	@	0	0			264G26&
				'I	'I	'I	−2	@	@	0	0	0	0			262G46&
						'I	−3	0	0	0	0	0	0			064G246&
'I	'I	'I	'I	'I	'I	'I	−1	0	0	@	@	@	@	@	@	682G2&
'I	'I	'I	'I	'I	'I	'I	−1	@	@	0	0	@	@	@	@	684G4&
~~'I~~	~~'I~~	~~'I~~	~~'I~~	~~'I~~	~~'I~~	~~'I~~	~~−1~~	~~@~~	~~@~~	~~@~~	~~@~~	~~0~~	~~0~~	~~@~~	~~@~~	~~686G6&~~
~~'I~~	~~'I~~	~~'I~~	~~'I~~	~~'I~~	~~'I~~	~~'I~~	~~−1~~	~~@~~	~~@~~	~~@~~	~~@~~	~~@~~	~~@~~	~~0~~	~~0~~	~~688G8&~~
		'I	'I	'I	'I	'I	−2	0	0	0	0	@	@	@	@	482G24&
		'I	'I	'I	'I	'I	−2	0	0	@	@	0	0	@	@	484G26&
		~~'I~~	~~'I~~	~~'I~~	~~'I~~	~~'I~~	~~−2~~	~~0~~	~~0~~	~~@~~	~~@~~	~~@~~	~~@~~	~~0~~	~~0~~	~~486G28&~~
		'I	'I	'I	'I	'I	−2	@	@	0	0	0	0	@	@	482G46&
		'I	'I	'I	'I	'I	−2	@	@	0	0	@	@	0	0	484G48&
		'I	'I	'I	'I	'I	−2	@	@	@	@	0	0	0	0	482G68&
				'I	'I	'I	−3	0	0	0	0	0	0	@	@	284G246&
				'I	'I	'I	−3	0	0	0	0	@	@	0	0	282G248&
				'I	'I	'I	−3	0	0	@	@	0	0	0	0	284G268&
				'I	'I	'I	−3	@	@	0	0	0	0	0	0	282G468&
						'I	−4	0	0	0	0	0	0	0	0	084G2468&

~ ~ ~

This subsection provides notation, for G-family solutions, for the number of QP-like harmonic oscillator pairs for which excitement is possible.

Table 3.3.2 pertains. For example, for 244G4&, excitement (for QP-like oscillators) is possible for just the P4L-and-P4R oscillator pair. Here, the sub-list % includes just the integer 4. Here, n"(%) = 1. For each mode of 244G4&, only n"(%) = 1 QP-like oscillator can be excited. More generally, for any one mode for any solution in Table 3.3.1, excitements impact exactly n"(%) QP-like oscillators.

Table 3.3.2 Notation, for G-family solutions, for the number of QP-like harmonic oscillator pairs for which excitement is possible

1. For a G-family solution with notation of the form epnG%&, ...

- 1.1. The symbol n"(%) denotes the number of elements in the sub-list %.
- 1.2. For each mode correlating with the solution, ...
  - 1.2.1. An excitement changes the N(P[positive integer]) value for precisely n"(%) values of positive integer.
- 1.3. The next equation pertains.

$$n''(\%) = -N(P0) \tag{3.2}$$

2. For each G-family solution that this monograph correlates with an elementary particle, the following inequalities pertain.

$$1 \leq n''(\%) \leq 4 \tag{3.3}$$

~ ~ ~

This subsection correlates G-family solutions with elementary particles.

Table 3.3.3 shows possibilities for correlations between candidates for elementary particles and G-family ground-state INTERN LADDER solutions for which $2 \leq \#P \leq 8$. The #E, #P, and solutions columns repeat information that is shown in or can be inferred from work above. Regarding correlating particles and solutions, we show correlations appropriate for the EACUNI set. (See Table 2.15.8.) People might say that the G-family includes no elementary particles for which Table 3.3.3 does not show a particle. The term %68odd pertains to each particle that the %68both column shows and the %68even column does not show.

Table 3.3.3 Elementary particles correlating with G-family INTERN LADDER solutions (EACUNI-set models)

#E	#P	Particle (%68both)	Particle (%68even)	Solution
0	2	2G2&	2G2&	022G2&
2	4			242G2&
2	4	4G4&	4G4&	244G4&
0	4	2G24&	2G24&	042G24&
4	6			462G2&
4	6			464G4&
4	6			~~466G6&~~
2	6			262G24&
2	6	4G26&		264G26&
2	6	2G46&		262G46&
0	6	4G246&		064G246&
6	8			682G2&
6	8			684G4&
6	8			~~686G6&~~
6	8			~~688G8&~~
4	8			482G24&

#E	#P	Particle (%68both)	Particle (%68even)	Solution
4	8			484G26&
4	8			~~486G28&~~
4	8			482G46&
4	8	4G48&		484G48&
4	8	2G68&	2G68&	482G68&
2	8			284G246&
2	8	2G248&		282G248&
2	8	4G268&	4G268&	284G268&
2	8	2G468&	2G468&	282G468&
0	8	4G2468&	4G2468&	084G2468&

Table 3.3.4 lists possible G-family elementary particles.

Table 3.3.4 Possible G-family elementary particles

Particle (%68both)	Particle (%68even)
2G2&	2G2&
4G4&	4G4&
2G24&	2G24&
4G26&	
2G46&	
4G246&	
4G48&	
2G68&	2G68&
2G248&	
4G268&	4G268&
2G468&	2G468&
4G2468&	4G2468&

In this monograph, subsequent work features the %68even set of G-family particles. That work deemphasizes items in the %68odd set (that is, items in the %68both set that do not belong to the %68even set). (See Table 2.15.9.)

~ ~ ~

This subsection discusses raising operators and lowering operators for G-family solutions.

Table 3.3.5 shows ground states for 042G24&. The modes have spins of 1, based on |S(P4n``) - S(P2n`)| = |2 - 1|, with S denoting spin and with n`` not equal to n`. For example, for 042G2R4L, n`` = L, S(P4L) = 2, n` = R, and S(P2R) = 1. We assume that combining 2 units of left-circular polarization with 1 unit of right-circular polarization results in 1 unit of left-circular polarization.

Table 3.3.5 042G24& modes and their ground-state INTERN LADDER solutions

E 6R	E 6L	E 4R	E 4L	E 2R	E 2L	E 0	P 0	P 2L	P 2R	P 4L	P 4R	P 6L	P 6R	P 8L	P 8R	Mode
						0	−2	@	0	0	@					042G2R4L
						0	−2	0	@	@	0					042G2L4R

Table 3.3.6 shows first excited states for 042G24& modes. Here, regarding Table 3.3.7, N(mode) = 1.

Table 3.3.6 042G24& modes and their first excited states (N(mode) = 1)

E 6R	E 6L	E 4R	E 4L	E 2R	E 2L	E 0	P 0	P 2L	P 2R	P 4L	P 4R	P 6L	P 6R	P 8L	P 8R	Mode
						2	−2	@	1	1	@					042G2R4L
						2	−2	1	@	@	1					042G2L4R

Throughout the G-family, Table 3.3.7 describes raising operators. Table 3.3.9 shows by example how this works for 042G2R4L. People can generalize from the example.

Table 3.3.7 Raising operator for G-family modes

1. The next equation shows the raising operator for G-family modes.

$$a^{+} \mid N(\text{mode}) > = (1 + N(\text{mode}))^{n'/2} \mid N(\text{mode}) + 1 > \quad (3.4)$$

1.1. Here, each of the following pertains.

$$n' = -N(P0) \quad (3.5)$$

n' = the number of QP-like oscillators for which 0 (and neither @ nor N(P0)) pertains for the ground state (3.6)

$$n' = n''(\%) \quad (3.7)$$

~ ~ ~

This subsection discusses the $(1 + N(\text{mode}))^{n'/2}$ factor in Table 3.3.7.

The mode 042G2R4L provides an example.

Table 3.3.8 shows the ground state (for which N = 0) and the first excited state (for which N = 1) for 042G2R4L. The table repeats information from Table 3.3.5 and Table 3.3.6.

Table 3.3.8 INTERN LADDER solutions for N = 0 and N = 1 for 042G2R4L

E	E	E	E	E	E	E	P	P	P	P	P	P	P	P	P	Excitation
6R	6L	4R	4L	2R	2L	0	0	2L	2R	4L	4R	6L	6R	8L	8R	
						0	−2	@	0	0	@					N = 0
						2	−2	@	1	1	@					N = 1

Table 3.3.9 shows spreading the E0 excitation between two oscillators - E0.2 and E0.1. The table also spreads the P0 quantum number between two oscillators - P0.1 and P0.2. The table shows four pairs of oscillators. One pair consists of E0.2 and E0.1. One pair consists of P0.1 and P0.2. One pair consists of P2L and P4R. One pair consists of P2R and P4L. Within each j-and-k pair, for N = 0 (that is, for the ground state of 042G2R4L), N(j) = N(k). Also, overall, Œ = 0.

Table 3.3.9 Alternative representation of N = 0 and N = 1 for 042G2R4L

E	E	P	P	P	P	P	P	P	P	P	P	Excitation
0.2	0.1	0.1	0.2	2L	2R	4L	4R	6L	6R	8L	8R	
0	0	−1	−1	@	0	0	@					N = 0
1	1	−1	−1	@	1	1	@					N = 1

Table 3.3.9 shows two linked systems, with each system having four oscillators. One system features oscillators E0.2, P0.2, P2R, and P4R. The other system features oscillators E0.1, P0.1, P2L, and P4L.

Based on this representation, P0 cannot be excited.

This example provides an n' = 2 illustration of Table 3.3.7.

For n' = 3, one spreads the quantum number for E0 equally across E0.3, E0.2, and E0.1. For n' = 3, one spreads the quantum number for P0 equally across P0.1, P0.2, and P0.3. And, so forth.

This work correlates with the concept that, for the G-family, for N(P0) = −n' ≤ −1, oscillator P0 cannot be excited. Work in the next subsection also correlates with the concept that, for the G-family, oscillator P0 cannot be excited.

~ ~ ~

This subsection discusses the concept of spatial dependences for interactions mediated by the G-family.

Table 3.3.10 pertains.

Table 3.3.10 Concepts regarding spatial dependences of interactions mediated by the G-family

1. People say that each of the electromagnetic interaction and the gravitational interaction has a spatial dependence of r^{-2}.

   1.1. Such a statement correlates with Newtonian physics and a Newtonian-physics characterization of interactions.
   1.2. People might say that, for Newtonian physics, an interaction propagates instantly indefinitely far from the object or objects that people would say generate the interaction.
2. For relativistic physics, people assume that electromagnetic interactions and gravitational interactions propagate (in a vacuum) at the speed, c, of light.
   2.1. In general relativity, people use a notion of curvature of space-time somewhat instead of a more-traditional-physics notion of a gravitational interaction.
3. Nevertheless, we find value (for purposes of this monograph) in using traditional terminology - such as spatial dependence of r^{-2}.
4. People might consider the notions that, ...
   4.1. Within and associated with a clump of stuff, energy and net charge may not change much.
   4.2. To the extent models account for the components of such a clump, ...
      4.2.1. At a distance, dynamics within the clump may have effects that people might characterize as other-than-first order.
   4.3. Work in this monograph correlates with a limit of c regarding the speed of propagation of effects.
      4.3.1. Such effects can correlate with happenings within a clump.

~ ~ ~

This subsection shows spatial dependences for interactions mediated by the G-family.

Table 3.3.11 pertains. For an epnG%& solution, −N(P0) equals the number of elements in % and equals n' in Table 3.3.7. A factor of r^{-2} correlates with each element in %. We think that work leading to and going beyond the n"(%) = −N(P0) = 2 case Table 3.3.9 shows dovetails with results Table 3.3.11 shows.

Table 3.3.11 Spatial dependences of interactions mediated by the G-family bosons

1. People might say that the spatial dependence of the interaction associated with a particle that correlates with a G-family solution is $r^{2\times N(P0)}$. Here, r denotes the distance between (in a sense of classical physics) the centers of property (such as charge {for 2G2& and the electromagnetic interaction} or mass or mass-energy {for 4G4& and gravity}). Here, N(P0) pertains to the G-family solution.

~ ~ ~

This subsection discusses a G-family interaction that interacts with elementary-fermion magnetic dipole moments.

In classical physics and traditional quantum physics, magnetic dipole moments contribute to magnetic fields.

MM1MS1-photons interact with 4-vector MM1MS1-currents for which the time-like components correlate with electric charge and the space-like components correlate with currents of electric charge. We expect that MM1MS1-photons adequately correlate with magnetic-field effects that correlate with fields generated by moving charges (including charges moving within objects).

So far, this leaves as open the question as to the extent to which MM1MS1-photons correlate with magnetic dipole moments of (at least) elementary-particle fermions (and, perhaps also, of elementary-particle bosons; and, perhaps also, of multi-particle objects that exhibit, for the particles, an overall spin-like state).

Table 3.3.12 lists aspects that we might expect regarding a boson that interacts with elementary-fermion magnetic dipole moments. We set these expectations based on successes regarding magnetic dipole moments of traditional physics. Here, we consider nominal magnetic dipole moments (that is, magnetic dipole moments other than anomalous magnetic dipole moments). Traditionally, people calculate nominal magnetic dipole moments for elementary fermions via the Dirac equation. Here, we are not considering anomalous magnetic dipole moments.

Table 3.3.12 Aspects that might correlate with a MM1MS1 elementary boson that interacts with elementary-fermion magnetic dipole moments

1. The boson has m' = 0.
2. The boson correlates with the G-family and not with the Y-family.
3. The boson has S = 1.
4. The boson correlates with a dipole-like force field.
    4.1. The force field carries information about a net size and a net spin-like state-centric orientation of the dipole moments that, in effect, create the field.
        4.1.1. Here, the term net correlates with a summation over contributions from individual elementary fermions.
        4.1.2. Here, for the impact of a single-elementary-particle magnetic dipole moment (or for the impact of a spin-like quantum-coupled magnetic dipole moment correlating with multiple elementary particles), ...
            4.1.2.1. The information the force field carries about the relevant magnetic dipole moment correlates with quantum spin-like states.

This monograph assumes that 2G24& intermediates interactions with elementary-particle magnetic dipole moments. (See Table 2.8.3.)

~ ~ ~

This subsection discusses the topic of correlating G-family interactions with electromagnetism.

Table 3.3.13 anticipates correlating G-family interactions with electromagnetism. (See Section 5.1.)

Table 3.3.13 Correlating G-family interactions with electromagnetism

1. In classical-physics electromagnetism and traditional quantum-physics electromagnetism, electromagnetism interacts similarly with each of ...
   - 1.1. Magnetic dipole moments that are generated by currents of moving charges.
   - 1.2. Magnetic dipole moments that are generated by the presence (and, in essence, spins) of (possibly non-moving) elementary particles.
2. People might say that, regarding traditional physics and thinking of ensembles (in the sense of each of ENS48 models, ENS6 models, and ENS1 models), electromagnetism that pertains to known elementary particles (and stuff made from those particles) pertains to one ensemble only.
3. We show a type of symmetries (oscillator swap symmetries) that, for 2G24&, allows for differentiation between the following two types of 2G24&-intermediated interactions (for which each type of interaction involves at least one particle in the ensemble with which people are familiar). (See Table 2.12.10.)
   - 3.1. The other particle (perhaps, a fermion) relevant to the interaction is part of the same ensemble. (This concept applies for ENS48 models, ENS6 models, and ENS1 models.)
   - 3.2. The other particle (perhaps, a fermion) relevant to the interaction is part of a different ensemble. (This concept applies for some ENS48 models and some ENS6 models, but not for ENS1 models. Applicability of this concept correlates with #E` < 6 for 2G24&. For cases in which the concept might apply, 2G24& might connect ordinary matter with one dark-matter ensemble. See, for example, Table 2.13.23. See the definition of DME-1 in Table 2.13.16.)
4. People might say that the following pertain.
   - 4.1. Photons correlate with a combination of MM1MS1-photons (or, 2G2&) and single-ensemble effects of 2G24&.
     - 4.1.1. For each of the three constructs (that is, for each of photons, 2G2&, and single-ensemble effects of 2G24&), ...
       - 4.1.1.1. A span of one ensemble pertains.
       - 4.1.1.2. S = 1 pertains.

4.2. A photon does not differentiate (at each of the two interaction vertices in which the photon is involved) between the (potentially two) types of sources (moving charges and elementary-particle spins) of magnetic dipole moment.
4.3. This interpretation leaves intact traditional physics, including the notion that traditional theory accurately correlates with anomalous magnetic dipole moments for electrons and muons.
4.4. For some ENS48 models and some ENS6 models, ...
    4.4.1. 2G24& can couple elementary-particle magnetic dipole moments between ordinary matter and one unit (the DME-1 unit) of five units of dark matter.
    4.4.2. Neither 2G2& nor 2G24& can couple ordinary-matter magnetic dipole moments correlating with currents of moving charges and dark-matter magnetic dipole moments correlating with currents of moving charges.

~ ~ ~

This subsection discusses some possibilities for detecting dark matter and/or cross-ensemble interactions meditated by 2G24&.

Table 3.3.14 pertains. Here, the result #E` < 6 must pertain regarding 2G24&.

Table 3.3.14 Some possibilities for detecting dark matter and/or cross-ensemble interactions meditated by 2G24& (some ENS48 models and some ENS6 models)

1. Possibly, people can do experiments to detect or rule out (to some degree of confidence) 2G24&-mediated interactions between ordinary-matter elementary-particle spin-like magnetic dipole moments and dark-matter elementary-particle spin-like magnetic dipole moments.
2. The following remarks pertain.
    2.1. Such an interaction does not pertain to the extent #E` = 6 pertains regarding 2G24&.
    2.2. Such an interaction would not pertain, to the extent oscillator swap symmetries (somehow) correlate with ruling out the interaction.
    2.3. Such an interaction does not pertain, to the extent ENS1 models pertain.
    2.4. This monograph does not firmly estimate a magnitude for such an interaction (say, between an ordinary-matter electron and {in some ENS48 models or some ENS6 models} a dark-matter electron analog to an electron).
    2.5. Possibly, experimental sensitivity can be enhanced to the extent experiments are designed to include coupling between spin-like magnetic dipole moments for (adequately dense) ordinary matter.

2.6. The likelihood of detecting such interactions in an experiment depends on (the density of DME-1 dark matter that is at the location of the experiment and) the amount of DME-1 dark matter that can effectively participate in the experiment.
    2.6.1. People conjecture that, for each of many galaxies that feature ordinary-matter stars, a significant fraction of the stuff in the galaxy is dark matter. (Note Section 4.7.)
    2.6.2. Possibly, ordinary matter and DME-1 dark matter anti-cluster. (See Section 4.3.)

~ ~ ~

This subsection discusses 2G24&-mediated interactions and particle properties related to those interactions.

Table 3.3.15 pertains.

Table 3.3.15 Interpretation of 2G24&-mediated interactions and particle properties related to those interactions

1. This work correlates with the concept that people can consider that elementary-particle 2G24&-related properties correlate with a 4-vector that (unlike just-charge-related 4-vectors and just-energy-related 4-vectors) carries information about quantum states correlating with elementary particle magnetic dipole moments.
2. For an elementary particle that is a rest with respect to an observer, ...
    2.1. A magnetic-dipole-moment-related 4-vector has QP-like values of zero.
    2.2. The QE-like value correlates with the magnitude of the magnetic dipole moment.
    2.3. For a single (quantum) base state of the elementary particle, the (classical-physics analog of a) spin vector orients along a QP-like axis.
3. For 2G24&, there is no vertex-strength variation based on generation. (Here, we ignore any differences between particles and their antiparticles.) The lack of variation correlates with #"G = 0. (See Table 2.13.23.)
    3.1. Across the 1C-subfamily, spins and charges are identical.
    3.2. Across the 1C-subfamily, masses vary.
    3.3. For 2G24&, ...
        3.3.1. 2 ∈ %.
            3.3.1.1. People might say that 2G24& couples to charge.
        3.3.2. 4 ∈ %.
            3.3.2.1. People might say that 2G24& couples to mass.
    3.4. Per results above (See. for example, Table 2.8.3 and Table 2.8.4.), ...

3.4.1. People might say that coupling of 2G24& to non-zero mass but not to generation-specific mass correlates with a symmetry that correlates with SU(2).
 3.4.1.1. People might say that the notion of 3 undistinguished generations correlates with SU(2).
3.5. People might say that the SU(2) symmetry that correlates with the 3 undistinguished generations correlates with aspects (that are somewhat like quantum spin and) that correlate with information that 2G24& transmits.

~ ~ ~

This subsection discusses 2G68&-intermediated interactions and particle properties related to those interactions.

Table 3.3.16 pertains.

Table 3.3.16 Interpretation of 2G68&-intermediated interactions and particle properties related to those interactions

1. For 2G68&, ...
   1.1. 2 $\notin$ %.
   1.2. 4 $\notin$ %.
2. People might say that 2G68& does not couple to charge (technically, charge currents) or mass (technically, energy currents).
3. People might say that 2G68& couples to spin. (See, for example, Table 3.13.2.)
4. People might say that 2G68& correlates with a type of interaction that can involve neutrinos.

~ ~ ~

This subsection discusses the concept, correlating with EACUNI-set models, of channels for interactions mediated by G-family bosons.

For EACUNI-set models, Table 3.3.17 provides a possible narrative about a role for channels. Other somewhat similar narratives could pertain. (See Table 2.13.12 and Table 3.9.4. Note, also, Table 3.3.9.)

Table 3.3.17 Possible mechanics related to channels (for EACUNI-set models and not for SOMMUL-set models)

1. People might think of the excitation of a G-family boson mode as including a step in which the QP-like oscillator pair for one channel excites −N(P0) times and a corresponding QE-like oscillator pair for the same channel excites −N(P0) times. In other steps, the QP-like excitement transfers to the appropriate (for the mode of the G-family member) Pll' oscillators (with each l satisfying 2 ≤ l ≤ #P and with each l' being either R or L). In a parallel step, the QE-like excitement transfers to oscillator E0.

~ ~ ~

This subsection discusses G-family rest masses.

Calculations based on Table 2.10.7 use only the {D + 2ν}(2,1,1) solution. For each member of the G-family, m' = 0.

Section 3.4 Elementary particles correlating with the WHO-families

Section 3.4 discusses all non-zero-mass elementary bosons. We show solutions correlating with the known non-zero-mass elementary bosons. We show solutions correlating with possible non-zero-mass elementary bosons for which σ = −1. We point to work below regarding masses of WHO-family elementary particles.

~ ~ ~

This subsection provides perspective about this section.

Traditional physics includes four fundamental interactions. One of those interactions is the weak interaction. People say that the W and Z bosons intermediate the weak interaction.

Traditional physics includes non-zero-mass elementary particles that people call the W, Z, and Higgs bosons. People observe that (in a frame of reference roughly correlating with the center of mass for the system of particles that create and destroy such a boson) such a boson lasts for little time and does not travel far. Typical distances are smaller than the radius of proton. Nevertheless, people consider these particles to be free-ranging.

Here, we correlate solutions with these particles and with possible other non-zero-mass elementary particles. The other particles would not be free-ranging.

~ ~ ~

This subsection shows ground states for W-, H-, and O-family bosons.

Table 3.4.1 lists ground-state INTERN LADDER solutions that correlate with non-zero-mass elementary bosons. (See Table 2.7.1.) Notation kW% correlates with the W-family. The symbol kH% that equals 0H0 correlates with the H-family. For the W- and H-families, % correlates with QP-like oscillators. Notation kO% correlates with the O-family. For the O-family, the % in kO% correlates with QE-like oscillators. (Table 2.2.11 shows equivalences between two sets of names for oscillators.) For each family, k = 2S. The number of particles per kX% set (for X = W, H, or O) is k + 1 = 2S + 1, per Table 2.6.5. (Here, we count a distinct antiparticle as if it differs from the corresponding particle.) Here, if N(P2L) = 0 for a particle, the row showing N(P2R) = 0 correlates with the particle's antiparticle. (Here, if N(P2R) = 0 for a particle, the row showing N(P2L) = 0 correlates with the particle's antiparticle.) Similar remarks pertain for N(E2R) and N(E2L). Similar remarks pertain for N(E4R) and N(E4L).

Table 3.4.1 Ground-state INTERN LADDER solutions for non-zero-mass elementary bosons

E 6R	E 6L	E 4R	E 4L	E 2R	E 2L	E 0	P 0	P 2L	P 2R	P 4L	P 4R	P 6L	P 6R	P 8L	P 8R	Particle
				'I	'I	'I	0	@	@							2W0
				'I	'I	'I	@	0	@							2W1
				'I	'I	'I	@	@	0							2W2
						'I	0									0H0
						0	'I									0O0
				0	@	@	'I	'I	'I							2O2
				@	0	@	'I	'I	'I							2O1
				@	@	0	'I	'I	'I							2O0
		0	@	@	@	@	'I	'I	'I	'I	'I					4O4
		@	0	@	@	@	'I	'I	'I	'I	'I					4O3
		@	@	0	@	@	'I	'I	'I	'I	'I					4O2
		@	@	@	0	@	'I	'I	'I	'I	'I					4O1
		@	@	@	@	0	'I	'I	'I	'I	'I					4O0

Table 3.4.2 associates some math solutions with known particles. Here, Q' denotes charge in units of $|q_e|$, in which q_e denotes the charge of an electron.

Table 3.4.2 W- and H-family particles

Particle	Symbol	Q'
Z boson	2W0	0
W^+ boson	2W1	+1
W^- boson	2W2	−1
Higgs boson	0H0	0

~ ~ ~

This subsection shows aspects of an interaction vertex involving a W boson.

For an interaction vertex in which a W⁻ and a MM1MS1-neutrino enter and an electron exits, Table 3.4.3 shows before and after states for the W boson. This example assumes that the W⁻ state starts with one (and not more than one) excitation.

Table 3.4.3 2W% states related to a specific interaction vertex

E	E	E	E	E	E	E	P	P	P	P	P	P	P	P	P	
6R	6L	4R	4L	2R	2L	0	0	2L	2R	4L	4R	6L	6R	8L	8R	Particle
				@	@	1	@	@	1							2W2 before
				@	@	0	@	@	0							2W2 after

~ ~ ~

This subsection discusses non-spin, non-mass properties of O-family bosons.

In Section 3.7, we show a means to compute charges for 2O-subfamily bosons. The method has as a basis the known charges for 2W (or, W-family) bosons.

No 4W subfamily exists. Possibly, some 4O-subfamily particles can transfer (at least, between spin-3/2 elementary fermions) a particle property. We do not discuss a description for such a property.

~ ~ ~

This subsection discusses W-, H-, and O-family rest masses.

We discuss W- and H-family rest masses in Section 3.7. We discuss possible O-family rest masses in Section 3.8.

Section 3.5 Elementary particles correlating with the CN- and QIRD-families

Section 3.5 discusses fermion elementary particles. We show FERTRA LADDER solutions correlating with all known fermion elementary particles. We show FERTRA LADDER solutions correlating with possible fermion elementary particles. People might say that FERTRA LADDER solutions correlate with abilities of elementary fermions to interact with elementary bosons correlating with W-family and the 2O- and 4O-subfamilies. People might say that MM1MS1-neutrinos are zero-mass Dirac fermions. People might say that we show a model correlating with handedness for leptons. We point to work below regarding relative masses of 1C-subfamily and 1Q-subfamily elementary particles.

~ ~ ~

This subsection provides perspective about this section.

Traditional physics includes particles people call leptons and particles people call quarks. People call these particles elementary fermions. Leptons include the electron, two other charged particles (the muon and the tauon), and neutrinos. Neutrinos have zero charge. Traditional models predict and people have found six quarks.

People speculate about some aspects of neutrinos. Are neutrinos Dirac particles? Specifically, does each neutrino have a distinct antiparticle? Or, are neutrinos Majorana particles? Specifically, is each neutrino its own antiparticle? And, do neutrinos have non-zero masses?

This section correlates some solutions with the known leptons and quarks. We also correlate some solutions with other possible elementary fermions. We discuss possible interactions between elementary fermions and non-zero-mass elementary bosons.

We discuss characteristics of MM1MS1-neutrinos. Some traditional physics posits that at least one neutrino has non-zero mass. We think that zero-mass MM1MS1-neutrinos can exhibit phenomena (namely, neutrino oscillations) that traditional physics correlates with a seeming necessity for a non-zero mass for at least one neutrino.

~ ~ ~

This subsection tabulates known and possible elementary fermions and discusses interactions between these particles and O- and W-family bosons.

Table 3.5.1 lists types of known and of some possible non-zero-mass elementary fermions (that is, members of the C-, Q-, and I-families). (See Table 2.6.5 and Table 2.7.3.)

Table 3.5.1 Types of C-, Q-, and I-family particles

S	Ω	D for fermion fields	2(2S+1)	D for fermion particles	Known types of particles	Possible other types of particles
1/2	3/4	1, 2, 3, or 4	4	3	1C	
1/2	−3/4	4	4	4	1Q	
3/2	−15/4	10	8	6		3Q, 3I

Table 3.5.2 lists elementary particles for the CN-families and the QIRD-families. Compared to Table 2.7.3 (which shows some FERTRA LADDER solutions), this list shows all possible N(..) = −1 ↔ N(..) = −2 exchanges within QE-like or QP-like oscillator pairs. (E0 and P0 do not form an eligible oscillator pair.) In the elementary particle column, a symbol shows the particle's subfamily, possible QE-like interaction vertices (between the open-parenthesis and the first semicolon), possible QP-like interaction vertices (between the two semicolons), and a symbol g (between the second semicolon

and the close-parenthesis) for designating the generation of the fermion. For each of 1C, 1N, 1Q, and (if the nature exhibits the subfamily) 1R, the number of generations is 3. (See Table 2.6.4.) For each of 3QIRD, the number of generations is #GEN(3/2). (See Table 2.6.4.) For each of 1C, 1Q, and one of 3Q or 3I, the number of particles and antiparticles does not exceed the limits (D for fermion fields and D for fermion particles) that Table 3.5.1 shows. We note a possible limit (regarding S = 3/2 non-zero mass elementary fermions) based on D for fermion particles as being 6 and, thereby, we assume that no more than one of 3Q and 3I correlates with nature.

Table 3.5.2 FERTRA LADDER solutions for CN-family particles and QIRD-family particles

	E	E	E	E	E	E	E	E	E	P	P	P	P	P	P	P	P	P	Elementary
σ	8R	8L	6R	6L	4R	4L	2R	2L	0	0	2L	2R	4L	4R	6L	6R	8L	8R	particle
+1					−1	−1	−1	−1	−1	−1	−1	−2							1C(;2R;g)
+1					−1	−1	−1	−1	−1	−1	−2	−1							1C(;2L;g)
+1					−1	−1	−1	−1	−2	−2	−1	−2							1N(0;0,2R;g)
+1					−1	−1	−1	−1	−2	−2	−2	−1							1N(0;0,2L;g)
+1	−1	−1	−1	−1	−1	−1	−1	−1	−1	−2	−1	−1	−1	−2					3N(;0,4R;g)
+1	−1	−1	−1	−1	−1	−1	−1	−1	−1	−2	−1	−1	−2	−1					3N(;0,4L;g)
−1							−2	−1	−1	−1	−1	−2							1Q(2R;2R;g)
−1							−1	−2	−1	−1	−1	−2							1Q(2L;2R;g)
−1							−2	−1	−1	−1	−2	−1							1Q(2R;2L;g)
−1							−1	−2	−1	−1	−2	−1							1Q(2L;2L;g)
−1							−2	−1	−2	−2	−1	−2							1R(2R,0;0,2R;g)
−1							−1	−2	−2	−2	−1	−2							1R(2L,0;0,2R;g)
−1							−2	−1	−2	−2	−2	−1							1R(2R,0;0,2L;g)
−1							−1	−2	−2	−2	−2	−1							1R(2L,0;0,2L;g)
−1					−2	−1	−1	−1	−1	−1	−1	−1	−1	−2					3Q(4R;4R;g)
−1					−1	−2	−1	−1	−1	−1	−1	−1	−1	−2					3Q(4L;4R;g)
−1					−2	−1	−1	−1	−1	−1	−1	−1	−2	−1					3Q(4R;4L;g)
−1					−1	−2	−1	−1	−1	−1	−1	−1	−2	−1					3Q(4L;4L;g)
−1					−2	−1	−2	−2	−1	−1	−2	−2	−1	−2					3I(4R,0;0,4R;g)
−1					−1	−2	−2	−2	−1	−1	−2	−2	−1	−2					3I(4L,0;0,4R;g)
−1					−2	−1	−2	−2	−1	−1	−2	−2	−2	−1					3I(4R,0;0,4L;g)
−1					−1	−2	−2	−2	−1	−1	−2	−2	−2	−1					3I(4L,0;0,4L;g)
−1					−2	−1	−1	−1	−2	−2	−1	−1	−1	−2					3R(4R,0;0,4R;g)
−1					−1	−2	−1	−1	−2	−2	−1	−1	−1	−2					3R(4L,0;0,4R;g)
−1					−2	−1	−1	−1	−2	−2	−1	−1	−2	−1					3R(4R,0;0,4L;g)
−1					−1	−2	−1	−1	−2	−2	−1	−1	−2	−1					3R(4L,0;0,4L;g)
−1					−2	−1	−2	−2	−2	−2	−2	−2	−1	−2					3D(4R,0;0,4R;g)
−1					−1	−2	−2	−2	−2	−2	−2	−2	−1	−2					3D(4L,0;0,4R;g)
−1					−2	−1	−2	−2	−2	−2	−2	−2	−2	−1					3D(4R,0;0,4L;g)
−1					−1	−2	−2	−2	−2	−2	−2	−2	−2	−1					3D(4L,0;0,4L;g)

In Table 3.5.2, for the 1C-, 1N-, 1Q-, and 1R-subfamilies, an entry for which N(Pk) = −2 and k = 1 or k = 2 correlates with the fermion's abilities to absorb one boson of the type 2Wk. For the 1N-, 1Q-, and 1R-subfamilies, an N(Ek) = −2 entry correlates with the fermion's ability to absorb a 2Ok boson. (See Table 2.2.11 regarding the notation E2, E1, P1, and P2.)

Per Table 2.13.42, we assume that 1CN and 1QR fermions can interact with 2WO bosons and not with 4O bosons. We assume 3QIRD fermions can interact with 4O bosons and not with 2WO bosons. Here, we assume 3N bosons cannot interact with 2WO bosons.

Table 3.5.3 highlights W- and O-family vertices for elementary fermions. The table correlates with vertices for which the fermion absorbs a boson. The table lists traditional names for known generation-1 elementary fermions. For elementary fermions, the table lists matter particles and antimatter particles. Because no 4W particles exist, the table has no P4L column and no P4R column. The term e-neutrino abbreviates electron-neutrino. The term p-neutrino abbreviates positron-neutrino. 2O2 particles have charge $-(1/3)|q_e|$. 2O1 particles have charge $+(1/3)|q_e|$. For each of the 1C-, 1N-, 1Q-, and 1R-subfamilies, the table shows interactions with O-family spin-1 bosons. For k = 2 or k = 1, an interaction (in which the 1Q particle absorbs a 2O particle) between a 1Q particle and a 2Ok particle converts an antimatter quark into a matter quark or converts a matter quark into an antimatter quark. MM1MS1-neutrinos (or, 1N particles) can interact with Z bosons and with 2O0 particles. Vertices involving 2O0 and Z particles can include particle-emission interactions. Regarding vertices involving 2O0 or Z particles, people might use the term elastic scattering. For the 3QIRD-subfamilies, the table shows interactions with O-family spin-2 bosons. Per remarks preceding Table 2.13.4, we think that, for FRERAN interactions (that Table 3.5.3 points to) involving 2W-subfamily particles, no change in fermion generation occurs.

Table 3.5.3 2W-subfamily, 2O-subfamily, and 4O-subfamily absorption interactions for elementary fermions

σ	E 4R	E 4L	E 2R	E 2L	E 0	P 0	P 2L	P 2R	Particle name (known, g = 1)	Elementary particle
+1								W^-	positron	1C(;2R;g)
+1							W^+		electron	1C(;2L;g)
+1					2O0	Z		W^-	e-neutrino	1N(0;0,2R;g)
+1					2O0	Z	W^+		p-neutrino	1N(0;0,2L;g)
+1										3N(;0,4R;g)
+1										3N(;0,4L;g)
−1			2O2					W^-	up	1Q(2R;2R;g)
−1				2O1				W^-	anti-down	1Q(2L;2R;g)
−1			2O2				W^+		down	1Q(2R;2L;g)

	E	E	E	E	E	P	P	P	Particle name	Elementary
σ	4R	4L	2R	2L	0	0	2L	2R	(known, g = 1)	particle
−1				201			W^+		anti-up	1Q(2L;2L;g)
−1			202		200	Z		W^-		1R(2R,0;0,2R;g)
−1				201	200	Z		W^-		1R(2L,0;0,2R;g)
−1			202		200	Z	W^+			1R(2R,0;0,2L;g)
−1				201	200	Z	W^+			1R(2L,0;0,2L;g)
−1	404									3Q(4R;4R;g)
−1		403								3Q(4L;4R;g)
−1	404									3Q(4R;4L;g)
−1		403								3Q(4L;4L;g)
−1	404		402	401						3I(4R,0;0,4R;g)
−1		403	402	401						3I(4L,0;0,4R;g)
−1	404		402	401						3I(4R,0;0,4L;g)
−1		403	402	401						3I(4L,0;0,4L;g)
−1	404				400					3R(4R,0;0,4R;g)
−1		403			400					3R(4L,0;0,4R;g)
−1	404				400					3R(4R,0;0,4L;g)
−1		403			400					3R(4L,0;0,4L;g)
−1	404		402	401	400					3D(4R,0;0,4R;g)
−1		403	402	401	400					3D(4L,0;0,4R;g)
−1	404		402	401	400					3D(4R,0;0,4L;g)
−1		403	402	401	400					3D(4L,0;0,4L;g)

~ ~ ~

This subsection discusses aspects regarding interaction vertices involving elementary fermions and WO-family bosons.

Table 3.5.2 and Table 3.5.3 point to possible interactions in which an elementary fermion absorbs a boson. Those tables do not explicitly point out properties of fermions that exit such interactions.

Table 3.5.4 pertains. People may find opportunities, regarding this table, for research.

Table 3.5.4 Aspects regarding an interaction between an extant elementary fermion and a WO-family boson

1. Table 3.5.2 (or Table 3.5.3) correlates with ...
   1.1. Absorption (by a fermion) of the boson.
   1.2. For other than interactions involving Z, 200, or 400 bosons, a conceivable change in property (such as charge) for the fermion.

        1.2.1. The term conceivable pertains for the 4O-subfamily. We are uncertain as to the extent interactions involving 4O-subfamily bosons would change something people would think of as an internal property (of a fermion).

2. Regarding such interactions, ...
    2.1. We correlate with interactions involving Z, 2O0, or 4O0 bosons the term elastic scattering.
    2.2. We correlate with some other interactions to which Table 3.5.3 alludes the term inelastic.
        2.2.1. For example, an interaction with a W^+ or W^- boson is inelastic.
3. For inelastic interactions involving 1CN-subfamily particles, ...
    3.1. The FERTRA LADDER solution correlating with the exiting elementary fermion differs from the FERTRA LADDER solution for the entering fermion via ...
        3.1.1. A change of value N(E0) involving one of the following ...
            3.1.1.1. From −1 to −2.
            3.1.1.2. From −2 to −1.
        3.1.2. A change of value N(P0) involving (the same) one of the following ...
            3.1.2.1. From −1 to −2.
            3.1.2.2. From −2 to −1.
        3.1.3. An exchange of N(P[#P − 1]) and N(P[#P]).
4. For inelastic interactions involving 1QR-subfamily particles, ...
    4.1. The FERTRA LADDER solution correlating with the exiting elementary fermion differs from the FERTRA LADDER solution for the entering fermion via exactly one of ...
        4.1.1. An exchange of N(E[#E]) and N(E[#E − 1]).
        4.1.2. An exchange of N(P[#P − 1]) and N(P[#P]).
5. For interactions involving 3QIRD-subfamily particles, ...
    5.1. Possibly, the term inelastic pertains to interactions with 4O4 or 4O3 bosons.
    5.2. Possibly, the term elastic pertains to interactions with 4O2 or 4O1 bosons.
6. For any interaction between an extant elementary fermion and a WO-family boson, ...
    6.1. Perhaps, conservation of generation (or lack of conservation of generation) correlates with results this monograph shows above. (See, for example, discussion regarding Table 2.13.4.)
7. For any interaction between an extant elementary fermion and a WO-family boson, ...
    7.1. We do not discuss (in the context of work in this table) the topic of color charge for the fermion.

Table 3.5.4 does not explicitly discuss interaction vertices that correlate with production of a matter-fermion-and-antimatter-fermion pair of particles. Perhaps, for such cases, people can develop results based on traditional thinking that correlates with techniques people associate with Feynman diagrams and on thinking the correlates with Table 3.5.4. For example, for a vertex correlated with creation of a pair of elementary fermion particles, perhaps the elementary fermion particles must be antiparticles to each other and (thereby) the generation for each fermion is the same as the generation for the other fermion.

Table 3.5.5 pertains.

Table 3.5.5 Concepts pertaining to fermion destruction operators and fermion creation operators correlating with interaction vertices that involve elementary fermions

1. People might say that ...
   1.1. Table 3.5.2 and Table 3.5.3 correlate, to some extent, with notions of destroying a fermion.
   1.2. Aspects of Table 3.5.4 correlate, to some extent, with notions of creating a fermion.
   1.3. Allowed combinations of such destruction operators and creation operators have parallels, to some extent, to pairs (each consisting of a color charge destruction operator and a color charge creation operator) that people correlate with models for interactions mediated by gluons. (Compare with, for example, Table 3.6.3.)

~ ~ ~

This subsection discusses a possible opportunity for research.

Of the known particles in Table 3.5.3, only MM1MS1-neutrinos would interact with the 2O0 boson. Table 3.5.6 pertains.

Table 3.5.6 Possible opportunities for research regarding neutrinos and the possible 2O0 boson

1. People might say that work in this monograph provides for (and does not rule out) possibilities that ...
   1.1. 2O0 particles interact with neutrinos.
   1.2. 2O0 interactions with a neutrino can change the generation (or, flavor) of the neutrino.
2. Propose theory that people can use as a basis for designing (or for showing a near-term lack of ability to design) experiments or observations for detecting or ruling out (to some confidence level) the existence of 2O0 bosons.

    2.1. See Section 3.8 for a possible mass for and other information regarding 2O0 particles.
3. Detect or rule out (to come confidence level) the existence of 2O0 bosons.

~ ~ ~

This subsection discusses incompatibilities between our models and the concept of Majorana MM1MS1-neutrinos.

People might interpret limits inherent in the first row of Table 3.5.1 as leaving room for 1C(;2L,2R;g) particles. These particles would have m' ≠ 0. These particles could absorb charge from either a W^+ or a W^-. Possibly, these particles would be Majorana MM1MS1-neutrinos with non-zero rest mass. However, for 1C(;2L,2R;g) solutions, S would equal #P/2 = 1. Traditional physics assumes that neutrinos have S = 1/2. This result rules out non-zero mass Majorana MM1MS1-neutrinos, at least the extent both the tradition that S must be 1/2 and models in this monograph correlate with nature. Similar considerations about S rule out zero-mass Majorana MM1MS1-neutrinos that might be part of the 1N-subfamily.

~ ~ ~

This subsection discusses some possible solutions that this monograph does not consider further.

Paralleling results for 1C particles, the number per generation of 3Q or 3I particles and antiparticles need not equal the relevant D (6, in this case) for fermion particles. (See the third row of Table 3.5.1.) Here, one can conceive of various solutions that might correlate with particles. For example, for 3Q, the two solutions per generation {3Q(2R;4R;g) ± 3Q(2L;4R;g)} and {3Q(2R;4L;g) ± 3Q(2L;4L;g)} might pertain. Here, each particle correlates with one of a symmetric or antisymmetric linear combination of two base-state solutions. (Given that the particles would be fermions, people might assume that the antisymmetric linear combinations pertain.) This monograph neither rules out nor further considers, for each generation, these two possible particles. This monograph neither rules out nor further considers other possibilities, such as 3Q(2R,2L;4R;g) and 3Q(2R,2L;4L;g).

~ ~ ~

This subsection discusses handedness for leptons.

Table 3.5.7 pertains.

Table 3.5.7 Handedness for leptons

1. For a 1C(;2j;g) lepton, the following pertain.
    1.1. People might say that j = R correlates with the term right-handed.
    1.2. People might say that j = L correlates with the term left-handed.

2. For a 1N(0;0,2j;g) lepton, the following pertain.
    2.1. People might say that j = R correlates with the term left-handed.
    2.2. People might say that j = L correlates with the term right-handed.

~ ~ ~

This subsection notes that our models are compatible with the concept that MM1MS1-neutrinos are zero-mass Dirac fermions.

Table 3.5.8 pertains.

Table 3.5.8 Terminology regarding 1N (or, MM1MS1-neutrinos) and 3N elementary particles

1. For 1N (or, MM1MS1-neutrino) particles, ...
    1.1. People might say that the particles are zero-mass Dirac fermions.
2. For 3N-subfamily elementary particles, ...
    2.1. People might say that the particles are zero-mass Dirac fermions.

~ ~ ~

This subsection discusses CN-family and QIRD-family rest masses.

For the N-, R-, and D-families, particle rest masses are zero. For the C-family and 1Q-subfamily, rest masses are non-zero and Section 3.9 shows formulas linking approximate rest masses. For the 3Q- and 3I-subfamilies, rest masses are non-zero.

Section 3.6 Elementary particles correlating with the Y-family

Section 3.6 discusses the Y-family. We show Y-family solutions. We note that 2Y-subfamily solutions do not correlate directly with gluons. We note that 2Y-subfamily solutions correlate with a basis for 8 gluons. We show that the Y-family provides a basis for possible spin-2 zero-mass bosons. We note that Y-family solutions correlate with 3 color charges for 1QR-subfamily elementary fermions and with 5 color charges for 3QIRD-subfamily elementary fermions. We note that 4Y-subfamily solutions correlate with a basis for 24 gluon-like particles. We discuss the notion that work above correlates with zero-mass for each gluon or gluon-like elementary particle.

~ ~ ~

This subsection provides perspective about this section.

Traditional physics includes four fundamental interactions. One of those interactions is the strong interaction.

Traditional physics includes particles people call gluons. People correlate gluons with the strong interaction. The strong interaction provides a force needed to bind quarks into protons, neutrons, and other composite particles. People consider gluons to be non-free-ranging elementary particles.

This section discusses solutions that correlate with gluons and that might correlate with some particles that would have similarities to gluons. These correlations between solutions and elementary particles are less direct than correlations between non-Y-family solutions and elementary particles.

~ ~ ~

This subsection discusses the 2Y-subfamily and correlates 2Y solutions with gluons.

Table 3.6.1 extends Table 2.3.14 to show members of the 2Y-subfamily. Here, 'I denotes −1. For the notation 2Yjk, N(Ej) = −2 and N(Ek) = 0.

Table 3.6.1 Ground-state INTERN LADDER solutions for 2Y

E	E	E	E	E	E	E	P	P	P	P	P	P	P	P	P	
6R	6L	4R	4L	2R	2L	0	0	2L	2R	4L	4R	6L	6R	8L	8R	Mode
				−2	−1	0	'I	'I	'I							2Y20
				0	−2	−1	'I	'I	'I							2Y12
				−1	0	−2	'I	'I	'I							2Y01
				−1	−2	0	'I	'I	'I							2Y10
				0	−1	−2	'I	'I	'I							2Y02
				−2	0	−1	'I	'I	'I							2Y21

Regarding the number of color charges, #CC = #E + 1 = 3.

For each 2Y solution, Table 3.6.2 pertains. Here, the symbols CCj denote color charges. The color charges relevant to 2Y are CC2, CC1, and CC0 (as in E2R, E2L, and E0, respectively).

Table 3.6.2 Characteristics of each 2Y solution

1. For exactly one value of j, N(Ej) = −2.
2. For exactly one value of k, N(Ek) = 0.
3. An excitement includes the following.
   3.1. A quark loses color charge CCj.
   3.2. N(Ej) becomes −1.
   3.3. N(Ek) becomes −1.

Such an excitement erases color charge CCj from a quark.

Similarly, a de-excitement from a state having N(E2R) = N(E2L) = N(E0) = −1 paints the quark with a color charge.

We consider two trios of 2Y-subfamily solutions. One of these trios consists of 2Y20, 2Y12, and 2Y01. Here, the cyclic order for N(E..) values is −2, −1, 0. This monograph calls this trio the ascending-order trio. Another trio consists of items 2Y10, 2Y02, and 2Y21. Here, the cyclic order for N(E..) values is 0, −1, −2. This monograph calls this trio the descending-order trio.

One trio of 2Y solutions correlates with gluons pertaining to quarks people consider to be matter. The other of these trios of 2Y solutions correlates with gluons pertaining to quarks people consider to be antimatter.

The following discussion assumes that the first trio corresponds to quarks people consider to be matter (and that the second trio corresponds to quarks people consider to be antimatter). (Possibly, the reversed pairing pertains.) Table 3.6.3 symbolizes a component for a gluon. The right element erases color charge CC1 from a quark. The left element paints the quark with color charge CC2.

Table 3.6.3 A component for a gluon

$| 2Y20 \rangle \langle 2Y12 |$

People sometimes denote the three color charges by r (for red), b (for blue), and g (for green). For such, this monograph uses l' to denote erasing color charge l. This monograph uses n to denote painting color charge n. This discussion assumes CC1 corresponds to r and CC2 corresponds to b. Table 3.6.4 restates Table 3.6.3.

Table 3.6.4 A restatement of a component for a gluon

br'

Table 3.6.5 shows two gluons for which the item in Table 3.6.4 comprises a component.

Table 3.6.5 Two gluons

$(rb' + br')/2^{1/2}$
$-i(rb' - br')/2^{1/2}$

Table 3.6.6 provides a way people symbolize another one of the eight gluons.

Table 3.6.6 Another gluon

$(rr' + bb' - 2gg')/6^{1/2}$

Table 3.6.7 pertains.

Table 3.6.7 A correlation between the r, b, g representation for color charge and the CC2, CC1, and CC0 representation for color charge

> 1. g (in the r, b, g representation for color charge) correlates with CC0 (in the CC2, CC1, and CC0 representation).

The number of spin-1 gluons is 8, which is the number of generators for SU(3). Here, 3 is the number of relevant color charges. Also, there are 8 anti-gluons.

~ ~ ~

This subsection introduces concepts regarding the 4Y-subfamily and correlates 4Y solutions with somewhat gluon-like particles.

Table 3.6.8 extends Table 2.3.14 to show some members of the 4Y-subfamily. Here, 'I denotes −1. For the notation 4Yjk, N(Ej) = −2 and N(Ek) = 0. Table 3.6.8 shows some (but not all) solutions that would correlate with ground states for 4Y. Other solutions would have values other than −1 for N(E4R) and/or for N(E4L).

Table 3.6.8 Some ground-state INTERN LADDER solutions for 4Y

E 6R	E 6L	E 4R	E 4L	E 2R	E 2L	E 0	P 0	P 2L	P 2R	P 4L	P 4R	P 6L	P 6R	P 8L	P 8R	Mode
		−1	−1	−2	−1	0	'I	'I	'I	'I	'I					4Y20
		−1	−1	0	−2	−1	'I	'I	'I	'I	'I					4Y12
		−1	−1	−1	0	−2	'I	'I	'I	'I	'I					4Y01
		−1	−1	−1	−2	0	'I	'I	'I	'I	'I					4Y10
		−1	−1	0	−1	−2	'I	'I	'I	'I	'I					4Y02
		−1	−1	−2	0	−1	'I	'I	'I	'I	'I					4Y21

Regarding the number of color-charge analogs, #CC = #E + 1 = 5.

Five color charges (or, color-charge analogs) pertain for 3QIRD fermions. Each such color charge would correlate with one of the oscillators E4R through E0.

The number of spin-2 matter gluon-like particles is 24, which is the number of generators for SU(5). Here, 5 is the number of relevant color charges. Also, there are 24 spin-2 antimatter gluon-like particles.

The value −2 can pertain to one of five N(Ej). Then, the value 0 can pertain to one of four remaining N(Ek). Then, the value −1 pertains to the 3 remaining N(Ek). Here, 20 = 5×4. The Y-family has 20 4Y (or, spin-2) solutions. The set of 20 solutions divides into 2 sets of 10 solutions. For one set of 10 solutions, the N(E..) = −2 value follows the N(E..) = 0 value by either one or three positions, when one considers the five QE-like N(E..) values as arrayed in a circle and one thinks of traversing the circle in a clockwise direction. For the other set of 10 solutions, the N(E..) = −2 value follows the N(E..) = 0 value by either two or four positions, when one considers the five QE-like N(E..) values as arrayed as before in a circle and one thinks of traversing the circle in a clockwise direction. One of the sets of 10 solutions correlates with spin-2 analogs to matter gluons. The other one of the 2 sets of 10 solutions correlates with spin-2 analogs to antimatter gluons.

~ ~ ~

This subsection discusses Y-family rest masses.

Calculations based on Table 2.10.7 use only the {D + 2v}(2,1,1) solution. For each member of the Y-family, m' = 0. We posit that m' = 0 for Y-family solutions correlates with m' = 0 for all gluons and all possible gluon-like elementary particles.

Section 3.7 W- and H-family masses and 2O-subfamily charges

Section 3.7 discusses masses for W- and H-family particles and charges for 2O-subfamily particles. People might say that MM1MS1 models correlate with the 2O2 boson having charge $-(1/3)|q_e|$ and the 2O1 boson having charge $+(1/3)|q_e|$. We show that ratios of integers correlate approximately with ratios of squares of rest masses for H- and W-family bosons. We show that the ratios $(m'(\text{Higgs}))^2 : (m'(Z))^2 : (m'(W))^2$ are approximately $17 : 9 : \approx 7\times(1 - (0.75)\times\alpha/(2\pi))$.

~ ~ ~

This subsection provides perspective about W-, H-, and O-family particles.

People have determined spins, masses, and charges for W- and H-family particles. As far as we know, the Standard Model does not contain theory that predicts ratios of masses for these particles.

Traditional physics does not include the O-family but does recognize possibilities for hypothetical particles people call leptoquarks.

~ ~ ~

This subsection computes charges for 2O-subfamily bosons.

The baryon asymmetry scenario that Section 4.5 features correlates with the magnitudes of the charges of the 2O2 and 2O1 particles being $|q_e|/3$.

People might say that Table 3.7.1 discusses and uses a method for computing charges for 2O bosons.

Table 3.7.1 Computation of charges for 2O bosons

1. Table 3.4.1 shows INTERN LADDER solutions for ground states of 2O% bosons. Based on Œ = 0, exciting an N(Ej) from 0 to 1 requires exciting an N(Pk) from 0 to 1.
   1.1. Here, we assume that there is no preferred value of Pk.
2. The amplitude this item shows pertains to the 2O1 particle. The expression shows an amplitude for a single excitement for oscillator E1. This work uses notation that features the construct | N(E2R) , N(E2L) , N(E0) , N(P0) , N(P2L) , N(P2R) >.

$$(1/3)^{1/2} \mid @ , 1 , @ , 1 , 0 , 0 > \\ + (1/3)^{1/2} \mid @ , 1 , @ , 0 , 1 , 0 > \\ + (1/3)^{1/2} \mid @ , 1 , @ , 0 , 0 , 1 > \qquad (3.8)$$

   2.1. Technically, each of the $(1/3)^{1/2}$ factors can be multiplied by any complex number having a magnitude of 1. Including such complex numbers would not impact results of work in this table.
3. We assume that the charge correlating with a 2O N(Ej) = 0 item in Table 3.4.1 correlates with the charge associated with the Pj oscillator. The charges associated with the other two values of Pk are not relevant.
4. The next result pertains regarding 2O1 particles. Here, as above, Q' denotes charge divided by $|q_e|$.

$$Q'(2O1) \\ = ((1/3)^{1/2} < @ , 1 , @ , 0 , 1 , 0 \mid) ((1/3)^{1/2} \mid @ , 1 , @ , 0 , 1 , 0 >) \\ \times Q'(2W1) + 0 \times Q'(2W2) + 0 \times Q'(2W0) \\ = (1/3)\ Q'(2W1) \qquad (3.9)$$

People might say that Table 3.7.2 summarizes charges of 2O-subfamily bosons.

Table 3.7.2 Charges for 2O-subfamily bosons

1. Q'(2Oj) = (1/3) Q'(2Wj), for j = 2, 1, and 0.
   1.1. 2O2 bosons have Q' = −1/3.
   1.2. 2O1 bosons have Q' = +1/3.
   1.3. 2O0 bosons have Q' = 0.

~ ~ ~

This subsection computes approximate masses for W- and H-family bosons.

Section 2.10 provides methodology that points to integers that approximately link masses of non-zero-mass elementary bosons for which $\sigma = +1$. We interpret experimental data as suggesting correlating 17 with the Higgs boson, 9 (= 10 − 1) with the Z boson, and 7 (= 10 − 1 − 2) with the W boson. (See Table 2.5.17, Table 2.10.7, and

Table 2.10.8.) The mass for the Z boson is more accurately known than are the masses for the other H- and W-family bosons.

Table 3.7.3 pertains. Here, $(\xi'/2) = (1/9)\,(m'(Z))^2\,c^4$, in which m'(Z) denotes the rest mass of a Z boson. Table 3.7.3 compares calculated and experimental masses. (See Table 8.1.1.) The calculated mass for the Higgs boson is within 1 standard deviation of the experimental mass. The calculated mass for the W boson is within 3 standard deviations of the experimental mass.

Table 3.7.3 Experimental masses and first-approximation calculated masses for H- and W-family bosons

Particle	Symbols for particles	$9(m')^2$ / $(m'(Z))^2$	Calculated mass (GeV/c^2)	Experimental mass (GeV/c^2)
Higgs	0H0	17	125.325	125.7±0.4
Z	2W0	9	91.1876	91.1876±0.0021
W	2W1, 2W2	7	80.420	80.385±0.015

~ ~ ~

This subsection presents a hypothesis regarding possible discrepancies, for the H- and W-families, regarding ratios of calculated masses and ratios of experimental masses.

Table 3.7.4 pertains. People might say that charge correlates with lower mass. The mass of a charged pion is less than the mass of zero-charge pion. The mass of a proton is less than the mass of a neutron. In each case, the difference is, at most, a few percent.

Table 3.7.4 Hypotheses regarding relative masses of H- and W-family bosons

1. From a standpoint of theoretical models, the best mass to use as a reference mass can be the mass of the Higgs boson.
   1.1. The Higgs boson has zero charge and zero spin.
2. To a first approximation, ratios of $(m')^2$ specified as follows pertain.
   2.1. The next formula pertains.
   $$(m')^2 = (I/17)\,(m'(\text{Higgs boson}))^2 \quad (3.10)$$
   2.2. The following numbers pertain.
   $$I = 17, \text{ for the Higgs boson} \quad (3.11)$$
   $$I = 9, \text{ for the Z boson} \quad (3.12)$$
   $$I = 7, \text{ for the W boson} \quad (3.13)$$
3. For W-family bosons, spin-related interactions (analogous to interactions based on quantum electromagnetism that correlates, for charged particles, with aspects people correlate with the fine-structure constant) based on 2G68& might correlate with actual masses differing from masses models might predict.

3.1. While experimental results shown in Table 3.7.3 for the Z boson may agree within 1 standard deviation (for the mass of the Higgs boson) with results from our model, further experimentation might lead to a discrepancy.

4. Within the family of W-family bosons, magnetic-dipole-moment-related interactions (analogous to interactions based on quantum electromagnetism that correlates, for charged particles, with aspects people correlate with the fine-structure constant) based on 2G24& might correlate with the actual mass possibly differing from the mass models might predict.
   4.1. Experimental results shown in Table 3.7.3 for the W boson agree within 3 standard deviations (for the mass of a W boson) with results from our model.
   4.2. People might say that this amount is too large to ignore.
   4.3. Elsewhere, we possibly correlate effects of 2G24& with the fine-structure constant. (See Table 3.10.2.)

We discuss the extent to which a contribution to the mass of the W boson might correlate with the fine-structure constant. Table 3.7.5 pertains.

Table 3.7.5 Estimate for an improved calculation of m'(W)

1. We start from the following formula.

$$(m'(W))^2 = (m'(Z))^2 ((9 - 2(1 + j\alpha))/9) \quad (3.14)$$

   1.1. Here, j is not known.
2. We use experimental data and the following formula.

$$j = (1/(2\alpha)) (7 - 9(m'(W))^2/(m'(Z))^2) \quad (3.15)$$

3. The following values of j pertain.
   3.1. For $j \approx 0.62$, ...
      3.1.1. The calculated m'(W) is approximately 1 standard deviation below the experimental value.
   3.2. For $j \approx 0.42$, ...
      3.2.1. The calculated m'(W) approximates the experimental value.
   3.3. For $j \approx 0.22$, ...
      3.3.1. The calculated m'(W) is approximately 1 standard deviation above the experimental value.

~ ~ ~

This subsection notes that work above provides an approximate result for the weak mixing angle.

People might say that Table 3.7.6 shows a definition of θ_W (the weak mixing angle) and the θ_W value work above calculates. Here, 0.83 approximates twice 0.42.

Table 3.7.6 θ_W (the weak mixing angle) and the θ_W value some work above estimates

$$\cos(\theta_W) = m'(\text{W boson})/m'(\text{Z boson}) \quad (3.16)$$
$$\approx (\,(1/9) \times (7 - 0.83\alpha)\,)^{1/2} \quad (3.17)$$
$$\approx (\,(1/9) \times 7(1 - 0.75\alpha/(2\pi))\,)^{1/2}$$
$$\approx (7/9)^{1/2} \times (1 - 0.37\alpha/(2\pi))$$

Section 3.8 Possibilities regarding O-family masses

Section 3.8 discusses models that produce algebraic expressions that might correlate with approximate masses of O-family particles. We show a process for developing models for masses of O-family particles. We show two such models. People might say that, based on one of the models we show, a candidate mass for the spin-0 (or, 0O0) O-family boson is 155 GeV/c^2. People might say that candidate spin-1 O-family bosons and candidate values for their masses (in units of GeV/c^2) are 2O0 with a value of 182 and 2O2 and 2O1 with values of 177. People might say that an O-family particle cannot be produced unless it is part of a pair (or triplet or ...) of O-family particles. People might say that we show candidate lower bounds for energies needed to produce O-family particles. We contrast some possibilities regarding O-family masses and information regarding masses of hypothetical leptoquarks.

~ ~ ~

This subsection discusses properties that people try to determine regarding elementary particles.

For elementary particles, spin, mass, and charge represent important quantities. People might say that, regarding experiments, determining mass is crucial. The higher the mass, the more powerful a particle accelerator it takes to produce the particle.

Before and when people first made likely sightings of the Higgs boson, people estimated or determined ranges of possible mass for the particle. As people performed more experiments, the possible range narrowed.

~ ~ ~

This subsection discusses uses for predictions regarding masses of elementary particles.

Predictions regarding one or more ranges of possible mass can influence choices of methods for trying to detect, make, of infer (the existence of) elementary particles.

Within a choice of method, predictions can influence people's decisions as to when, regarding the evolution of techniques, trying to detect, make, or infer particles is feasible.

When a method starts to provide promising results, specific predictions regarding masses can prove useful.

~ ~ ~

This subsection provides perspective about relationships, for similar particles, between charge and mass.

People might spot a possible relationship between charge and mass. For each of various sets of similar particles, a charged particle has less rest mass than does the particle with no charge. For example, a charged pion has less mass than does a non-charged pion. Or, a proton has less mass than does a neutron.

~ ~ ~

This subsection characterizes work in this section.

People might say that, regarding the O-family, work above in this monograph provides precise characterization of spins for all candidate particles. People might say that Section 3.7 shows a method for calculating charges for 20 particles. People might say that Section 4.5 suggests phenomena (regarding matter/antimatter imbalance) that, if they occurred (some billions of years ago), may have depended on those values for charge.

This monograph does not offer a confirmed model for masses of O-family bosons. This section discusses concepts. We base the concepts on methodology Table 2.10.7 and Table 2.10.8 show, patterns pertaining to W- and H-family masses, and integers that Table 2.5.17 shows. We think that the squares of masses of O-family bosons may approximate algebraic combinations of integers in the D + 2ν column in Table 2.5.17. Here, an algebraic combination would produce a positive integer and would include plus at least one relatively (for Table 2.5.17) large integer and possibly include minus some other relatively (for Table 2.5.17) small integers and possibly include plus some relatively (for Table 2.5.17) small integers.

We discuss such models in the hopes people can use the models in endeavors to estimate approximate masses, design and conduct experiments, and analyze results of experiments. We hope that people also can use the models, at least after experiments determine adequately bounded ranges for some masses, to predict approximate, if not nearly exact, masses.

~ ~ ~

This subsection discusses possibilities regarding estimating approximate O-family masses.

Table 3.8.1 provides a process for developing candidate masses of O-family bosons. Other somewhat similar processes could pertain. For example, one might assume that the mass of 000 equals the mass of 0H0 and that the masses of 20 particles correlate (at least somewhat) with masses of 2W particles.

Table 3.8.1 A process for developing candidate masses for 0-family bosons.

1. Correlate values of D (in Table 2.5.16) with WHO-family subfamilies.
2. Note the following correlations.

$$D = 1 \text{ for 2W. (Here, } \sigma = +1.) \quad (3.18)$$

$$D = 3 \text{ for 0H. (Here, } \sigma = +1.) \quad (3.19)$$

$$D = 3 \text{ for 00. (Here, } \sigma = -1.) \quad (3.20)$$

$$D = 5 \text{ for 20. (Here, } \sigma = -1.) \quad (3.21)$$

$$D = 9 \text{ for 40. (Here, } \sigma = -1.) \quad (3.22)$$

3. Assume that Table 2.10.7 and Table 2.10.8 pertain.
4. Assume that Table 2.5.18 pertains.
5. Determine a means to limit the number of Table 2.5.17 values of D + 2ν that pertain for each particle and determine a means to assign values of D + 2ν for each need to use of such a number.
   - 5.1. For example, in Table 3.8.2, we ...
     - 5.1.1. Allow for use of S + 1 so-called family-related values.
       - 5.1.1.1. One use correlates with (a positive-value) use of a non-zero D + 2ν number.
         - 5.1.1.1.1. For example, assume that this use correlates with the projection correlating with the oscillator pair E0-and-P0.
       - 5.1.1.2. Any other uses correlate with use of the negative of a non-zero D + 2ν number.
         - 5.1.1.2.1. For example, assume that each of these uses correlates with a projection into D* = 2 dimensions.
     - 5.1.2. Allow for use of S so-called subfamily-related values.
       - 5.1.2.1. Any use correlates with use of one of ...
         - 5.1.2.1.1. The zero D + 2ν number.
         - 5.1.2.1.2. The negative of a D + 2ν number.
       - 5.1.2.2. For example, assume that each of these uses correlates with a projection into D* = 2 dimensions.
6. Take into account that notion that calculations above in this table ...
   - 6.1. Use some algebraic terms correlating with o(n"(%) = 1). (See Table 3.3.2.)

6.2. Do not include o(n"(%)) ≥ 2 effects that might correlate with G-family bosons for which the sub-list % has at least two elements. (See Table 2.3.13 and Table 3.7.4.)

6.3. Might need to include o(n"(%) ≥ 2) effects in order to better approximate actual masses.

~ ~ ~

This subsection discusses one set of possibilities regarding approximate O-family masses.

Table 3.8.2 shows possible results from applying concepts that correlate with Table 3.8.1. (See Table 3.8.1 for explanations of the leftmost five columns.) For the 2W- and 0H-subfamilies, numbers in the leftmost five columns correlate with work that Section 3.7 shows. In column headings, we abbreviate subfamily-related as sub. In column headings, we abbreviate family-related as fam. Numbers in the family S ≥ 0 column follow an order that Table 2.5.17 shows for ν = −1. The sixth column shows the sum of numbers in the leftmost five columns. The m' column shows calculated approximate masses. (Compare with results and discussion in Section 3.7.) We line out the 4W-subfamily, per Table 2.3.1. People might say that results for 2O2, 2O1, 4O2, 4O1, 4O4, and 4O3 involve more speculation that do results for 2O0 and 4O0.

Table 3.8.2 H-family mass, approximate W-family masses, and possible approximate O-family masses

Sub. S ≥ 2	Sub. S ≥ 1	Fam. S ≥ 0	Fam. S ≥ 1	Fam. S ≥ 2	Sum (also: $9(m')^2 / (m'(Z))^2$)	m' (GeV/c^2)	Bosons (or math solutions)
~~−1~~	~~−2~~	~~5~~	~~−2~~	~~−1~~	~~−1~~		~~4W3, 4W4~~
~~0~~	~~−2~~	~~5~~	~~−2~~	~~−1~~	~~0~~		~~4W1, 4W2~~
~~0~~	~~0~~	~~5~~	~~−2~~	~~−1~~	~~2~~		~~4W0~~
	−2	10	−1		7	80.4	2W1, 2W2
	0	10	−1		9	91.2	2W0
		17			17	125.3	0H0
		26			26	155.0	0O0
	0	37	−1		36	182.4	2O0
	−2	37	−1		34	177.2	2O2, 2O1
0	0	50	−2	−1	47	208.4	4O0
0	−2	50	−2	−1	45	203.9	4O2, 4O1
−1	−2	50	−2	−1	44	201.6	4O4, 4O3

~ ~ ~

This subsection discusses minimum energies needed to create free-ranging (or, composite) particles that include O-family bosons.

Experimentally, approaches to producing free-ranging particles that exhibit O-family bosons might feature trying to convert mesons (that is, 2 × 1Q+2Y) to 2 × 1Q+(0,2)O or trying to convert baryons (that is, 3 × 1Q+2Y) to 3 × 1Q+(0,2)O. Here, the notation (0,2) denotes 0 or 2. (We think that 2 might pertain and that 0 might not pertain.) Converting mesons might require lower energy than converting baryons. Converting baryons might be experimentally more attractive, based on lifetimes and availability of these particles.

Table 3.8.3 shows possible minimal energies needed to create free-ranging particles involving integer numbers of WHO-family bosons. W- and H-family bosons are free-ranging. Regarding the O-family, we consider that each such corresponding composite likely would involve at least two or three O-family bosons. We do not estimate binding energies that might also pertain to the composites. O-family numbers the table shows represent results from one of various possible ways to estimate O-family masses. Other ways could correlate with, for the O-family, different masses and different lower energy bounds.

Table 3.8.3 Possible minimum energies to create W-, H-, and O-family bosons, assuming O-family bosons are created as parts of composite particles

m' (GeV/c^2)	Bosons (or math solutions)	Lower energy bound (GeV), assuming 1 non-zero mass elementary boson	Lower energy bound (GeV), assuming 2 non-zero mass elementary bosons	Lower energy bound (GeV), assuming 3 non-zero mass elementary bosons
80.4	2W1, 2W2	80.4		
91.2	2W0	91.2		
125.3	0H0	125.3		
155.0	0O0		310.0	465.0
182.4	2O0		364.8	547.1
177.2	2O2, 2O1		354.5	531.7
208.4	4O0		416.8	625.2
203.9	4O2, 4O1		407.8	611.7
201.6	4O4, 4O3		403.2	604.9

~ ~ ~

This subsection shows an alternative set of possibilities regarding O-family masses.

Here, we assume that, for O-family members, each contribution (to which Table 3.8.2 alludes) to the square of an approximate mass is nonnegative. The respective sums (correlating with the sixth column in Table 3.8.2) become 26 (= 26), 38 (=37 +1), 40 (=37 + 3), 53 (= 50 + 3), 55 (= 50 + 5), and 56 (= 50 + 6). Table 3.8.4 pertains for these estimates. Perhaps, the estimates in Table 3.8.3 are minimums and the estimates in Table 3.8.4 are maximums. In either case, we ignore o(n"(%) ≥ 2) effects. In either case, we ignore possible needs to consider binding energies for composite particles or for compound particles. In either case, the lower energy bounds are lower bounds.

Table 3.8.4 Alternative possible minimum energies to create W-, H-, and 0-family bosons, assuming 0-family bosons are created as parts of composite particles

m' (GeV/c^2)	Bosons (or math solutions)	Lower energy bound (GeV), assuming 1 non-zero mass elementary boson	Lower energy bound (GeV), assuming 2 non-zero mass elementary bosons	Lower energy bound (GeV), assuming 3 non-zero mass elementary bosons
80.4	2W1, 2W2	80.4		
91.2	2W0	91.2		
125.3	0H0	125.3		
155.0	000		310.0	465.0
187.4	200		374.7	562.1
192.2	202, 201		384.5	576.7
221.3	400		442.6	663.9
225.4	402, 401		450.8	676.3
227.5	404, 403		454.9	682.4

~ ~ ~

This subsection provides experimental data regarding hypothetical leptoquarks.

People use the term leptoquark when discussing various concepts for bosons and interactions people do not include in the Standard Model. Perhaps, some of those concepts dovetail with aspects of the 0-family and of 0-family-related phenomena.

Experimental results provide lower bounds for masses of various hypothesized leptoquarks. K.A. Olive (2014; LEPTOQUARKS) notes a range of reported lower bounds (with minimum confidence levels of 95%). (The reference also cites one instance of a lower bound of 100 GeV/c^2.) Table 3.8.5 shows some lower bounds people have estimated for rest masses for hypothetical leptoquarks.

Table 3.8.5 Estimated lower bounds on rest masses for leptoquarks

1. 226 - 785 GeV/c^2.
2. 100 GeV/c^2.

This monograph does not speculate regarding relations between thresholds related to leptoquarks and thresholds related to possible composite particles that would include 0-family bosons.

~ ~ ~

This subsection discusses another topic regarding O-family properties.

This monograph does not discuss the topic of 4O-subfamily properties other than spin and mass.

~ ~ ~

Reference 4 K.A. Olive (2014; LEPTOQUARKS)

K.A. Olive et al. (2014) (See Reference 18). Specifically, LEPTOQUARKS, Updated August 2013 by S. Rolli (US Department of Energy) and M. Tanabashi (Nagoya U.).

Section 3.9 Masses and charges of C-family and 1Q-subfamily particles

Section 3.9 discusses a formula that approximates masses of six quarks and three charged leptons. People might say that a ratio of strengths of electromagnetism and gravity correlates with the ratio of masses of the tauon and the electron. People might say that the mass of a tauon may be $1.77685(6)\times10^3$ MeV/c^2. Here, the standard deviation is less than half the standard deviation correlating with data from experiments. People might say that this result provides impetus to improve the accuracies of measurements of the tauon mass and the gravitational constant. People might say that the formula that approximates masses of six quarks and three charged leptons involves no more than 6 parameters.

~ ~ ~

This subsection provides perspective about masses for quarks and charged leptons.

Masses for charged leptons are well-established. These particles have $\sigma = +1$. For quarks, $\sigma = -1$. Masses of quarks are less accurately known.

~ ~ ~

This subsection suggests two integer indices for a mass-centric table for known charged elementary fermions.

Table 3.9.1 shows an orderly array of approximate particle masses for quarks and charged leptons. For each particle, an item shows (in successive rows) $\log_{10}(m'/m_e)$, charge in units of Q', and particle name. The table does not show M' = 2 and M' = 3 columns. As with the M' = 1 column, for n = 2 or 3, each M' = |n| column would show antiparticles correlating with the particles the M' = −|n| column shows. The charge of an antiparticle is the negative of the charge of the particle. The mass of an antiparticle equals the mass of the particle. For each column pertaining to quarks (that is, columns for which |M'| = 2 or |M'| = 1), the particle charge varies with M".

Table 3.9.1 A mass-centric array for 1C- and 1Q- elementary particles

M" \ M'	−3	−2	−1	0	1	...
0	0.00	0.61	0.97		0.97	
0	(−1)	(+2/3)	(−1/3)		(+1/3)	
0	electron	up	down		anti-down	
1		2.26	3.40		3.40	
1		(−1/3)	(+2/3)		(−2/3)	
1		strange	charm		anti-charm	
2	2.32	3.93	5.51		5.51	
2	(−1)	(−1/3)	(+2/3)		(−2/3)	
2	muon	bottom	top		anti-top	
3	3.54					
3	(−1)					
3	tauon					

~ ~ ~

This subsection discusses a numerical relationship between a ratio of strengths for electromagnetism and gravity and a ratio of masses for two 1C-subfamily particles.

This work points to a Feynman diagram interpretation of the ratio of strengths for electromagnetism and gravity.

Table 3.9.2 characterizes the relative strengths of electromagnetism and gravity. In the first item in the table, the numerator pertains for electromagnetic interactions between two electrons. In the first item in the table, the denominator pertains for gravitational interactions between (the same two) two electrons. Here, q_e denotes the charge of an electron, $1/(4\pi\varepsilon_0)$ denotes the Coulomb constant, G_N denotes the gravitational constant, and m_e denotes the mass of an electron. In the table, m'(tauon) denotes the mass of a tauon. The table defines β and β'. (Regarding β, see discussion preceding Table 2.13.11 or see Table 3.9.4.) This work bases numbers (here and below) on data Section 8.1 shows.

Table 3.9.2 Notation and some data regarding a relationship among the strength of electromagnetism, the strength of gravity, the mass of a tauon, and the mass of an electron

$$\{(q_e)^2/(4\pi\varepsilon_0)\} / \{G_N(m_e)^2\} \approx 4.166\times10^{42} \quad (3.23)$$

$$(4/3)(\beta^6)^2 = \{(q_e)^2/(4\pi\varepsilon_0)\} / \{G_N(m_e)^2\} \quad (3.24)$$

$$\beta' = m'(\text{tauon}) / m_e \quad (3.25)$$

Table 3.9.3 estimates β and β'. For β, this work estimates a standard deviation based on the standard deviation for G_N. (The relative uncertainty for G_N much exceeds the relative uncertainty for the other constants in the right side of the second item in Table 3.9.2.) For β', this work bases the standard deviation on experimental results. People might say that β' = β. This monograph assumes that β' = β.

Table 3.9.3 Numeric results for and an assumed equality between β and β'

$$\beta \approx 3.47721(12)\times 10^3 \quad (3.26)$$
$$\beta' \approx 3.47715(31)\times 10^3 \quad (3.27)$$
$$\beta' = \beta \quad (3.28)$$

People might say that Table 3.9.4 interprets elements of the left side of the second item in Table 3.9.2. (See, also, discussion before Table 2.13.11 and/or see, also, Table 3.9.20, Table 3.9.21, and Table 3.13.4.)

Table 3.9.4 Vertex strengths for electromagnetism and gravity

1. In the expression $(4/3)(\beta^6)^2 = \{(q_e)^2/(4\pi\varepsilon_0)\} / \{G_N(m_e)^2\}$, the leftmost exponent 2 represents the number of vertices in a Feynman diagram, β^6 represents the ratio of strengths per channel for electromagnetism and gravity (for interactions between two electrons), 4 represents the number of channels for a MM1MS1-photon, and 3 represents the number of channels for a graviton.

For G-family EACUNI-set models (and when considering INTERN LADDER solutions), the channel that pertains for MM1MS1-photons and does not pertain for gravitons correlates with oscillators E2R, E2L, P4L, and P4R. For FRERAN SPATIM LADDER solutions, the channel that pertains for MM1MS1-photons and does not pertain for gravitons correlates with oscillators E8, E7, P4L, and P4R. (See Table 2.13.12.)

For G-family SOMMUL-set models, numbers of channels correlate with numbers of solutions.

~ ~ ~

This subsection discusses a tauon mass that is consistent with and would have more accuracy than experimental results.

Table 3.9.5 restates results from Table 3.9.2 and Table 3.9.3. People might say that Table 3.9.5 predicts a rest mass for tauons.

Table 3.9.5 Possibly predicted rest mass and experimental rest mass for tauons

1. Formula for calculating a result.

$$m'(\text{tauon})/m_e = \beta' = \beta \quad (3.29)$$

$= \exp((1/12) \log\{ \{3/4\} \{(q_e)^2/(4\pi\varepsilon_0)\} / \{G_N(m_e)^2\} \})$

2. Calculated result.

   $m'(\text{tauon}) \approx 1.77685(6)\times10^3 \text{ MeV}/c^2$ (calculated) (3.30)

   2.1. The largest contributor to the standard deviation comes from measurements of G_N.

3. Experimental results.

   $m'(\text{tauon}) \approx 1.77682(16)\times10^3 \text{ MeV}/c^2$ (experimental) (3.31)

   3.1. Section 8.1 shows this result and a reference for the result.

Table 3.9.6 points to possible opportunities for research. Possibly, at any time, at least one of the two possibilities exists.

Table 3.9.6 Possible opportunities for research, based on m'(tauon) and G_N

1. Measure m'(tauon) accurately enough to predict G_N to more than known accuracy.
2. Measure G_N accurately enough to predict m'(tauon) to more than known accuracy.

~ ~ ~

This subsection develops a formula that approximately fits the masses of all charged leptons and all quarks.

Table 3.9.7 pertains to 1C particles and 1Q particles. Here, L denotes a principal quantum number for spin-like systems. Here, M denotes a secondary quantum number. Here, M can equal $-L$, $-L + 1$, ..., $L - 1$, or L. (People might correlate L and M with a possible presence of exponential functions that might correlate with $\delta(r)$ pertaining for integrals over the range $-\infty < r < \infty$ (See Table 2.5.10.), while a traditional-physics-theory $\Psi(r)$ pertains for the range $0 \le r < \infty$ (See Table 2.5.13.). Possibly, some such exponential functions correlate with angular coordinates and extensions of $\Psi(r)$ to negative values of r. For example, for a radial coordinate θ, if the function $\cos(\theta)$ pertains for $r > 0$, the function $\cosh(\theta)$ might pertain for $r < 0$. This work interprets Table 3.9.1 as correlating with supporting, but not necessarily requiring, these concepts.)

Table 3.9.7 Integers L and M pertaining to 1C and 1Q rest masses

1. For each one of three generations, ...
   1.1. The following maximum numbers of relevant solutions pertain. (See Table 2.6.5.)
      1.1.1. 3, for $S = 1/2$ with $\Omega = +3/4$.
      1.1.2. 4, for $S = 1/2$ with $\Omega = -3/4$.
   1.2. For the C-family, 2 solutions correspond to 2 of the 3 possible members of an $L = 1$ set. (The $M = 0$ member does not pertain.)

1.3. For the 1Q-subfamily, combinations of the 4 solutions correspond to 4 of the 5 members of an L = 2 set. (The M = 0 member does not pertain.)

In Table 3.9.1, M" and M' are integer indices. The table shows an orderly array of approximate particle rest masses for charged leptons and quarks. Here, M correlates with (but does not necessarily equal) M'.

Table 3.9.8 correlates with the last two main sub-items in Table 3.9.7 and with Table 3.9.1. This work considers all three generations. The various values of L` correlate with notions above of L.

Table 3.9.8 Integers L` pertaining to 1C and 1Q rest masses

1. For each of the rows (in Table 3.9.1) for which M" = 0, 2, or 3, the following pertain.
   - 1.1. The row includes an instance characterized by L` = 1.
     - 1.1.1. M' = −3, 0, and +3 characterize this instance.
     - 1.1.2. A matter-lepton-and-antimatter-lepton pair correlates with M' = −3 and M' = +3.
     - 1.1.3. No particle correlates with M' = 0 for that row.
2. For each of the rows (in Table 3.9.1) for which M" = 0, 1, or 2, the following pertain.
   - 2.1. The row includes an instance characterized by L` = 2.
     - 2.1.1. M' = −2, −1, 0, +1, and +2 characterize this instance.
     - 2.1.2. A matter-quark-and-antimatter-quark pair correlates with M' = −2 and M' = +2.
     - 2.1.3. A matter-quark-and-antimatter-quark pair correlates with M' = −1 and M' = +1.
     - 2.1.4. No particle correlates with M' = 0.
3. For each of the rows (in Table 3.9.1) for which M" = 0 or 2, the following pertain.
   - 3.1. The row includes an instance characterized by L` = 3.
     - 3.1.1. M' = −3, −2, −1, 0, +1, +2, and +3 characterize this instance.
     - 3.1.2. A matter-lepton-and-antimatter-lepton pair correlates with M' = −3 and M' = +3.
     - 3.1.3. A matter-quark-and-antimatter-quark pair correlates with M' = −2 and M' = +2.
     - 3.1.4. A matter-quark-and-antimatter-quark pair correlates with M' = −1 and M' = +1.
     - 3.1.5. No particle correlates with M' = 0.

Table 3.9.9 prepares for developing a formula that approximates rest masses.

Table 3.9.9 Preparation for developing a formula that approximates the rest masses of charged leptons and of quarks

1. This item defines notation regarding Table 3.9.1.
   1.1. m(M",M') denotes a calculated m' correlating with a position in Table 3.9.1, which includes positions for S = 1/2 charged elementary fermions.
2. This item pertains to trends in Table 3.9.1.
   2.1. For charged leptons (either M' = −3 or M' = +3), people can benefit by correlating the range $-1 \le M'' \le 3$ with an L`` = 2 system.
3. For the L`` = 2 system with |M'| = 3 and varying M", ...
   3.1. The range $-2 \le M'' - 1 \le +2$ pertains.
   3.2. Regarding the next item, this work assumes that no solution with d(0) = −d(2) appeals. Such a solution would correspond to a trigonometric function that is anti-symmetric with respect to M" − 1. The next item features an expression that corresponds to a symmetric trigonometric function.
   3.3. For the L`` = 2 system (with |M'| = 3 and varying M") that includes charged leptons and for some number ζ", the following formula pertains.

$$m(M'',-3) \propto e^{M''\zeta''}(1 + d(M'')), \text{ in which ...} \quad (3.32)$$

$$-1 \le M'' \le 3$$

$$d(0) = d(2)$$

$$d(-1) = d(1) = d(3) = 0$$

Table 3.9.10 develops a formula to approximate the rest masses of charged leptons and of quarks.

Table 3.9.10 Steps that develop a formula that approximates the rest masses of charged leptons and of quarks

1. This step develops results for M' = − 3 charged leptons and relevant values of M". This work uses experimental masses for the electron and muon to calculate ζ". Then, this work uses an experimentally-acceptable calculated mass for the tauon to calculate m`. (See the calculated tauon mass that Table 3.9.5 states.) Then, the work calculates d(0).

$$m(M'',-3) = m` \times \exp((M'' + 1)\zeta'') \times (1 + d(M'')) \quad (3.33)$$

$$\zeta'' = (1/2) \log(m'(\text{muon}) / m_e) \approx 2.665799 \quad (3.34)$$

$$m` = m'(\text{tauon}) / \exp(4\zeta'') \approx 4.155987 \times 10^{-2}\ \text{MeV}/c^2 \quad (3.35)$$

$$1 + d(0) = m_e / (m` \exp(\zeta'')) \quad (3.36)$$

$$d(2) = d(0) \approx -0.144926 \quad (3.37)$$

$$d(-1) = d(1) = d(3) = 0 \tag{3.38}$$

2. This step shows an item that provides a calculated number that does not necessarily directly correlate with a particle.

$$m(1,-3) \approx 8.59326 \text{ MeV}/c^2 \tag{3.39}$$

3. This step correlates with a possible role for trigonometric functions. (See, also, Table 3.9.11.)

$$d(M'') \approx d'' (1/2) (\cos(M''\pi) + 1) \tag{3.40}$$

$$d'' = d(0) = d(2) \approx -0.144926 \tag{3.41}$$

4. This work assumes that the next item in this table provides a useful set of m(M",0). Here, α denotes the fine-structure constant. No known particles correlate with the combination $M'' \geq 0$ and $M' = 0$.

$$m(M'', 0) \approx m(M'', -3) \times \exp(\, (1/4) \log(1/\alpha'')\, 3(1 + M'')\,) \tag{3.42}$$

$$\alpha'' = \alpha\, (1 + \alpha') \tag{3.43}$$

$$\alpha' = 0 \tag{3.44}$$

5. This step shows numbers.

$$m(0, 0) \approx 2.05\times10^{1} \text{ MeV}/c^2 \tag{3.45}$$

$$m(1, 0) \approx 1.38\times10^{4} \text{ MeV}/c^2 \tag{3.46}$$

$$m(2, 0) \approx 6.79\times10^{6} \text{ MeV}/c^2 \tag{3.47}$$

$$m(3, 0) \approx 4.57\times10^{9} \text{ MeV}/c^2 \tag{3.48}$$

6. This step finishes defining the formula correlating with approximate masses for quarks and charged leptons. This work specifies d(M",M') adequate to fit known experimental results. (For each M", this work could use a trigonometric function to specify d(M", M').)

$$m(M'', M') \approx m(M'', 0) \times \exp(\, (1/4) \log(\alpha'')\, (1 + M'')\, |M'|\,) \times (1 + d(M'',M')) \tag{3.49}$$

$$d(M'', \pm2) = -d(M'', \pm1) \approx d`(-d``)^{M''}, \text{ for } 0 \leq M'' \leq 2 \text{ and } 2 \geq |M'| \geq 1 \tag{3.50}$$

$$d(M'', M') = 0, \text{ otherwise} \tag{3.51}$$

$$d` \approx 0.2 \tag{3.52}$$

$$d`` \approx 0.4 \tag{3.53}$$

7. This step shows relationships that pertain.

$$-d(0, -1) = d(0, -2) \sim 0.2 \tag{3.54}$$

$$-d(1, -1) = d(1, -2) \sim -0.08 \tag{3.55}$$

$$-d(2, -1) = d(2, -2) \sim 0.032 \tag{3.56}$$

$$d(M'', M') = 0, \text{ otherwise within } 0 \leq M'' \leq 3 \text{ and } 3 \geq |M'| \geq 0 \tag{3.57}$$

Table 3.9.11 pertains. We do not speculate further regarding possible significance of $\tan^{-1}(1 + d'') \approx 2^{-1/2}$.

Table 3.9.11 A possible hint toward additional significance related to d" and some applications of trigonometric functions

1. To what extent might people find significance in the following approximate relationship?

$$(\tan^{-1}(1 + d'') - 2^{-1/2})/2^{-1/2} \approx 4.6\times10^{-4} \quad (3.58)$$

Table 3.9.12 and Table 3.9.13 compare calculated numbers with experimental numbers. The tables show masses in units of MeV/c^2. For each particle, the bottom number (calc) comes from this monograph's calculations.

Table 3.9.12 compares calculated masses with experimental masses for charged leptons. (Table 8.1.1 provides experimental masses.)

Table 3.9.12 Experimental and calculated masses for charged leptons

	M'	-3	...
M"			
0	exp	0.510998928±0.000000011	
0	calc	0.510998928 MeV/c^2	
1	exp		
1	calc		
2	exp	105.6583715±0.0000035	
2	calc	105.6583715	
3	exp	1776.82±0.16	
3	calc	1776.85(6)	

Table 3.9.13 compares calculated masses with experimental masses for quarks. (K.A. Olive et al. (2014) provides experimental masses for quarks.) Except regarding M" = 2 quarks, the upper number (exp) comes from experiments. For each M" = 2 quark, K.A. Olive et al. (2014) provides at least two ranges for quark masses. For M" = 2, M' = −2, the symbol m_b(MS) {with an over-bar on the MS} correlates with the result the table lists first. For M" = 2, M' = −2, the symbol m_b(1S) correlates with the result the table lists second. For M" = 2, M' = −1, the first result the table lists correlates with the phrase direct measurements, the second result correlates with the phrase MS {with an over bar on the MS} from cross-section measurements, and the third result correlates with the phrase Pole from cross-section measurements.

Table 3.9.13 Experimental and calculated masses for quarks

	M'	...	−2	−1
M"				
0	exp		$2.3^{+0.7}_{-0.5}$	$4.8^{+0.5}_{-0.3}$
0	calc		2.10 MeV/c^2	4.79
1	exp		95±5	$(1.275\pm0.025)\times10^3$
1	calc		92.5	1.272×10^3

M"		M' ...	−2	−1
2	exp		$(4.18\pm0.03)\times10^3$ $(4.66\pm0.03)\times10^3$	$(173.21\pm0.51\pm0.71)\times10^3$ $(160^{+5}_{-4})\times10^3$ $(176.7^{+4.0}_{-3.4})\times10^3$
2	calc		4.367×10^3	164.1×10^3
3	exp			
3	calc			

Possibly, m(M",M') correlates with each of the 9 masses for quarks and charged leptons.

Either of the two items Table 3.9.14 shows lists numbers this work uses to fit the data.

Table 3.9.14 Alternative sets, each of six numbers, sufficient to approximate the rest masses of the three charged leptons and six quarks

1. m_e, m'(muon), m'(tauon), α', d\`, and d\`\`.
2. m\`, ζ", d(0), α', d\`, and d\`\`.

Table 3.9.15 shows a relationship that correlates with work above. This work assumes d(−1,±3) = d(−1,0) = 0.

Table 3.9.15 Relationships between m\` and some results from a formula that approximates the rest masses of three charged leptons and the six quarks

$$m(-1,-3) = m(-1,0) = m(-1,+3) = m` \quad (3.59)$$

~ ~ ~

This subsection shows a possible correlation between ratios of masses of the three 1C- or 1Q-subfamily elementary particles for which M" = 0 and results pertaining for D* = 2 and ν = −1/2.

Table 3.9.16 pertains. (See Table 3.9.10.)

Table 3.9.16 An approximate relationship between |M'| and m(M" = 0, M')

$$m(M" = 0, M') \sim m(M" = 0, M') \times ((\alpha")^{-1/4(1+M")})^{-|M'|}, \text{ for } 3 \geq |M'| \geq 1 \text{ and } M" = 0. \quad (3.60)$$

People might say that Table 3.9.17 shows items that correlate values of |M'| with values of D + 2ν correlating with D* = 2 and ν = −1/2. For example, for Ω = 1/4, D + 2ν = 1. (See Table 2.5.17.)

Table 3.9.17 Possible relationships between $|M'|$ and values of $D + 2\nu$ correlating with $D^* = 2$ and $\nu = -1/2$

$\|M'\|$	Values of $D + 2\nu$, correlating with $D^* = 2$ and $\nu = -1/2$
1	$D + 2\nu$ for $\Omega = 1/4$
2	$D + 2\nu$ for $\Omega = -1/4$
3	($D + 2\nu$ for $\Omega = 1/4$) + ($D + 2\nu$ for $\Omega = -1/4$)

~ ~ ~

This subsection points to an involvement of square roots of masses.

For charged leptons, Table 3.9.18 pertains. (See Table 3.9.10.) The formula involves integer powers of the square roots of two masses.

Table 3.9.18 An approximate relationship between square roots of masses and masses

$$m(M'', M' = \pm 3) \approx m` \times ((m'(\text{muon})/m_e)^{(1/2)})^{(M'' + 1)} \times (1 + d(M'')). \quad (3.61)$$

~ ~ ~

This subsection discusses the Koide formula.

Table 3.9.19 shows the Koide formula and results this monograph calculates based on the monograph's calculated value for m'(tauon).

Table 3.9.19 The Koide formula and results this monograph calculates

1. The Koide formula is the following formula.

$$(m_e + m'(\text{muon}) + m'(\text{tauon})) / (m_e^{1/2} + (m'(\text{muon}))^{1/2} + (m'(\text{tauon}))^{1/2})^2 \approx 2/3 \quad (3.62)$$

2. Results reflecting nominal numbers this work uses, including the calculated value for m'(tauon), are the following.

$$(m_e + m'(\text{muon}) + m'(\text{tauon})) / (m_e^{1/2} + (m'(\text{muon}))^{1/2} + (m'(\text{tauon}))^{1/2})^2 \approx 0.66666(\sim 4) \quad (3.63)$$

   2.1. The uncertainty-range may not be accurate.

~ ~ ~

This subsection discusses, for charged leptons, relative masses and relative vertex strengths per channel.

Table 3.9.20 correlates with Table 3.9.3. In particular, k = 5 for tauons.

Table 3.9.20 Some G-family interactions and their relative strengths

Interaction	Channels	Electron relative vertex strength per channel	Electron k for β^{-k}	Tauon relative vertex strength per channel	Tauon k for β^{-k}
2G2&	4	$1 = \beta^{-0}$	0	$1 = \beta^{-0}$	0
4G4&	3	β^{-6}	6	β^{-5}	5

Table 3.9.21 includes information about the mass of a muon.

Table 3.9.21 Some information about a pattern pertaining to masses of charged leptons

Particle	Symbol for mass m(M",M')	mass/m_e	M"
electron	m(0,−3) = m_e	$\beta^{0/3}$	0
-			1
muon	m(2,−3) = m'(muon)	$\sim\beta^{(2/3)\,(1\,-\,0.02)}$ or $\sim 0.9\,\beta^{2/3}$	2
tauon	m(3,−3) = m'(tauon)	$\beta^{3/3}$	3

~ ~ ~

Reference 5 K.A. Olive et al. (2014)

K.A. Olive et al. (Particle Data Group), *Chin. Phys. C*, 38, 090001 (2014). (http://pdg.lbl.gov/2014/html/authors_2014.html)

Section 3.10 Some interactions between elementary bosons and fermions

Section 3.10 provides perspective regarding interactions between elementary fermions and elementary bosons. We list elementary bosons for which conservation of fermion generation pertains regarding FRERAN interactions between spin-1/2 fermions and elementary bosons. We discuss aspects pertaining to COMPAR interactions.

~ ~ ~

This subsection discusses some aspects of interactions between elementary bosons and elementary fermions.

We discuss aspects related to #"G = 1 and to conservation of fermion generation at vertices.

Table 3.10.1 pertains. Above, we do not find a G-family boson for which #"G would equal 3.

Table 3.10.1 Elementary bosons for which #"G = 1 (assuming that G-family %68odd solutions do not pertain)

1. Elementary bosons for which #"G = 1 or 3.
   1.1. 2W in FRERAN environments. (See Table 2.13.5.)
   1.2. 2YO. (See Table 2.13.8.)
   1.3. G-family. (See Table 2.8.3 and Table 3.3.4.)
      1.3.1. 4G4&.
      1.3.2. 4G268&.
      1.3.3. 2G468&.
2. G-family elementary bosons for which #"G ≠ 1 and #"G ≠ 3. (See Table 2.8.3 and Table 3.3.4.)
   2.1. 2G2&.
   2.2. 2G24&.
   2.3. 2G68&.
   2.4. 4G2468&.

People might say that #"G = 1 correlates, for an interaction vertex involving a spin-1/2 elementary fermion, with the vertex exhibiting conservation of generation for the fermion. For a vertex correlating with one entering fermion and one exiting fermion, conservation of generation correlates with the generation of the fermion that exits the vertex being the same as the generation of the fermion that enters the vertex. For a vertex correlating with creation of a matter-fermion-and-antimatter-fermion pair, conservation of generation correlates with the generation of the antimatter fermion being the same as the generation of the matter fermion.

People might say that, as far as people know, such conservation of generation pertains to interactions between leptons and 2W. We correlate FRERAN models with these interactions.

People might say that, as far as people know, such conservation of generation pertains to interactions between quarks and 2W, except for some COMPAR interactions that we term COMPAR-fermion-vertex interactions.

An example of a COMPAR-fermion-vertex interaction is one in which each of a pair of 2W interacts (in effect) with the same two quarks. Such interactions involving two 2W particles and two quarks can lead to changes in quark generations (and to violations of CP-symmetry). We discuss such COMPAR-fermion-vertex interactions in Section 2.13 (See, for example, discussion preceding Table 2.13.36.) and Table 3.10.2.

People might say that, as far as people know, conservation of generation pertains to interactions between quarks and 2Y.

~ ~ ~

This subsection introduces the topic of correlations between the presence of gravity and the phenomenon of neutrino oscillations.

People might say that (assuming EACUNI-set models pertain) Table 3.10.1 correlates with no possibility that gravity directly correlates with MM1MS1-neutrino oscillations. We discuss neutrino oscillations in Section 3.11.

~ ~ ~

This subsection discusses possible aspects of interactions that involve multiple vertices.

Table 3.10.2 provides questions that might correlate with opportunities for research. (See, for example, Table 2.9.4 and Table 2.9.6.)

Table 3.10.2 Possible relationships between R_0 (and similar lengths) and some strengths of interactions

1. To what extent does the magnitude of R_0 for charged 2W-subfamily particles ...
    - 1.1. Correlate with the magnitude of CP-violation that people correlate with, for example, interactions between the two quarks in a kaon?
        - 1.1.1. People might say that, at least to a first approximation, CP-violation in a kaon correlates with COMPAR-fermion-vertex interactions involving two 2W particles.
    - 1.2. Correlate with measurements regarding neutral B meson flavor oscillation?
2. Regarding the G-family member 2G24&, ...
    - 2.1. A formula of the form ...
        - 2.1.1. ... $(|q_e|^2/((4\pi\varepsilon_0)r^2)) \times (R^2/r^2) \times$ (correlation of two somewhat spin-like states) ...
        - 2.1.2. ... Pertains to an interaction between two elementary particles for which each elementary particle has charge $|q_e|$, with ...
            - 2.1.2.1. R being independent of the masses of the two interacting charged particles.
    - 2.2. To what extent does R correlate with a known property of nature?
3. Regarding the G-family member 2G68&, ...
    - 3.1. A formula of the form ...
        - 3.1.1. ... $(\hbar c/r^2) \times (R^2/r^2) \times$ (correlation of two spin-like states) ...
        - 3.1.2. ... Pertains to an interaction between two elementary particles for which each elementary particle has spin $\hbar/2$, with ...
            - 3.1.2.1. R being independent of the charges and masses of the two interacting charged particles.

| 3.2. To what extent does R correlate with a known property of nature?

Section 3.11 MM1MS1-neutrino oscillations

Section 3.11 correlates MM1MS1-neutrino oscillations (or, flavor mixing) with various phenomena. People might say that models for MM1MS1-neutrinos correlate with the passing through ordinary matter of MM1MS1-neutrinos contributing to MM1MS1-neutrino oscillations. People might say that gravity does not directly contribute to MM1MS1-neutrino mixing. We show the possibility that 2O0 bosons can contribute to MM1MS1-neutrino mixing.

~ ~ ~

This subsection discusses neutrino oscillations (or, flavor mixing).

People observe that neutrinos change flavor. For example, the mixture of flavors for solar neutrinos observed on earth does not match the mixture that nuclear reactions in the sun likely produce.

Traditionally, people say that gravity catalyzes neutrino mixing and that such mixing can occur only to the extent that at least one of the three flavors of neutrinos has non-zero rest mass.

~ ~ ~

This subsection discusses neutrino masses.

Perhaps, no measurements exist that would indicate that neutrinos move, in a near-vacuum, with speeds less than c, the speed of light. People might say that, to the extent all neutrinos move (in a near-vacuum) at the speed of light, each flavor of neutrino would have zero-mass.

People infer, from cosmological data, upper bounds on the sum (over three flavors) of neutrino masses. The number $\Sigma(m')_{\nu}$ provides an estimated upper bound of a fraction of one eV/c^2. (See Table 8.1.1.)

Possibly, people's statements that at least one flavor of neutrino needs to have non-zero mass correlate with attempted applications of models that do not adequately well encompass quantum gravity, gravity, and/or aspects of neutrinos. Perhaps, people might consider that a model that does not adequately incorporate gravity may not adequately address relevant aspects of mass.

We think that the notion that each MM1MS1-neutrino has zero mass may not present significant problems.

~ ~ ~

This subsection discusses possibilities for MM1MS1-neutrino flavor mixing (or, MM1MS1-neutrino oscillations).

Table 3.11.1 summarizes some results. Subsequent subsections provide some details. Regarding possibilities for neutrino oscillations correlating with interactions mediated by the G-family, the table shows results for G-family bosons with spatial dependence of interaction of either r^{-2} or r^{-4}. The table does not show results for G-family bosons with spatial dependence of interaction of either r^{-6} or r^{-8}. (See Table 2.13.23.) We do not show results regarding %68odd G-family particles. (See Table 3.3.4.) Per Section 3.10, we assume interactions between MM1MS1-neutrinos and gravity do not directly correlate with MM1MS1-neutrino oscillations.

Table 3.11.1 Some phenomena that correlate with MM1MS1-neutrino oscillations (assuming that the %68even set pertains)

1. Interactions (with MM1MS1-neutrinos) mediated by 2G68&.
   - 1.1. For 2G68&, MM1MS1-neutrino oscillations can correlate with ...
     - 1.1.1. Interactions with non-zero spin OME elementary particles.
     - 1.1.2. Interactions with non-zero spin DME-1 elementary particles, assuming ENS48 models or ENS6 models pertain.
2. Interactions (with MM1MS1-neutrinos) mediated by 2OO.
   - 2.1. MM1MS1-neutrino oscillations can correlate with ...
     - 2.1.1. Interactions with OME elementary particles that interact with 2OO particles.
       - 2.1.1.1. Such particles ...
         - 2.1.1.1.1. Include MM1MS1-neutrinos.
         - 2.1.1.1.2. Might include 1QR-subfamily elementary particles and/or 2YO-subfamily elementary particles.

The first item in Table 3.11.1 correlates (at least in part and substantially) with MM1MS1-neutrinos passing through ordinary matter.

~ ~ ~

This subsection correlates possibilities for MM1MS1-neutrino flavor mixing (or, MM1MS1-neutrino oscillations) with MM1MS1-neutrinos interacting, via 2G68&, with stuff.

Work above indicates that any MM1MS1-neutrino can interact via 2G68&. Such interactions would not differentiate between MM1MS1-neutrino generations. (See, for example, Table 2.8.3 and Table 2.8.4.) Such interactions would not necessarily conserve MM1MS1-neutrino generation. MM1MS1-neutrino oscillations can occur.

People might say that 2G68& can participate in an interaction vertex that involves any elementary particle with non-zero spin.

People might say that MM1MS1-neutrino flavor mixing occurs (at least) when a MM1MS1-neutrino moves through almost any OME-related stuff. People might say that this phenomena does not correlate with whether MM1MS1-neutrinos have non-zero masses.

People might say that, for cases correlating with ENS48 models or ENS6 models, MM1MS1-neutrino flavor mixing occurs also when a MM1MS1-neutrino moves through almost any DME-1-related stuff. (See, for example, Table 2.13.23.)

~ ~ ~

This subsection notes that the presence or creation of 2O0 particles may contribute to MM1MS1-neutrino flavor mixing (or, MM1MS1-neutrino oscillations).

People might say that Table 3.5.3 correlates with possibilities that MM1MS1-neutrinos interact with 2O0 bosons.

People might say that Table 3.5.3 correlates with phenomena people call elastic scattering of Z bosons by MM1MS1-neutrinos. People might say that the term elastic correlates with no change in generation (or, flavor) for the MM1MS1-neutrino. We think that conservation of generation correlates with the appearance of exactly one "G symbol in the Table 2.13.5 row regarding the 2W-subfamily. The appearance of that one "G symbol correlates with (1;3)> symmetries pertaining to the 2W-subfamily.

The symmetries (1;3)> do not necessarily pertain to interactions involving a single 2O0 particle. COMPAR symmetries might pertain. (See, for example, Table 2.12.6.)

People might say that scattering by 2O0 bosons can result in MM1MS1-neutrino flavor mixing (or, MM1MS1-neutrino oscillations). (See, for example, Table 2.12.6.)

Section 3.12 The fine-structure constant

Section 3.12 lists traditional and possible appearances, in physics models, of the fine-structure constant.

~ ~ ~

This subsection discusses the fine-structure constant.

Physics models have various uses for the fine-structure constant, α. The formula $\alpha = ((q_e)^2/(4\pi\varepsilon_0))\times(1/(\hbar c))$ pertains. The number is measured to a high degree of accuracy. Approximately, $\alpha \approx 7.297\times10^{-3}$.

~ ~ ~

This subsection notes possible appearances of the fine-structure constant in physics models.

Table 3.12.1 pertains. People might say that some of these appearances correlate with interactions.

Table 3.12.1 Possible appearances of the fine-structure constant, α, in physics models

1. In the following formula.
   $$\alpha = ((q_e)^2/(4\pi\varepsilon_0))\times(1/(\hbar c)) \quad (3.64)$$
2. As a contributor toward the ground-state energy for a hydrogen atom (in the limit that the mass of the nucleus is large compared to the mass of an electron). (See Table 6.2.2.)
3. As a contributor toward computing some anomalous magnetic dipole moments.
   3.1. This statement reflects work in traditional quantum physics.
4. As a number in a formula approximately linking the masses of quarks and charged leptons. (See Section 3.9.)
5. Possibly, as a contributor toward computing, from the mass of a Z boson, the mass of a W boson. (See Section 3.7.)
6. Possibly, as a contributor toward computing interaction strengths relevant for 2G24&. (See Table 3.10.2.)
7. Possibly, as a contributor in a possible decomposition into two components of the measured mass of the electron. (See Table 3.13.5.)

Section 3.13 Other aspects regarding particles, properties, and interactions

Section 3.13 discusses results that may be coincidences and/or may be pointers to opportunities for further research. We show a basis for a series that, for the electron, features charge (or, q_e), mass (or, m_e), and nominal magnetic dipole moment (or, $(g_S)\hbar/2$ or $2S\hbar$). We show a possible approximate relationship between the mass of a Higgs boson and the mass of an electron. We discuss some possible significance of β regarding ratios of G-family vertex strengths. We discuss possible significance for the notion that $\beta^{1/3} \approx e^e$.

~ ~ ~

This subsection discusses concepts regarding patterns and approximations. Sometimes, modeling can advance based on identifying patterns.

In this section, we look for elements in the series such that, for each element, the power associated with a factor is 0. We try to correlate that series with models and nature.

Sometimes, people benefit by trying to develop models that might correlate with closeness of some ratios of physics numbers to seemingly possibly relevant algebraic numbers.

In this section, we discuss possible interconnections among the sizes of physics constants.

~ ~ ~

This subsection discusses possible interpretations of aspects of a series of related formulas that correlate with lengths that include R_0, the Planck length, and the radii of black holes.

Table 3.13.1 pertains. (See Section 2.9.)

Table 3.13.1 Notes regarding R_j for which a particle property has (in Table 2.9.3) an exponent $k = 0$

1. This monograph attaches possible significance to R_j for which a particle property has an exponent $k = 0$ (in Table 2.9.3).
   - 1.1. G_N is not a particle property.
     - 1.1.1. We note the first item in Table 3.9.2.
     - 1.1.2. Regarding R_0, we assume we can consider q_e to be a particle property for which $|q_e|^0$ pertains.
2. Also, c is not a particle property.

Table 3.13.2 pertains. The integer j' correlates this table with Table 2.9.3. In Table 3.13.2, the leftmost three columns pertain to electrons. Here, g_S denotes magnetic dipole moment. For electrons, $g_S = 2$. The rightmost column provides names for physical properties.

Table 3.13.2 The series (for electrons) charge, mass, and magnetic dipole moment

j'	Factor (for an electron) for which k = 0	Size of the property (for an electron)	Traditional name of the property
2	q_e	q_e	Charge
4	m_e	m_e	Mass
6	$\hbar$	$(g_S)\hbar/2$ or $2S\hbar$	Magnetic dipole moment
8	-	-	-

~ ~ ~

This subsection notes a possible approximate relationship between the mass of a Higgs boson and the mass of an electron.

Table 3.13.3 defines β" and shows some numeric relationships. This monograph leaves unanswered the question in the last item in the table.

Table 3.13.3 A possible link between masses of elementary bosons and the strength of electromagnetism

1. Define β" via the following formula.
   $$(4/3)(\beta'')^{12} = \{(q_e)^2/(4\pi\varepsilon_0)\} \,/\, \{G_N(m'(\text{Higgs boson}))^2\} \quad (3.65)$$
2. Calculate β".
   $$\beta'' \approx 4.3928\times10^2 \quad (3.66)$$
3. Note the following approximation.
   $$(\beta/\beta'')^{1/3} \approx 2 \times (1 - 3.5\times10^{-3}) \approx 2 \quad (3.67)$$
4. To what extent can theory link masses of non-zero-mass elementary fermions to masses of non-zero-mass elementary bosons?

~ ~ ~

This subsection discusses some possible significance of β regarding ratios of G-family vertex strengths.

Table 3.9.3, Table 3.9.20 and Table 3.9.21 discuss, for charged leptons, relative masses and relative vertex strengths per channel.

Table 3.13.4 pertains. The functions m(...,−3) extend functions Section 3.9 defines and uses. This monograph leaves unanswered the question in the last item in the table.

Table 3.13.4 Possible ratios of G-family vertex strengths, beyond for just 2G2&, 4G4&, and interactions involving ordinary matter

1. For integers j such that $0 \le j \le 6$, define the following.
   $$m(0 + 3j,-3) = \beta^j\, m(0,-3) \quad (3.68)$$
   $$m(2 + 3j,-3) = \beta^j\, m(2,-3) \quad (3.69)$$
   $$m(3 + 3j,-3) = \beta^j\, m(3,-3) \quad (3.70)$$
2. Note that, if (for $j > 0$) the various $m(l + 3j,-3)$ represented particle masses, ...
   - 2.1. For the particle for which $l = 3$ and $j = 5$, ...
     - 2.1.1. The mass would be $\sim 9.03\times10^{20}$ MeV/c^2.
     - 2.1.2. The vertex strength per channel of 4G4& would equal the strength per channel of 2G2&.
   - 2.2. For the particle for which $l = 0$ and $j = 6$, ...
     - 2.2.1. The mass would be $\sim 9.03\times10^{20}$ MeV/c^2.
     - 2.2.2. The vertex strength per channel of 4G4& would equal the strength per channel of 2G2&.

3. Similarly, ...
   3.1. We could define ...
      3.1.1. Charges ...

$$q(0-3j,-3) = q(2-3j,-3) = q^*(3-3j,-3) \tag{3.71}$$

$$= \beta^{-j}\, q(0,-3) \tag{3.72}$$

         3.1.1.1. Here, we use the symbol q^* to differentiate (for example, for $j = 1$) $q^*(3 - 3j,-3)$ from the electron's charge of $q(0,-3)$.
      3.1.2. Masses ...

$$m(0-3j,-3) = m(0,-3) \tag{3.73}$$

$$m(2-3j,-3) = m(2,-3) \tag{3.74}$$

$$m(3-3j,-3) = m(3,-3) \tag{3.75}$$

   3.2. Here, for $q^*(0-6,-3)$ and $m(0-6,-3)$, the vertex strength per channel for 2G2& would equal the vertex strength per channel of 4G4&.
4. Possibly, a range of 6 integers for j correlates (in ENS48 models and ENS6 models), at least mathematically, with ...
   4.1. The existence of 6 OMDME (or, ordinary-matter or dark-matter ensembles).
5. For ENS48 models for which #E\` < 6 for 2G24& and for ENS6 models for which #E\` < 6 for 2G24&, ...
   5.1. Consider the following two 2G24&-mediated interactions ...
      5.1.1. An interaction between two electrons.
      5.1.2. An interaction between an electron and the lowest-mass charged lepton in the one dark-matter ensemble for which such an interaction is possible. (For #E\` < 6, the span of an instance of 2G24& is 2 ensembles.)
   5.2. To what extents does (for those ENS48 models and those ENS6 models) a ratio of vertex strengths for those two interactions correlate with ...
      5.2.1. $1 : \beta^{-1}$?
      5.2.2. A function of α (or, the fine-structure constant)?

~ ~ ~

This subsection discusses possible significance of β.

Table 3.13.5 pertains. This monograph leaves unanswered the question in the last item in the table.

Table 3.13.5 A possible link between β and a number (or between various masses and interaction strengths)

1. $\beta^{1/3} = e^e(1 - x'')$, with $x'' \approx 3\times10^{-4}$.
    1.1. Here, the following approximation pertains.
    $$(\pi/2)/\cos^{-1}(1 - x'') \approx 65 \qquad (3.76)$$
2. $\beta^{1/3} = e^{e(1 - x'')}$, with $x'' \approx 1\times10^{-4}$.
    2.1. Here, the following approximation pertains.
    $$(\pi/2)/\cos^{-1}(1 - x'') \approx 107 \qquad (3.77)$$
3. For the following formula, various results below pertain.
    $$(4/3)e^{36e} = \{(q_e)^2/(4\pi\varepsilon_0)\} / \{G_N(m^*)^2\} \qquad (3.78)$$
    3.1. $m_e \approx (1 + 0.709109\alpha)\, m^*$.
    3.2. More specifically, ...
        3.2.1. For a value of G_N that exceeds the current nominal value by 1 standard deviation, ...
            3.2.1.1. $m_e \approx (1 + 0.717365\alpha)\, m^*$.
        3.2.2. For a value of G_N that is less than the current nominal value by 1 standard deviation, ...
            3.2.2.1. $m_e \approx (1 + 0.700853\alpha)\, m^*$.
    3.3. Perhaps the factor multiplying α correlates with $2^{-1/2}$.
        3.3.1. To the extent that factor is $2^{-1/2}$, ...
        $$G_N \approx 6.67365\times10^{-11}\ \mathrm{m^3\,kg^{-1}\,s^{-2}}. \qquad (3.79)$$
    3.4. An experimental value is the following. (Compare with the experimental result Table 8.1.1 states.)
    $$6.67384(80)\times10^{-11} \qquad (3.80)$$
4. To what extent can people attach significance to the closeness of $\beta^{1/3} \approx e^e$?

~ ~ ~

This subsection discusses possible significance of m*.

Suppose that the concept of m* is physics-relevant. (See Table 3.13.5.) Table 3.13.6 describes points that people may find useful. Or, people may find that some of the consequences are sufficiently implausible that people do not want to consider the concept of m* further.

Table 3.13.6 Possible consequences, to the extent m* has physics-relevance

1. The measured mass of the electron includes components correlating with each of ...
    1.1. m*.
    1.2. A number such as $m_e - m^*$ or $(m_e^2 - m^{*2})^{1/2}$, with the relevant number correlating with ...

1.2.1. A measure of aspects of the cloud of virtual particles that accompanies an elementary particle having rest mass m^* and charge q_e.

2. The following result provides a useful possible prediction.

$$G_N \approx 6.67365 \times 10^{-11}\ \mathrm{m^3\ kg^{-1}\ s^{-2}} \tag{3.81}$$

3. Other.

Chapter 4 From particles to cosmology and astrophysics

Chapter 4 (From particles to cosmology and astrophysics) starts with the known and candidate elementary and composite particles that the previous chapter discusses and shows possibilities for correlating with or anticipating data about cosmology or astrophysics. We show models that possibly correlate with various cosmological phenomena and with various astrophysical phenomena. People might say that some models point to how to close some gaps between physics data and results from traditional theory.

~ ~ ~

People might say that results in Chapter 4 (From particles to cosmology and astrophysics) might correlate with or provide insight about the following aspects of physics. Mechanisms governing the rate of expansion of the universe. Dark energy. Dark matter. Stuff other than ordinary matter, dark matter, and dark-energy stuff. Ratios of densities of dark matter to densities of ordinary matter. Clustering and anti-clustering mechanisms, regarding ordinary matter and dark matter. Objects, such as galaxies, that include both ordinary matter and dark matter. A mechanism possibly leading to imbalance between matter and antimatter. Mechanisms, correlating with quantum phenomena, that may correlate with phenomena people describe via general relativity. The galaxy rotation problem. Phenomena regarding the spacecraft flyby anomaly. Mechanisms leading to quasars. Cosmic microwave background cooling. Some phases, for quark-based plasmas.

Section 4.1 Introduction

Section 4.1 discusses correlations between values of #ENS and aspects of cosmology and astrophysics.

~ ~ ~

This subsection provides perspective regarding discussion in this monograph about cosmology and astrophysics and regarding a parameter, #ENS, correlating with MM1MS1 models.

T.J. Buckholtz, *Models for Physics of the Very Small and Very Large*,
Atlantis Studies in Mathematics for Engineering and Science 14,
DOI 10.2991/978-94-6239-166-6_4

People might say that work in this monograph features cosmology and astrophysics because, in part, doing so discusses possible tests for and possible application of models pertaining to elementary particles.

People might say that aspects of this monograph's discussions about cosmology and astrophysics couple closely with elementary particles and their properties. People might say that #ENS is not a property of elementary particles.

Table 4.1.1 pertains regarding correlations between #ENS and aspects of cosmology and astrophysics.

Table 4.1.1 Correlations between #ENS and aspects of cosmology and astrophysics

1. Regarding the rate of expansion of the universe (See Section 4.2.), ...
    1.1. People might say that (until further analysis of extant data or analyses of future data) ...
        1.1.1. People might want to explore definitions for DIU (or, directly inferable universe).
        1.1.2. People might not need to much explore a value of #ENS.
    1.2. People might say that (eventually) ...
        1.2.1. A value of #ENS influences the definition of DIU.
2. Regarding dark matter, dark-energy stuff, and objects that include ordinary matter and dark matter (See Section 4.3 and Section 4.4.), ...
    2.1. People might say that assuming (for modeling now) and determining (eventually) a value of #ENS is important.
3. Regarding other topics this chapter discusses (See Section 4.5 through Section 4.11.), ...
    3.1. People might not need to much explore a value of #ENS.

Section 4.2 The rate of expansion of the universe

Section 4.2 discusses interactions governing the rate of expansion of the universe. People might say that interactions mediated by G-family particles provide for changes in the rate of expansion of the universe. People might say that possible observations or theories that some people correlate with term inflationary epoch would (to the extent the observations or theories pertain to nature) correlate with phenomena related to 3QIRD or 1R fermions.

~ ~ ~

This subsection discusses concepts related to the term expansion of the universe.

People interpret observations as correlating with the moving apart of the largest clusters of stuff of which people know. These clusters are larger than galactic superclusters. On smaller scales, such as at the sizes of galactic superclusters or within galactic superclusters, sizes currently do not vary.

People interpret observations and theory as correlating with the concept that all of the stuff people currently observe was once contained within a quite small volume. People correlate the term big bang with that era.

When discussing the evolution in time of such stuff from very small to very large, people discuss concepts of expansion of the universe and of a rate of expansion of the universe. People say that the universe expands. People say that the rate changes over time.

Possibly, people should exert caution regarding using the term expansion of the universe. Observations pertain directly only to visible objects. Physics theory correlates with the concept that less than 5 percent of the stuff in the so-called observable universe correlates with visible objects. (Perhaps, the term indirectly observable or the term directly inferable would be more appropriate than the term observable.) And, perhaps, the so-called observable part of the universe exists within a much broader universe (that, perhaps, includes an infinite amount of stuff).

We use the acronym DIU to denote directly inferable universe.

People might say that the potentially visible stuff and the other DIU stuff undergo expansion in adequately similar manners that the DIU exhibits just one notion of expansion and, at least regarding inferences people can make, just one rate of expansion. The rate can change as the universe evolves.

Work in this monograph correlates with an assumption that discussing just one rate of expansion of the DIU suffices for purposes of this monograph.

~ ~ ~

This subsection discusses concepts related to the term expansion of space-time.

Traditional discussion may posit a so-called expansion of space-time. We think that such discussion correlates with use of the term expansion of the DIU.

We think that discussion of the expansion of the DIU need not necessary correlate with discussion of an expansion of space-time.

People might say that our work does not depend on assuming that space-time has properties (such as a size that expands or a non-zero curvature).

~ ~ ~

This subsection discusses a thought experiment regarding interactions between two objects.

Think about two clumps of stuff. These clumps have similar size and similar amounts of material. The clumps neighbor each other. Think about forces (or, interactions) that affect the two clumps.

Table 4.2.1 lists interactions that dominate within each clump and between the two clumps. This table reflects observations that people characterize by saying that the universe expands. During such expansion, effects of an r^{-2j} interaction (within or between the two clumps) evolve from being more than effects of an $r^{-2(j-1)}$ interaction to being less than effects of an $r^{-2(j-1)}$ interaction. The table defines four eras. (See Table 2.8.3.) Table 4.2.1 does not show %68odd particles.

Table 4.2.1 Eras pertaining to interactions between two clumps and relevant %68even particles

1. During era FE1, interactions with r^{-8} spatial dependence dominate.
   1.1. These interactions correlate with 4G2468&.
2. During era FE2, interactions with r^{-6} spatial dependence dominate.
   2.1. These interactions include ones that correlate with 2G468& and/or 4G268&.
3. During era FE3, interactions with r^{-4} spatial dependence dominate.
   3.1. These interactions include ones that correlate with 2G24& and/or 2G68&.
4. During era FE4, interactions with r^{-2} spatial dependence dominate.
   4.1. These interactions include ones that correlate with 4G4& and 2G2&.

~ ~ ~

This subsection correlates, for various sized objects, various FEk eras with data about expansion of the DIU.

Today, observed atoms, planets, stars, galaxies, and galactic superclusters exhibit era FE4 behavior. At and beyond a size that is larger than sizes of galactic superclusters, people observe FE3 behavior.

People deduce rates of expansion of the DIU from observations of photons. People express approximate time (after the big bang) of emission (or, creation) of photons in terms of the redshift, z, that people correlate with the photons. Redshift numbers are nonnegative. Redshift $z \approx 0$ pertains to photons emitted recently. A value of z correlates with a time before the time people associate with a smaller value of z.

Table 4.2.2 correlates some redshifts with times after the big bang. (We performed calculations using a tool N. Gnedin (2014) provides. We used calculations that assume the DIU is flat.)

Table 4.2.2 Some redshifts and the related times after the big bang

z	Time after the big bang (years)
2.3	2.9×10^{9}
0.7	7.3×10^{9}
0.46	8.9×10^{9}
0	13.8×10^{9}

Based on N. G. Busca (2013) and A. Riess (2004), we think that three eras exist. Throughout these eras, the rate of expansion is greater than zero. Table 4.2.3 pertains.

Table 4.2.3 Three eras relevant to the rate of expansion of the DIU

1. For z > some number greater than 2.3, the observed rate of expansion increases.
   1.1. This monograph calls this era ZE1.
2. For ~2.3 > z > ~0.7, the observed rate of expansion decreases.
   2.1. This monograph calls this era ZE2.
3. For 0.46±0.13 > z > 0, the observed rate of expansion increases.
   3.1. This monograph calls this era ZE3.

~ ~ ~

This subsection provides a model for the expansion of the DIU.
This monograph assumes that Table 4.2.4 pertains.

Table 4.2.4 Model for the rate of expansion of the DIU

1. For observed astrophysical objects of above some size, era FEk correlates with era ZEn, for $1 \le k = n \le 3$.

Table 4.2.5 pertains. This table correlates interactions mediated by G-family particles Table 3.3.3 lists with work above in this section. Table 4.2.5 lists all %68even interactions except 2G2& and 4G4&.

Table 4.2.5 G-family-mediated interactions that govern the rate of expansion of the DIU (assuming %68odd particles do not pertain)

%68even bosons	Interaction	Effects
4G2468&	r^{-8}	The net effects repel astrophysical objects from each other.
2G468& and 4G268&	r^{-6}	The net effects attract astrophysical objects to each other.
2G24& and 2G68&	r^{-4}	The net effects repel astrophysical objects from each other.

Table 4.2.6 suggests possible opportunities for research.

Table 4.2.6 An example of possible opportunities for research based (at least in part) on observations of large-scale phenomena

1. Estimate ...
   1.1. The relative extents to which each of 2G24& and 2G68& dominates the present growth in the observed rate of expansion of the DIU.

~ ~ ~

This subsection discusses possible rates of expansion for possible objects that would be larger than objects people observe.

People do not know the extent of the natural universe. Perhaps stuff exists beyond the DIU. Perhaps enough stuff exists that some very large objects today experience era ZE2 behavior or era ZE1 behavior.

~ ~ ~

This subsection provides context for discussing some possible phenomena and theory that some people correlate with the term inflationary epoch.

Some people interpret some observations or theories as correlating with a rapid increase in size of the DIU between approximately 10^{-36} seconds after the beginning of the big bang and approximately 10^{-33} to 10^{-32} seconds after the beginning of the big bang. Some people refer to this time period via the term inflationary epoch. Some people say that (in some sense) space-time expanded at faster than the speed of light.

~ ~ ~

This subsection provides a model correlating with possible phenomena people might correlate with the term inflationary epoch.

Possibly the end of the inflationary epoch has parallels to the end of the era in which quarks and gluons formed nucleons. However, differences exist.

People say that a quark-and-gluon plasma existed until the formation of individual nucleons. Regarding some of the time leading up to the formation of individual nucleons, people use the term quark epoch. People estimate that nucleons formed between approximately 10^{-6} seconds and 1 second after the big bang. People call some of the time after the formation of nucleons the hadron epoch.

We think that quarks and gluons formed nucleons under conditions for which n × 1Q+2Y composite particles form. We think that, previously, 1Q+2O or 1Q+2Y phenomena would pertain to a plasma (or, sea) of 1Q fermions. (See Section 4.5 and Section 4.11.)

Some inflationary-epoch phenomena might feature combinations of 3X, 4Y-related, and 4O particles. (We use the term 4Y-related because each 4Y solution does not correlate directly with one elementary particle.) Here, X could be any one of Q, I, R, or D. Perhaps, more than one such value of X pertains. For each of 3X, 4O, and 4Y-related particles, $\sigma = -1$. We think that 3X particles can exist in at least two of various contexts. One context is a plasma of 3X in which 4O particles provide the most important boson effects. Another context is a plasma of 3X in which 4Y-related particles provide the most important boson effects. Another perhaps possible context is primordial black holes consisting of 3X bound by bosons. Here, the bosons could include one of 4O particles and 4Y-related particles and might include 4G268& and/or 2G468&. Primordial black holes would have $\sigma = +1$.

To the extent that such inflationary-epoch concepts correlate with nature, this work suggests that, possibly, for at least one X (or, and least one of X = Q, X = I, X = R, or X = D), at least one phase change occurred regarding contexts for 3X. Possibly, an initial phase features 3X and 4O. Possibly, a later phase features 3X and 4Y. Possibly, the end of the epoch correlates with the formation of primordial black holes and/or with matter/antimatter annihilation of 3X particles.

To the extent nature includes 1R particles, perhaps parallel concepts pertain for combinations of 1R, 2O, and 2Y-related particles.

Before the inflationary epoch, 4G2468& dominates the expansion of the DIU. To the extent something special pertains during the inflationary epoch, this work suggests that, during inflationary epoch, at least one of the phase transitions this work suggests above occurred.

People might say that, regarding during such an inflationary epoch, to the extent people attempt to use a classical-physics interpretation for $\sigma = -1$ phenomena, the DIU expands at faster than the speed of light. We suggest that people should be cautious regarding stating or using such an interpretation.

~ ~ ~

Reference 6 N. Gnedin (2014)

N. Gnedin, Cosmological Calculator for the Flat Universe (2014). (http://home.fnal.gov/~gnedin/cc/)

Reference 7 N. G. Busca (2013)

N. G. Busca, et. al., Baryon Oscillations in the Lyα forest of BOSS quasars, *Astronomy & Astrophysics* 552 A96, April 2013. (arXiv:1211.2616 [astro-ph.CO])

Reference 8 A. Riess (2004)

A. Riess, et. al., Type Ia Supernova Discoveries at z > 1 from the *Hubble Space Telescope*: Evidence for Past Deceleration and Constraints on Dark Energy Evolution, *The Astrophysical Journal*, 607, 665 (2004). (doi:10.1086/383612) (http://iopscience.iop.org/0004-637X/607/2/665)

Section 4.3 Dark energy and dark matter

Section 4.3 discusses two uses of the term dark energy. People might say that some G-family particles correlate with a force-like (or, pressure-like) use of the term. We provide perspective regarding models that may correlate with densities of the DIU (or, the directly inferable universe) for ordinary matter, dark matter, and dark-energy stuff. People might say that, in ENS48 models, dark-energy stuff features 7 peers of the stuff people might associate with the combination of ordinary (or, baryonic) matter and dark matter. People might say that, for ENS48 models and ENS6 models, this work provides an explanation for the inferred ratio of density of dark matter to density of ordinary matter being equal to or greater than 5 : 1. People might say that, for ENS48 models, this work provides an explanation for the inferred ratio of density of dark-energy stuff to density of dark matter plus ordinary (or, baryonic) matter being less than 7 : 1. People might say that ENS48 models and ENS6 models point to possibly observable phenomena that would arise from clustering within a unit of stuff that features ordinary matter and dark matter, clustering within each of seven units of dark-energy stuff (ENS48 models, but not ENS6 models), anti-clustering between each pair among the just mentioned eight units (ENS48 models, but not ENS6 models), anti-clustering between ordinary matter and one of five units of dark matter, anti-clustering between two of the other four units of dark matter, anti-clustering between the remaining two units of dark matter, and clustering within each of six units - one of ordinary matter and five of dark matter. We discuss ENS1 models. We discuss possible correlations between concepts people have proposed regarding dark matter and results correlating with models this monograph discusses. We discuss concepts for experiments or observations to possibly attempt regarding the nature of dark matter and/or the nature or dark-energy stuff. People might say that we discuss, assuming ENS48 models correlate with nature, a way to envision dark matter and dark-energy stuff. People might say that we discuss, assuming ENS6 models correlate with nature, a way to envision dark matter.

~ ~ ~

This subsection provides perspective about this section.

This section discusses dark energy and dark matter. Traditional discussions of these topics tend to feature the topic of dark matter and then, possibly, the topic of dark energy. For ENS48 models, we think that it may be more useful to start with a holistic approach, then discuss dark energy, and then discuss dark matter. We use this notion to structure this section.

The section follows the following agenda. Provide some perspective. Discuss pressure-centric uses of the term dark energy. Provide perspective about stuff. Discuss stuff-centric aspects that people correlate with the term dark energy. Discuss, from the perspectives of models we feature, dark matter. Discuss some hypothetical aspects that people propose regarding dark matter.

~ ~ ~

This subsection discusses terminology related to dark energy.

People make two not-necessarily compatible uses of the term dark energy.

One use of the term dark energy correlates with contributions to the total density of the DIU (or, in traditional terms, of the universe). Here, people might correlate (to some extent) density with the term energy or with the term energy per unit of spatial volume. People interpret data as suggesting that dark energy makes up more than about two-thirds the total density. People correlate the remaining density with stuff people call ordinary matter and dark matter. People say that ordinary matter and dark matter interact via gravity. People say that ordinary matter plus dark matter need not interact with dark energy via gravity.

One use of the term dark energy correlates with mechanisms contributing to increases in the observed rate of expansion of the DIU. People discuss a concept of a pressure.

~ ~ ~

This subsection discusses correlations between work in this monograph and the pressure-centric use of the term dark energy.

People might say that models in this monograph correlate the concept that (at least) the five G-family bosons Table 4.2.5 lists correlate with pressure-centric (or, force-centric) uses of the term dark energy.

~ ~ ~

This subsection suggests perspective regarding stuff-centric aspects of dark energy and regarding aspects of dark matter.

Table 4.3.1 pertains.

Table 4.3.1 Perspective regarding stuff-centric aspects of dark energy and regarding aspects of dark matter

1. We anticipate that advances in physics can lead to definitions of dark-energy stuff and dark matter (or to definitions of other terms) that will be more precise and more useful than traditional uses of such terms.
    - 1.1. Presumably, the definitions will meaningfully span more advanced (than today's physics) data and theory.
    - 1.2. Presumably, more advanced theory will correlate with better understanding of the implications of the patterns in CMB (or, cosmic microwave background radiation) observations that people correlate with the existence of dark energy and with the existence of dark matter.
2. People might say that, to the extent that ENS48 models pertain to nature, concepts that Table 2.13.15, Table 2.13.16, and Table 2.13.17 discuss provide a useful cataloging of stuff.
3. People might say that, to the extent that ENS6 models pertain to nature, ...
    - 3.1. With respect to concepts that Table 2.13.15, Table 2.13.16, and Table 2.13.17 discuss, ...
        - 3.1.1. The OME pertains.
        - 3.1.2. 5 DME pertain.
        - 3.1.3. DEE do not pertain.
4. People might say that, to the extent that ENS1 models pertain to nature, ...

4.1. With respect to concepts that Table 2.13.15, Table 2.13.16, and Table 2.13.17 discuss, ...
 4.1.1. The OME pertains.
 4.1.2. DME do not pertain.
 4.1.3. DEE do not pertain.

~ ~ ~

This subsection summarizes a possibly useful look, via ENS48 models, at the topics of densities of the DIU of dark-energy stuff, dark matter, and ordinary matter.

People might say that Table 4.3.2 pertains. (See Table 2.13.15.) Table 4.3.3, Table 4.3.6, and discussion before Table 4.3.6 provide more detail and provide references regarding observations.

Table 4.3.2 The relative densities of DEE, DME, OME, and other stuff and the relative densities of dark-energy stuff, dark matter, and ordinary matter (ENS48 models)

1. Regarding stuff that nature includes, the following pertain.
 1.1. Possibly, nature created and maintains the 48 ensembles approximately equally.
  1.1.1. The 48 ensembles include ...
   1.1.1.1. 1 ensemble that we label (relative to observations people make) the OME.
   1.1.1.2. 5 ensembles that we label (relative to observations people make) DME.
   1.1.1.3. 42 ensembles that we label (relative to observations people make) DEE.
 1.2. To the extent nature created and maintained the 48 ensembles approximately equally, ...
  1.2.1. Letting AD denote actual density and considering adequately large segments of the DIU, people might assume that the following integers correlate with ratios of densities ...
   1.2.1.1. 1 - ADOME.
   1.2.1.2. 5 - ADDME.
   1.2.1.3. 42 - ADDEE.
 1.3. People might say that, regarding the various G#XY (that Table 2.13.15 shows, with # = 2, 6, or 48; X = OM, DM, or DE; and Y = OM, DM, or DE), ...
  1.3.1. Table 2.13.15 defines these categories relative to observations people make.
  1.3.2. Cases for which X = OM and Y = OM correlate with the OME.

1.3.3. Cases for which X = DM and Y = DM correlate with DME.
1.3.4. Cases for which X = DE and Y = DE correlate with DEE.
1.3.5. Cases for which X ≠ OM and Y ≠ OM do not correlate with OME.
1.3.6. Cases for which X ≠ DM and Y ≠ DM do not correlate with DME.
1.3.7. Cases for which X ≠ DE and Y ≠ DE do correlate with DEE.

1.4. People might say that the above classification correlates with the following sets of stuff.
1.4.1. OMS ...
1.4.1.1. Denotes ordinary-matter stuff.
1.4.1.2. Includes OME stuff.
1.4.1.2.1. Possibly, excludes OME SEDMS. (See Table 4.3.14.)
1.4.1.2.2. Possibly, excludes OME SEDES. (See Table 4.3.13.)
1.4.1.3. Includes G#XY stuff that correlates with the OME.
1.4.2. DMS ...
1.4.2.1. Denotes dark-matter stuff.
1.4.2.2. Includes DME stuff.
1.4.2.2.1. Possibly, excludes DME SEDES. (See Table 4.3.13.)
1.4.2.3. Possibly, includes OME SEDMS. (See Table 4.3.14.)
1.4.2.4. Includes G#XY stuff that correlates with DME.
1.4.3. One can define OMDMS to be the combination of OMS, DMS, G2OMDM, G6OMDM, and G48OMDM.
1.4.4. DES ...
1.4.4.1. Denotes dark-energy stuff.
1.4.4.2. Includes DEE stuff.
1.4.4.3. Possibly, includes OMDME SEDES. (See Table 4.3.13.)
1.4.4.4. Includes G#XY stuff that correlates with DEE.
1.4.5. OTS ...
1.4.5.1. Denotes other stuff.
1.4.5.2. Includes G#XY stuff that does not correlate with either of OMDMS and DES.

2. Regarding stuff that people can infer the existence of, the following pertain.
    2.1. The notion of infer includes concepts such as ...
        2.1.1. People can observe directly.
        2.1.2. People can infer from combinations of data and models.
    2.2. PIOMS ...
        2.2.1. Denotes possibly inferable ordinary-matter stuff.
        2.2.2. Includes OMS.
    2.3. PIDMS ...
        2.3.1. Denotes possibly inferable dark-matter stuff.
        2.3.2. Includes DMS.
    2.4. PIOMDMS ...
        2.4.1. Denotes possibly inferable ordinary-matter-and-dark-matter stuff.
        2.4.2. Includes OMDMS.
    2.5. PIDES ...
        2.5.1. Denotes possibly inferable dark-energy stuff.
        2.5.2. Includes DES.
    2.6. PIOTS ...
        2.6.1. Denotes possibly inferable other stuff.
        2.6.2. Includes OTS.
3. People might say that the following pertains.
    3.1. To the extent nature created ensembles equally, the densities of SEDES and SEDMS are adequately small, and the densities of the various G#XY are adequately small, ...
        3.2. Ratios of possibly inferable densities of the DIU would approximate ratios of the following integers. (Here, the letters PI abbreviate the term possibly inferable.)
            3.2.1. PIOMS = 1 - Ordinary matter.
            3.2.2. PIDMS = 5 - Dark matter.
            3.2.3. PIDES = 42 - Dark-energy stuff.
        3.3. One can define PIOMDMS to be PIOMS + PIDMS.
        3.4. To the extent nature created ensembles equally, ratios of inferred densities of the DIU would be ratios of the following numbers. (Here, IN denotes inferred density.)
            3.4.1. INOMS = 1 - Ordinary matter.
            3.4.2. INDMS ≥ 5 - Dark matter.
            3.4.3. INDES ≥ 42 - Dark-energy stuff.
        3.5. One can define INOMDMS to be INOMS + INDMS.
4. People infer ratios of densities.

- 4.1. People make such inferences based on observations of CMB (or, cosmic microwave background) radiation.

5. Inferred approximate ratios for the present DIU include ...
   - 5.1. 1 : 5.3 - INOMS : INDMS (or, ordinary matter : dark matter).
     - 5.1.1. A possible prediction (from above in this table) could be ...

$$1:5 \tag{4.1}$$

   - 5.2. 1 : 2.2 - INOMDMS : INDES (or, ordinary matter plus dark matter : dark-energy stuff).
     - 5.2.1. A possible prediction (from above in this table) could be ...

$$1:7 \tag{4.2}$$

6. Subsequent subsections provide means to reconcile ...
   - 6.1. The possible discrepancy regarding the ratio of inferred ordinary-matter density (or, INOMS) to inferred dark-matter density (or, INDMS).
     - 6.1.1. Possibly, a difference between the inferred ratio (1 : ~5.3) and 1 : 5 correlates with ...
       - 6.1.1.1. Densities of some G-family bosons.
       - 6.1.1.2. Densities of OME SEDMS.
   - 6.2. The seeming discrepancy regarding the ratio of inferred ordinary-matter-plus-dark-matter density (or, INOMDMS) to inferred dark-energy-stuff density (or, INDES).
     - 6.2.1. Possibly, this difference (between an inferred ratio of 1 : ~2.2 and a possible ratio of 1 : 7) correlates with slowness of the impact of effects off dark-energy stuff on phenomena that people measure.

~ ~ ~

This subsection summarizes a possibly useful look, via ENS6 models, at the topics of densities of the DIU of dark-energy stuff, dark matter, and ordinary matter.

People might say that concepts in Table 4.3.2 pertain, with the following provisos. Each aspect that correlates with DEE does not pertain. Only 6 ensembles (the OME and DME), not 48 ensembles, pertain. Ratios based on 42 ensembles does not pertain. The ratio of 1 : 7 does not pertain.

~ ~ ~

This subsection notes inferred ratios of densities of the DIU.

Table 4.3.3 pertains.

Table 4.3.3 Concepts pertaining to inferred OMDMS and DES densities of the DIU

1. People report the following contributions to the inferred total density of the DIU.
   1.1. Ordinary matter (or, baryonic matter), not including photons and neutrinos.
   1.2. Photons.
   1.3. Neutrinos.
   1.4. Dark matter.
   1.5. Dark energy.
2. People might say that, for the purposes of this monograph, no harm comes from assuming that inferences about OMS include inferences about ordinary matter, photons, and neutrinos.
3. People might say that, for the purposes of this monograph, no harm comes from assuming that inferences about OMDMS include inferences about ordinary matter, photons, neutrinos, and dark matter.
4. People suggest that, over the history of the universe, a ratio of the inferred density of the DIU for DES to inferred density of the DIU for OMDMS has grown from less than 1 : 99. (We interpret results that T. Ferris (2015) states.)
5. People suggest, for the current state of the DIU, a ratio of ~2.2 : 1 for the inferred density of DES to the inferred density of OMDMS. (See, respectively, Ω_Λ and $\Omega_m = \Omega_{cdm} + \Omega_b$ in Table 8.1.1.)

~ ~ ~

This subsection reviews aspects of how people estimate ratios of the inferred density of DES to the inferred density of OMDMS.

People infer ratios of contributions to the density of the DIU of ordinary (or, baryonic) matter, dark matter, and dark energy. People base estimates on data about CMB (that is, cosmic microwave background radiation). Traditional physics considers that observed CMB photons have existed since near the time of the big bang. Physics considers that before $10^{5.6}$ years after the big bang, such photons interacted significantly with ionized plasma. Around $10^{5.6}$ years after the big bang, the plasma ceased to be ionized. Much CMB travels today.

Data suggest late-time (or, secondary) anisotropies. Late-time refers to any time after the above-mentioned plasma ceased to be ionized. Relevant processes for generating non-uniformities in CMB may feature photon scattering by free electrons (Thompson scattering), scattering by clouds of high-energy electrons (Compton scattering by hot electron gases), and frequency-shifting because of changing gravitational fields (integrated Sachs-Wolfe effect). (Here, we summarize statements in J. Beringer (2012).)

Analyses of CMB data determine, in effect, aspects of clumping. This monograph uses the term clumping to denote non-homogeneity in the distribution of energy or matter. Within OMDMS, clumps correlate with nucleons atomic nuclei, atoms, planets, stars, solar systems, clouds of electrons, and so forth. OMDMS clumps also include galaxies, galactic clusters, and so forth.

~ ~ ~

This subsection discusses assumptions regarding phenomena leading to ratios of inferred DES to inferred OMDMS.

This monograph assumes that phenomena leading toward inferred DES densities reflect influences of DES clumping on OMDMS clumping.

People might say that, for the purposes of discussing ENS48 models, this monograph deemphasizes possibilities for the existence of SEDES (or, single-ensemble dark-energy stuff). (See Table 4.3.13.) Regarding ENS48 models, existence of SEDES could correlate with actual ratios of DES to OMDMS that would exceed 7.

People might say that, for the purposes of discussing ENS6 models, this monograph requires the existence of SEDES (or, single-ensemble dark-energy stuff). (See Table 4.3.13.)

~ ~ ~

This subsection discusses, for ENS48 models, ratios of inferred DES to inferred OMDMS.

For ENS48 models, this monograph assumes that DES influences on OMS clumping occur via interactions that bridge OMDMS and DES. Relevant interactions correlate with just 4G2468&.

At some time early in the evolution of the universe, presumably, effects of DES clumping on OMDMS clumping are small. Based on data regarding CMB photons via which people infer densities for that time, people detect little or no effects of DES.

Impact of DES on CMB comes from interactions between OMDMS and DES. Over time, such effects grow.

Table 4.3.4 shows possible reasons why, for ENS48 models, the inferred ratio of DES to OMDMS need not approximate or exceed 7. Above, this monograph discusses the first candidate reason. This monograph does not further emphasize the other candidate reasons.

Table 4.3.4 Candidate reasons why the ratio of inferred DES to inferred OMDMS need not approximate or exceed 7 (ENS48 models)

1. It takes time for the presence of DES to impact phenomena through which people infer the existence of DES.
    - 1.1. It takes time for impacts of clumping outside of OMDMS to impact clumping within OMDMS.
    - 1.2. It takes time for clumping within OMDMS to impact observed CMB.

2. Nature might not have populated the peers consisting of OMDME and each of 7 similar DEE units (with each unit featuring 6 DEE) roughly equally.
   2.1. Perhaps, no data that implies lack of equality.
3. Some properties of elementary particles in peers of OMDMS might not equal properties of their counterpart particles in OMDMS.
   3.1. Perhaps, no data that implies such inequalities exists.
4. Other.
   4.1. For example, see Table 4.3.5.

People might say that an inferred ratio of ~2.2 : 1 need not be inconsistent with an actual ratio of ensembles of 7 : 1.

~ ~ ~

This subsection discusses, for ENS6 models, ratios of inferred DES to inferred OMDMS.

For ENS6 models, this monograph does not propose a way to model an inferred ratio of density of DES to density of OMDMS.

~ ~ ~

This subsection notes other types of phenomena that might impact ratios of inferred DES density to inferred OMDMS density.

Table 4.3.5 pertains.

Table 4.3.5 Other factors that might affect the ratio of inferred DES to inferred OMDMS

1. DES converts to OMDMS.
   1.1. For example, SEDES converts to SEDMS.
2. OMDMS converts to DES.
   2.1. For example, SEDMS converts to SEDES.
3. Nature creates stuff (after early in the big bang) within the DIU.
4. Other.

~ ~ ~

This subsection discusses concepts and data people have regarding dark matter.

People have yet to settle on a definition of dark matter. Generally, people say that dark matter denotes any stuff (except Standard-Model stuff) that interacts with Standard-Model stuff via gravity and that does not much or at all interact directly with Standard-Model stuff via electromagnetism.

People think dark matter may provide mass needed to bind the material within a galaxy. (See K.A. Olive (2014; 25). See, also, Section 4.4 and Section 4.7.)

People infer dark-matter densities of the universe for various times during the evolution of the universe. (See K.A. Olive (2014; 25).) Table 4.3.6 pertains. (See T. Ferris (2015). See, also, Ω_{cdm} and Ω_b in Table 8.1.1.)

Table 4.3.6 Inferences pertaining to dark-matter and ordinary-matter densities of the universe

1. For data pertaining to and after about 100 million years after the big bang, the ratio of the inferred density of dark matter to the inferred density of Standard Model stuff equals or exceeds about 5.
2. Inferences about the current state of the universe suggest that ~5.3 times as much dark matter as ordinary matter exists.

~ ~ ~

This subsection discusses, for ENS48 models and ENS6 models, ratios of inferred DMS to inferred OMS.

People might say that Table 4.3.7 shows possible reasons why, for ENS48 models and ENS6 models, the inferred ratio of DMS to OMS need not equal 5. Above, this monograph discusses some of these candidate reasons.

Table 4.3.7 Candidate reasons why the ratio of inferred DMS to inferred OMS need not equal 5 (ENS48 models and ENS6 models)

1. DMS includes more than 5 times as much G#XY stuff as does OMS.
   1.1. People might say that Table 4.3.2 correlates with this possibility.
2. The OME includes some stuff (perhaps, SEDMS {See Table 4.3.14.}) that (regarding the physics of OMS, DMS, and clumping) exhibits behavior that correlates with DMS.
   2.1. People might say that discussion, regarding ENS1 models, below discusses aspects of this possibility. (See Table 4.3.14.)
3. Nature might not have populated the peers consisting of the OME and each of 5 similar units (with each of those 5 units featuring one DME) roughly equally.
   3.1. Perhaps, no data that implies lack of equality.
4. Some properties of elementary particles in DME peers of the OME might not equal properties of their counterpart particles in the OME.
   4.1. Perhaps, no data that implies such inequalities exists.
5. Some OMS evolves into SEDMS.
6. Some SEDMS evolves into OMS.

People might say that the inferred ratio of ~5.3 : 1 need not be inconsistent with an actual ratio of ensembles of 5 : 1.

~ ~ ~

This subsection summarizes a possibly useful look, assuming that ENS48 models correlate with nature, at the nature of dark-energy stuff and at the nature of dark matter.

Table 4.3.8 pertains.

Table 4.3.8 Process for envisioning dark-energy stuff and dark matter (ENS48 models)

1. Review concepts that Table 4.3.1 and Table 4.3.2 state specifically regarding ENS48 models.
2. Determine the extent to assume that such work correlates with nature.
3. To the extent that work correlates with nature, ...
    - 3.1. Consider similarities among the 48 ensembles.
    - 3.2. Consider that each ensemble would exhibit elementary particles and physics that would be similar to elementary particles and physics that each other ensemble exhibits.
    - 3.3. Realize that the OME (or, the ordinary-matter ensemble) exhibits life as people know it.
    - 3.4. Realize the each other ensemble could include adequately physics-savvy beings.
        - 3.4.1. Here, the term adequately physics-savvy includes the notion of knowing of the existence of constructs that the beings might call (from the perspective of the beings and their ensemble) ordinary matter, dark matter, and dark-energy stuff.
    - 3.5. Realize that the relationship of being (or not being) each other's dark matter is reciprocal between ensembles.
    - 3.6. Realize that the relationship of being (or not being) each other's dark-energy stuff is reciprocal between ensembles.
    - 3.7. Look around.
    - 3.8. Realize that what you see ...
        - 3.8.1. Would be considered to be dark matter by adequately physics-savvy beings in each of 5 ensembles.
            - 3.8.1.1. (Also, perhaps, note that you would consider those beings to be part of what you consider to be dark matter.)
        - 3.8.2. Would be considered to be dark-energy stuff by adequately physics-savvy beings in each of 42 ensembles.
            - 3.8.2.1. (Also, perhaps, note that you would consider those beings to be part of what you consider to be dark-energy stuff.)
        - 3.8.3. Would be similar to and would behave somewhat similarly to corresponding aspects of each of ...

3.8.3.1. The 5 ensembles people would consider to be dark matter.
3.8.3.2. The 42 ensembles people would consider to be dark-energy stuff.

~ ~ ~

This subsection summarizes a possibly useful look, assuming that ENS6 models correlate with nature, at the nature of dark matter.

Concepts in Table 4.3.8 that pertain to ordinary matter and dark matter pertain. Concepts in Table 4.3.8 that pertain to dark-energy stuff or to dark-energy ensembles do not pertain.

~ ~ ~

This subsection discusses, regarding ENS48" models, possible large-scale clustering and anti-clustering phenomena.

Table 4.3.9 points to interactions that might have led to observable large-scale non-uniformities in the currently observable (spatial) distributions of ordinary matter, dark matter, and dark-energy stuff. The interaction column shows the spatial dependence of the interaction. The instances and span (per instance) columns show information from Table 2.13.23. The effect and particles columns show information from Table 4.2.5.

Table 4.3.9 Spatial dependences and spans for interactions affecting throughout much of the evolution of the universe (present-day) large-scale objects (assuming ENS48" models and %68even models)

Interaction	Instances	Span	Effect	%68even particles
r^{-8}	1	48	Repel	4G2468&
r^{-6}	8	6	Attract	2G468&, 4G268&
r^{-4}	24	2	Repel	2G24&, 2G68&
r^{-2}	8	6	Attract	4G4&
r^{-2}	48	1	Attract/repel	2G2&

People might say that Table 4.3.10 discusses clustering effects and anti-clustering (or abating of clustering) effects that might have significant impact during the evolution of the universe. Generally, each subsequent item in the table would have significant impact (on present-day large objects) after the previous items in the table have started to have impact (on the same large objects). Here, for purposes of clarity and simplicity, we use notation such as OME and DME (and not such as OMS and DMS.)

Table 4.3.10 Possible clustering and anti-clustering of dark-energy stuff and of dark matter (ENS48" models)

1. Based on r^{-6} attraction, ...
   1.1. Clusters form within the combination of the OME and DME.
   1.2. Clusters form within each of 7 units of DEE.
2. Based on r^{-8} repulsion, ...
   2.1. Some anti-clustering occurs between the eight units of which ...
      2.1.1. One unit is OMDME (or, the combination of the OME and DME).
      2.1.2. The other seven units are the 7 units of DEE.
   2.2. Some anti-clustering occurs within clusters that form within ...
      2.2.1. OMDME.
      2.2.2. Each of 7 units of DEE.
3. Based on r^{-4} repulsion, ...
   3.1. Anti-clustering occurs within ...
      3.1.1. The combination of the OME and DME-1.
      3.1.2. Two of the other 4 units of DME.
         3.1.2.1. We use the terms DME-B and DME-C to denote these units.
      3.1.3. The other 2 units of DME.
         3.1.3.1. We use the terms DME-D and DME-E to denote these units.
      3.1.4. Each of 21 mutually distinct units, each consisting of 2 DEE.
4. Based on r^{-2} attraction, ...
   4.1. Some large-scale clustering occurs within ...
      4.1.1. OMDME.
      4.1.2. Each of 7 units of DEE.

~ ~ ~

This subsection discusses, regarding some ENS6 models, possible large-scale clustering and anti-clustering phenomena.

Concepts that Table 4.3.9 and Table 4.3.10 discuss pertain for some ENS6 models, with the following provisos. Statements involving DEE do not pertain. The span of 4G2468& can be (correlating with #E` for 4G2468&) 1, 2, or 6 ensembles and is not 48 ensembles. The span of 2G24& can be (correlating with #E` for 2G24&) 1 ensemble or 2 ensembles (the OME and DME-1). For each particle Table 4.3.9 shows, the number of instances is 6 divided by the span for the ENS6 model.

~ ~ ~

This subsection notes types of experiments or observations people might attempt regarding dark matter or dark-energy stuff.

People might say that Table 4.3.11 provides examples of types of observations (that people might attempt to make) that could correlate with phenomena Table 4.3.10 suggests. Other types of observations may be possible.

Table 4.3.11 Observations to possibly attempt regarding the distributions of ordinary matter, dark matter, and dark-energy stuff

1. To what extent do dark matter clusters correlate with gaps between ordinary-matter clusters?
2. To what extent does the density of dark-energy stuff vary by region in the cosmos?
3. To what extent does anti-clustering within the combination of the OME and DME-1 abate over time?
4. To what extent is each (of many) observable or inferable galaxies predominantly either mostly ordinary-matter stuff or mostly dark-matter stuff?

People might say that Table 4.3.12 pertains.

Table 4.3.12 Experiments or observations possibly to attempt regarding the nature of dark matter and/or the nature of dark-energy stuff

1. Look for interactions between ordinary-matter elementary-particle magnetic dipole moments and dark-matter elementary-particle magnetic dipole moments. (See Table 3.3.13.)
   1.1. Here, 2G24& might intermediate relevant interactions.
2. Look for interactions between ordinary-matter elementary-particle spins and dark-matter elementary-particle spins. (See, for example, the span Table 2.13.23 shows for 2G68&.)
   2.1. Here, 2G68& might intermediate relevant interactions.
3. Determine various aspects related to dark-matter galaxies. (See Section 4.4. Compare with Table 4.3.11.)

~ ~ ~

This subsection begins exploration of dark-energy stuff and dark matter for ENS1 models.

For ENS1 models, DME and DEE are not physics-relevant.

Traditional physics correlates the notion of dark-energy stuff with a notion of stuff that interacts with gravity either not at all or not very much. For at least ENS1 models, we call this stuff SEDES. SEDES abbreviates the phrase single-ensemble dark-energy stuff. People might say that Table 4.3.13 pertains.

Table 4.3.13 Possible examples of SEDES (single-ensemble dark-energy stuff)

1. n × 3X+4Y, ...

1.1. For $n \geq 2$.
1.2. For X = R or D.
1.3. Which ...
1.3.1. Would have $\sigma = +1$ and (1;3)> symmetries.
1.3.2. Might not interact much with gravity.
2. n × 1R+4Y, ...
2.1. For $n \geq 2$.
2.2. Which ...
2.2.1. Would have $\sigma = +1$ and (1;3)> symmetries.
2.2.2. Might not interact much with gravity.
3. 3N-subfamily elementary particles.
4. Some G-family bosons.
4.1. For ENS48 models, these might include G48OMOM.
4.2. For ENS6 models, these might include G6OMOM.
4.3. For ENS1 models, these might include 4G2468&.
5. Some $\sigma = -1$ pre-composite particles.

~ ~ ~

This subsection extends our use of the term DES to pertain to ENS1 models.

Above, we define DES for ENS48 models and for ENS6 models.

Here, we extend use of DES to encompass ENS1 models.

People might say that, in ENS1 models, DES is, generally, stuff that people today would consider not to be part of the Standard Model, that does not interact much via gravity with Standard Model stuff, and that possibly does not interact much via electromagnetism with Standard Model stuff. Regarding ENS1 models, DES features SEDES.

~ ~ ~

This subsection discusses, for ENS1 models, ratios of inferred DES to inferred OMDMS.

People might say that, as yet, ENS1 models do not correlate with means to estimate an actual ratio of DES density to OMDMS density. People might say that, as yet, ENS1 models do not correlate with means to estimate a rate at which effects of DES clumping would impact CMB.

Thus, the inferred ratio of ~2.2 : 1 need not be inconsistent with work in this monograph regarding ENS1 models.

~ ~ ~

This subsection continues exploration of dark matter for ENS1 models.

Traditional physics correlates the notion of dark matter with stuff that interacts directly with gravity and does not interact with light either at all or very much. For at least ENS1 models, we call this stuff SEDMS. SEDMS abbreviates the phrase single-ensemble dark-matter stuff. People might say that Table 4.3.14 pertains.

Table 4.3.14 Possible examples of SEDMS (single-ensemble dark-matter stuff)

1. n × 3X+4YO, ...
    - 1.1. For $n \geq 2$.
    - 1.2. For X = Q or I.
    - 1.3. Which ...
        - 1.3.1. Would have $\sigma = +1$ and (1;3)> symmetries.
        - 1.3.2. Might interact with gravity.
        - 1.3.3. Might not interact much (if at all) with electromagnetism.
2. Some G-family bosons.
3. Some $\sigma = -1$ pre-composite particles.

~ ~ ~

This subsection extends our use of the term DMS to pertain to ENS1 models.

Above, we define DMS for ENS48 models and for ENS6 models.

Here, we extend use of DMS to encompass ENS1 models.

People might say that, in ENS1 models, DMS is, generally, stuff that people today would consider not to be part of the Standard Model, that interacts via gravity with Standard Model stuff, and that does not interact much via electromagnetism with Standard Model stuff. Regarding ENS1 models, DMS features SEDMS.

~ ~ ~

This subsection discusses, for ENS1 models, ratios of inferred DMS to inferred OMS.

People might say that, as yet, ENS1 models do not correlate with means to estimate an actual ratio of DMS density to OMS density.

Thus, the inferred ratio of ~5.3 : 1 need not be inconsistent with work in this monograph regarding ENS1 models.

~ ~ ~

This subsection discusses hypothetical candidates that people propose as possibilities for dark matter.

People consider how to explain traditionally unexplained experimental and observational data. Based on such thinking, people hypothesize candidates for constituents of dark matter. K.A. Olive (2014; 25) lists some candidates that people discuss. Table 4.3.15 pertains.

Table 4.3.15 Some candidates that people discuss for dark matter

1. Primordial black holes.
2. Axions.

3. Sterile neutrinos.
4. Weakly interacting massive particles (WIMPs).

~ ~ ~

This subsection provides perspective regarding subsequent subsections in this section.

Subsections below attempt to explain phenomena some people have correlated with the term dark matter. (See Table 4.3.15.) People think of (hypothetical) natural mechanisms that might correlate with a mismatch between data and traditional theory. This monograph tries to match possible particles (to which this monograph points) with the hypothetical natural mechanisms people propose. People might say that descriptions this monograph provides are, in some ways, more definitive than other descriptions of some such hypothetical natural mechanisms. People might say that this monograph does not take a definitive stand about the existence of constructs Table 4.3.15 lists.

~ ~ ~

This subsection notes possible correlations between various combinations of $\sigma = -1$ particles and primordial black holes.

Section 4.2 notes that some of the following combinations of $\sigma = -1$ particles might have produced primordial black holes - combinations of 3QIRD and 4YO or of 1R and 2YO. Possibly, clusters of 0O could also form primordial black holes.

~ ~ ~

This subsection discusses possible correlations between 2O particles and axions, between 2G24& and axions, and between 4G2468& and axions.

People conjecture that axions may correlate with violation of CP-symmetry. People conjecture that axions played a key role in the formation of baryon asymmetry. K.A. Olive (2014; 25) states that axions may have zero mass "at temperatures well above the QCD phase transition" and may have a mass at temperatures (in units of energy) less than or equal to approximately 1 GeV.

This monograph correlates baryon asymmetry with lasing of charged 2O bosons. (See Section 4.5.) People might interpret such interactions as violating CP-symmetry. This monograph suggests non-zero masses for 2O bosons.

This monograph does not suggest elementary particles that we think would change rest energy based on temperature.

For each G-family boson, m' = 0 pertains. Perhaps an nG%& for which 4 is a member of % would appear (based on traditional thinking) to have non-zero rest mass regarding some interaction vertices.

For example, perhaps, in interactions, a 4G2468& can actually (or would appear to) decay (at an interaction vertex or within a cluster of related interaction vertices) into (at least) a 2G2& (MM1MS1-photon) and a 4G4& (graviton). People might correlate perceived m' ≠ 0 (for an axion) with the presence of the graviton.

Assuming #E` < 6 for 4G2468&, regarding ENS48 models and ENS6 models, perhaps, when intermediating an interaction between an ordinary-matter-ensemble fermion and an other-ensemble fermion, 4G2468& could provide for interactions that people would interpret as violations of CP-symmetry.

Assuming #E` < 6 for 2G24&, for 2G24&, perhaps, considerations similar to ones just stated regarding 4G2468& pertain.

~ ~ ~

This subsection discusses possible sterile neutrinos.

Members (if such elementary particles exist) of the 3N-subfamily may correlate with some notions of sterile neutrinos. The particles might not interact with any non-zero-spin W- or 0-family bosons. Or, possibly the particles interact with the Z boson. (See discussion before Table 3.5.3.) The particles might not interact directly with either photons or gravitons. However, some concepts of sterile neutrinos correlate with spin-1/2 particles. 3N-subfamily particles would be spin-3/2 particles.

~ ~ ~

This subsection discusses possible correlations between clusters of 0-family particles and WIMPs.

People use the acronym WIMP to abbreviate the phrase weakly interacting massive particle. People conjecture that WIMPs might have been created early in the evolution of the universe. People conjecture that the universe cooled in way that WIMP states still may be populated. K.A. Olive (2014; 25) notes that people conjecture that WIMP rest masses might be in a range from 10 GeV/c^2 to a few TeV/c^2. People conjecture that WIMPs have cross sections of approximately weak strength. People conjecture that WIMPs interact with photons.

Above, this work discusses possible composite particles of the forms n × 00, n × 1Q+2O, n × 3Q+4O, n × 3I+4O, n × 1R+2O, n × 3R+4O, and n × 3D+4O. Section 3.8 discusses possible masses for 0-family particles and possible minimum threshold energies for such composite particles. The masses and minimum threshold energies fall within the conjectured range for WIMPs. The composites n × 1Q+2O and n × 1R+2O would interact with MM1MS1-photons.

~ ~ ~

Reference 9 T. Ferris (2015)

Timothy Ferris, A First Glimpse of the Hidden Cosmos, *National Geographic*, Vol. 227, No. 1, January 2015, pp. 108-123.
(http://ngm.nationalgeographic.com/2015/01/hidden-cosmos/ferris-text)

Reference 10 J. Beringer (2012)
J. Beringer et al. (Particle Data Group), *Phys. Rev. D86*, 010001 (2012). (http://pdg.lbl.gov/2012/reviews/rpp2012-rev-cosmic-microwave-background.pdf)
Reference 11 K.A. Olive (2014; 25)
K.A. Olive et al. (2014) (See Reference 18). Specifically: 25. DARKMATTER, Revised September 2013 by M. Drees (Bonn University) and G. Gerbier (Saclay, CEA). (http://pdg.lbl.gov/2014/reviews/rpp2014-rev-dark-matter.pdf)

Section 4.4 Objects containing ordinary matter and dark matter

Section 4.4 discusses objects that might contain significant amounts of each of ordinary matter and dark matter. People might say that ENS48 models and ENS6 models correlate with clustering and anti-clustering phenomena that correlate with possibilities that a galaxy or a globular cluster can contain a somewhat (but statistically not completely) arbitrary ratio of ordinary matter to dark matter.

~ ~ ~

This subsection provides perspective about the topic of dark-matter galaxies and dark-matter globular clusters.

Based on various observations and theories, people speculate that the universe may include objects, from which people and equipment do not detect much if any electromagnetic radiation, that otherwise have approximately galactic (or, globular-cluster) characteristics.

Discussion of such so-called dark-matter galaxies and globular clusters includes results published during 2015. (See Jin Koda, et. al. (2015); Sukanya Chakrabarti, et. al. (2015); C. Yozin (2015); Elizabeth Howell (2015); and Evan N. Kirby, et. al. (2015).)

Possibly, galaxies can contain various proportions of ordinary matter and dark matter. In a galaxy with mostly dark matter, possibly some ordinary-matter objects can form. (Evan N. Kirby, et. al. (2015) describes a dark-matter dwarf galaxy that includes ordinary-matter stars.)

For the remainder of this section, we deemphasize use of the term globular cluster. Generally, use of the word galaxy correlates with a phrase such as galaxy and/or globular cluster.

~ ~ ~

This subsection provides perspective about the topic of van der Waals forces.

Generally, van der Waals forces are electromagnetic forces between molecules or between various parts of single molecules, for which the forces do not correlate with covalent bonds and do not correlate with attraction or repulsion because objects are ions (as opposed to being charge-neutral objects).

~ ~ ~

This subsection discusses mechanisms that could have led to the formation of objects that contain both ordinary matter and dark matter.

Here, we work in a context of ENS48 models or of ENS6 models.

We consider clumping of stuff that would be characterized as having today sizes typical of atoms and molecules. People might say that clumping of such stuff exhibits the same four eras as does clumping of large objects. (See Table 4.2.1.) We assume that era FE4 includes an interval after the first formation of atoms and molecules. Between nearby atoms and small molecules, electromagnetic interactions dwarf gravitational interactions. We assume that, for a portion of that interval, (for interactions between atomic nuclei, free leptons, atoms, and small molecules) electromagnetic interactions (including covalent bonding and van der Waals forces) dominate gravity.

People might say that covalent bonding provides one form of relevant-scale clumping. The one OME clumps with itself. Stuff correlating with each one of the five DME clumps with itself.

People might say that, to the extent applicable van der Waals forces provide net attraction, these forces also help stuff correlating with the one OME clump and help stuff correlating with each one of the five DME clump.

People might say that, to the extent such clumps coalesce on adequately large scales before (relevant-sized clumps become generally charge-neutral and ...) gravitational interactions dominate electromagnetic interactions, objects form for which each object contains stuff correlating mostly or perhaps nearly exclusively with one ensemble. Such objects could include molecules, gas-and/or-dust clouds of various masses (or sizes), and so forth.

People might say that, when gravitational interactions dominate at relevant size scales, such objects clump to form larger objects such as nascent galaxies. People might say that such objects and some of their components (such as nascent solar systems) can contain somewhat arbitrary mixtures of stuff correlating with the OME and stuff correlating with each of the five DME.

~ ~ ~

This subsection discusses other effects that might pertain to the formation of objects that contain both ordinary matter and dark matter.

Here, we work in a context of ENS48 models or of ENS6 models.

People might say that effects correlating with Table 4.3.10 pertain. Anti-clustering based on r^{-4} repulsion could start before clustering based on electromagnetism and/or on gravity. A nascent galaxy might feature 3 ensembles and deemphasize 3 ensembles.

For example, for a galaxy with much stuff that correlates with the OME, there might not be as much stuff that correlates with DME-1. Similarly, that galaxy might have more stuff that correlates with (say) DME-B than stuff that correlates with (say) DME-C. Similarly, that galaxy might have more stuff that correlates with (say) DME-D than stuff that correlates with (say) DME-E.

People might say that discussion above provides possibilities that, for example, our solar system may contain dark-matter asteroids or other dark-matter objects.

~ ~ ~

This subsection discusses, in a context of ENS1 models, objects that contain significant amounts of dark matter.

ENS1 models do not correlate with a dark-matter analog to the electromagnetic interaction. ENS1 models do not necessarily correlate with dark-matter direct analogs to atoms. People might say that people should wait, until people better understand dark matter, before considering the extent to try to use ENS1 models meaningfully to model objects that feature dark matter.

~ ~ ~

Reference 12 Jin Koda, et. al. (2015)

Jin Koda, Masafumi Yagi, Hitomi Yamanoi, and Yutaka Komiyama, Approximately a thousand ultra diffuse galaxies in the Coma cluster, *ApJ Letters*, Volume 807, Number 1, 2015. http://iopscience.iop.org/article/10.1088/2041-8205/807/1/L2

Reference 13 Sukanya Chakrabarti, et. al. (2015)

Sukanya Chakrabarti, Roberto Saito, Alice Quillen, Felipe Gran, Christopher Klein, and Leo Blitz, Clustered Cepheid Variables 90 kiloparsec from the Galactic Center, *Astrophysical Journal Letters*, 1502, 1358 (2015).

Reference 14 C. Yozin (2015)

C. Yozin and K. Bekki, The quenching and survival of ultra diffuse galaxies in the Coma cluster, *MNRAS* 452, 937–943 (2015).

Reference 15 Elizabeth Howell (2015)

Elizabeth Howell, Dark Matter Can Interact With Itself, Galaxy Collisions Show, Space.com (April 15, 2015). http://www.space.com/29115-dark-matter-interactions-galaxy-collisions.html

Reference 16 Evan N. Kirby, et. al. (2015)

Kirby, Evan N. and Cohen, Judith G. and Simon, Joshua D. and Guhathakurta, Puragra, Triangulum II: Possibly a Very Dense Ultra-faint Dwarf Galaxy, *Astrophysical Journal*, 814 (1), Art. No. L7, 2015.

Section 4.5 Baryon asymmetry

Section 4.5 discusses the possibility that spin-1 0-family bosons catalyzed the currently existing matter/antimatter imbalance.

~ ~ ~

This subsection discusses the possibility that nature previously exhibited less baryon asymmetry (or, matter/antimatter imbalance) than nature currently exhibits.

People discuss the possibility that the early universe contained equal amounts of matter and antimatter. Conceptually, for example, for each electron, there would be positron. For each positron, there would be an electron. For each matter quark, there would be an antimatter quark having the negative of the charge of the matter quark. For each antimatter quark, there would be a matter quark having the negative of the charge of the antimatter quark. Perhaps, some theories and models correlate with such circumstances.

Generally, perhaps, for some era in the past, a ratio of matter-charged-leptons-and-matter-quarks (1C matter plus 1Q matter coupled with, say, 2O matter or 2Y-related matter-gluons) to antimatter-charged-leptons-and-antimatter-quarks (1C antimatter plus 1Q antimatter coupled with, say, 2O antimatter or 2Y-related antimatter-gluons) was less than it is now. If so, the question arises as to when and how the balance changed.

Perhaps, such discussion could be improved. For example, as reprised above, the discussion does not consider neutrinos. We are not aware of any estimates of matter/antimatter balance or imbalance regarding neutrinos. People might say that current imbalance regarding known charged elementary fermions might correlate with an imbalance regarding 1N matter and 1N antimatter. Possibly, such a concept correlates with existence of more 1N matter than 1N antimatter. Or, possibly, such a concept correlates with enough more 1N antimatter than 1N matter that an overall count of elementary fermions correlates with balance between matter and antimatter.

~ ~ ~

This subsection points to a mechanism that could have led to a change from less asymmetry to today's amount of asymmetry.

Here, we focus on asymmetry regarding charged elementary fermions.

Interactions between quarks and 2O% bosons (for % = 2 or 1) convert antimatter quarks into matter quarks and vice versa. For example, an anti-down quark has Q' = +1/3 (or, a charge of $+(1/3)|q_e|$). If an anti-down quark absorbs a 2O1 boson (for which Q' = +1/3), an up quark (with Q' = +2/3) results. This reaction converts antimatter (an anti-down quark) into matter (an up quark).

Lasing of 2O% bosons could have led to a transformation from equal amounts of matter and antimatter to the imbalance people correlate with today's universe. Presumably, even a small asymmetry regarding direction (more conversion of antimatter quarks to matter quarks than vice versa) could have grown. Reactions that convert antimatter charged leptons (such as positrons) into antimatter MM1MS1-neutrinos could have occurred concurrently with the interactions that led to changes in ratios between matter quarks and antimatter quarks. Reactions that convert matter MM1MS1-neutrinos into matter charged leptons (such as electrons) could have occurred concurrently with the interactions that led to changes in ratios between matter quarks and antimatter quarks.

Such a transformation would have occurred before stuff became sufficiently dispersed that 2O2 and 2O1 bosons would have limited impact.

People might say that the transformation occurred before or during the hadron epoch. People place this epoch as lasting from about 10^{-6} second until about 1 second after the big bang. People correlate with the beginning of the epoch the existence of a plasma (or, sea) of quarks. By the end of the epoch, nucleons formed. Perhaps, that epoch also included a transformation from 1Q+2O physics to 1Q+2Y physics. (See, also, remarks pertaining to Table 2.13.41.)

~ ~ ~

This subsection discusses a possible aspect of nature that people might try to detect or infer.

The above discussion suggests that today's ordinary matter may include more anti-neutrinos than neutrinos. Table 4.5.1 pertains.

Table 4.5.1 Possibility for experiments or inference regarding balance between matter neutrinos and antimatter neutrinos

1. People might say that trying to determine the amount of imbalance between matter 1N-subfamily particles and antimatter 1N-subfamily particles is worthwhile.

~ ~ ~

This subsection discusses possible synchronicity of conversion of antimatter quarks to matter quarks.

To the extent ENS48 models pertain, each instance of 1Q-subfamily elementary particles would associate with a specific ensemble. People might say that such association allows for some non-synchronicity, among the 48 ensembles, in achieving each ensemble's respective present baryon asymmetry. People might say that such association allows for some non-uniformity, across the 48 ensembles, in each ensemble's respective present baryon asymmetry. (Perhaps, no data exists by which people can estimate baryon asymmetries for any of the 47 non-ordinary-matter ensembles.)

To the extent ENS6 models pertain, similar results could pertain regarding the 6 (not 48) ensembles.

People might say that, to the extent ENS1 models pertain, baryon-asymmetry forms somewhat simultaneously (in terms of time after the big bang) in the universe's one ensemble.

Section 4.6 Phenomena that people model via general relativity

Section 4.6 discusses quantum-based models for phenomena that people model via general relativity. People might say that MM1MS1 models correlate with gravitational redshift and blueshift, gravitational bending of trajectories of light, and shifts of perihelia. We discuss a means by which black holes can lose energy.

~ ~ ~

This subsection notes some successes and aspects of models based on general relativity.

Early successes based on general relativity include explaining gravitational redshifts and blueshifts as photons pass by objects and predicting and explaining a continuing shift in the perihelion of the planet Mercury. Here, redshift refers to (relative to an observer) decreasing the energy of a photon. (The term redshift correlates with the notion that the wavelength of the photon increases.) Blueshift refers to (relative to an observer) increasing the energy of a photon. (The term blueshift correlates with the notion that the wavelength of the photon decreases.)

In general-relativity models, trajectories of zero-mass (and relatively small-mass objects) correlate with geodesics of space-time. The geodesics correlate with the existence and motion of (relatively) large-mass objects.

~ ~ ~

This subsection notes possible motivations for exploring quantum-based alternative models for phenomena people model via general relativity.

People correlate aspects of general relativity with curvature. For example, people say that trajectories of small-mass objects correlate with geodesics that correlate with curved (or, not flat) space-time. People say that such curvature correlates with gravity. People might say that, however, in ENS48 models, it would be difficult (or even impossible) to meaningfully correlate space-time curvature with gravity. (See Section 6.4. People might say that curvature correlating with any of the seven instances of 4G4& other than the instance correlating with OMDME would correlate with inappropriate motion of small-mass OMS, DMS, or OMDMS objects.) People might say that ENS48 models for which #b" = 3 for all G-family elementary particles would correlate with space-time having zero curvature.

People might say that people should look for models that are not as asymmetric (as is general relativity) regarding high-mass objects and low-mass objects.

~ ~ ~

This subsection introduces a basis for thought experiments that we discuss in this section.

This work considers some thought experiments regarding particles or other objects moving under the influence of gravity associated with an object this monograph calls the hub. This work sets up the thought experiments so that there is a reference frame in which the hub remains stationary.

People might say that we try, in this section, to use the somewhat least sophisticated physics models that might meaningfully pertain.

Imagine two focal points, point-1 and point-2, positioned so that the hub bisects a straight line between the two focal points. Imagine a plane that includes the three points. All the phenomena this work discusses occur in this plane. Imagine ellipse-1, which has focal points at point-1 and the hub. Imagine ellipse-2, which has focal points at the hub and point-2. Imagine that four identical objects orbit the hub. Two objects have orbits that track ellipse-1. These two objects orbit in opposite directions. The other two objects have orbits that track ellipse-2. These two objects orbit in opposite directions. In the frame of reference in which the hub is stationary, all four objects arrive simultaneously at the perihelia for their respective orbits. This discussion ignores collisions between objects. For these circumstances, momentum and angular momentum balance adequately so that that the hub remains stationary in the frame of reference. This monograph considers, for each of some examples, models regarding one orbiting object for which the orbit follows ellipse-1.

~ ~ ~

This subsection correlates some phenomena people model via general relativity with models based on this monograph's approach.

Consider an example in which the hub is a non-rotating black hole, the orbiting object is a MM1MS1-photon and point-1 is at the center of the hub. Here, the orbit for the MM1MS1-photon is a circle. Here, we assume that the orbit occurs within the event horizon for the black hole. The MM1MS1-photon has rest mass m' = 0. Gravitons (or, 4G4&) associated with the hub do not interact directly with the MM1MS1-photon. The non-zero energy and momentum of the MM1MS1-photon correlate with a cloud of virtual particles associated with the MM1MS1-photon. (See Table 2.7.4.) Gravity associated with the hub interacts with the virtual particles in that cloud. (Perhaps, more precisely regarding gravitational phenomena, because gravitons have m' = 0, one might consider that a key quantum-mechanical aspect includes the notion that the cloud of virtual particles associated with a graviton interacts with the cloud of virtual particles associated with the photon.) Gravitational attraction keeps the MM1MS1-photon in its orbit. Each of the MM1MS1-photon's observed energy and observed magnitude of momentum is proportional to E. In the chosen frame of reference, the expression E describes the perceived energy associated with the MM1MS1-photon. In the chosen frame of reference, the expression P = E/c describes the perceived magnitude of momentum associated with the MM1MS1-photon.

Consider another example in which the orbiting object is a MM1MS1-photon and orbit occurs within the event horizon of a hub that is a black hole. This time, point-1 does not coincide with the center of the hub. The orbit is an ellipse that is not a circle. Here, the MM1MS1-photon's E varies, depending on where the MM1MS1-photon is in its orbit. The MM1MS1-photon's E is greatest at the orbit's perihelion (that is, at the point at which the orbit is closest to the hub). The MM1MS1-photon's E is least at the orbit's aphelion (that is, at the point at which the orbit is farthest from the hub).

~ ~ ~

This subsection discusses thermal radiation pertaining to black holes.

Some people conjecture that a black hole can radiate photons. Perhaps, eventually, a black hole could shed much or all of its energy.

Work above in this section discusses examples regarding orbits of MM1MS1-photons within black-hole event horizons. At least a small part of the cloud of virtual particles associated with a MM1MS1-photon's non-zero E exists outside the event horizon. The cloud of virtual particles associated with the MM1MS1-photon can interact with objects outside the event horizon. These interactions can reduce the MM1MS1-photon's E. People might say that these interactions provide bases for black-hole thermal radiation and/or black-hole dissipation.

~ ~ ~

This subsection discusses gravitational redshift and gravitational blueshift and discusses gravitational alteration of paths of photons.

Consider an example in which the orbiting object is a MM1MS1-photon. Here, the hub is a star. Here, point-1 lies far from the star. Here, the orbit passes near, but does not touch, the star. Results paralleling those (regarding variations in E) from a previous example correlate with blueshift as the MM1MS1-photon moves toward the star and with redshift as the MM1MS1-photon moves away from the star. People might say that Table 4.6.1 pertains.

Table 4.6.1 Physics of paths of light and of redshifts and blueshifts

1. For describing the bending of a path of light by gravity, people can use a model based on a MM1MS1-photon's E (instead of a model based on geodesics).

2. For describing the red-shifting or blue-shifting of light by gravity, people can use a model based on a MM1MS1-photon's E (instead of a model based on geodesics).

~ ~ ~

This subsection discusses planetary perihelion shifts.

Consider an example in which the orbiting object has non-zero m' that is adequately small. (We elaborate on a concept of adequately small here and in the next paragraphs.) We assume that the object has sufficiently small kinetic energy that we can use some aspects of non-relativistic kinematics. Here, point-1 is suitable for thinking of the object as being a planet orbiting a star (that the hub represents) in a non-circular elliptical orbit. Assume that neither object rotates.

For models that this monograph features, the QE-like component of the orbiting object's energy-momentum 4-vector varies throughout the orbit. (The QP-like components also vary.) Let E denote that QE-like component. Except at the aphelion, E > E(aphelion). Based on assumptions that there are three other objects similar to the object on which this example focuses, the example ensures that the hub does not move. Also, we assume no object rotates. Based on the condition that E for the object is always adequately small, this example approximates as zero the effects on the object of gravitation associated with the other three similar objects. Thus, gravitational somewhat analogs to electromagnetism's magnetic field do not significantly influence the orbiting object's behavior. (See Table 2.15.10.)

We continue this discussion, based in part on classical, pre-relativistic physics models.

For the sake of discussion, assume that the object's orbit remains unchanged (compared to a classical-physics, pre-relativity calculation based on a constant value of m'). Assume that, at each point in the orbit, the object's kinetic energy remains unchanged (compared to a pre-relativity calculation). Then, at each point in the orbit, the kinetic energy can be expressed as $(1/2) \times (f \times m') \times (f^{-(1/2)} \times v)^2$, in which $f > 1$. (Here, and below in this example, the symbol × denotes a multiplication of numbers and does not represent a vector cross product.)

At each point in the orbit, let r denote the magnitude of the lever-arm for which $m' \times v \times r$ computes the angular momentum of the object (with respect to the hub) in the pre-relativity calculation. (At the perihelion and aphelion, r equals the distance between the object and the hub. At other points in the orbit, the velocity vector is not perpendicular to the radius vector and so the magnitude of the cross product of the two vectors is smaller than the product of the magnitudes of the two vectors. Thus, at these other points, the lever-arm value of r is less than the magnitude of the radius vector.)

At each point in the orbit, the angular momentum of the object is $(f \times m') \times (f^{-(1/2)} \times v) \times r$. This equals $f^{1/2} \times m' \times v \times r$. At each point, this angular momentum exceeds (based on the factor $f^{1/2}$, which exceeds 1 and which depends on the point) the angular momentum from the pre-relativity calculation.

An observer would observe more angular momentum than the pre-relativity calculation shows. People might say that the observer could correlate the extra angular momentum with observations of a perihelion shift that continues to occur orbit after orbit.

~ ~ ~

This subsection discusses other phenomena people model via general relativity.

People might say that people can extend such types of examples to consider other effects - such as some aspects of time dilation - that people model via general relativity. For some of these effects, at least one of the interacting objects rotates and the 4G4& somewhat analog to electromagnetism's magnetic field pertains. (See Table 2.15.10.)

~ ~ ~

This subsection posits possible opportunities for research.

People might say that Table 4.6.2 pertains.

Table 4.6.2 Possible research opportunities related to general relativity and alternative models

1. For each of various phenomena people model via general relativity, determine the extent to which MM1MS1 models provide useful alternative models.

Section 4.7 The galaxy rotation problem

Section 4.7 discusses the galaxy rotation problem. People might say that, depending on models people use, a gravitational somewhat analog to electromagnetism's magnetic field may point to an overlooked phenomenon that helps contain material within galaxies.

~ ~ ~

This subsection introduces the galaxy rotation problem.

People say that visible galaxies rotate too rapidly to retain visible material, if the only significant containment interaction is traditional gravity and the galaxy does not include dark matter. People say that dark matter may contribute enough mass and traditional gravitational interaction to account for the retention of material. People speculate as to the extent that other effects (and possibly less dark matter) may help account for the retention of material.

~ ~ ~

This subsection provides additional perspective.

We are not certain as to the extents traditional models for trajectories of objects within galaxies take into account effects of rotation of galactic cores, effects of motion of objects (outside of cores) on other objects, and/or effects people model via general relativity.

Perhaps, such traditional models do not adequately account for effects correlating with the gravitational somewhat analog to electromagnetism's magnetic field.

~ ~ ~

This subsection discusses the possibility that the gravitational somewhat analog to electromagnetism's magnetic field helps contain material within galaxies.

Work in previous sections correlates with the concepts that galaxies can contain mixtures of ordinary matter and dark matter and that each of ordinary matter and dark matter contributes to gravitational binding for the galaxy.

People might say that the 4G4& somewhat analog to the magnetic field people associate with electromagnetism can provide force that helps contain objects within galaxies. (See Table 2.15.10.) Think of a (geometric) sphere, centered at the center of a galaxy. The radius of the sphere can be any length that is not too small compared to the radius of (some concept of) the core of the galaxy and that is smaller than (some concept of) the radius of the galaxy. The somewhat analog field generated by the material inside the sphere affects the (somewhat disk of) material outside the sphere. Each of the inside stuff and the outside stuff rotates in the same direction. Possibly (at least by analogy to charge currents and electromagnetism), the force is attractive. If so, the force helps keep stuff outside the sphere from moving into orbits with larger radii or from leaving the galaxy. (Presumably, effects of the outer stuff on the inner stuff occur and are not as significant as the effects of the inner stuff on the outer stuff.)

For ENS48 models and ENS6 models, Section 4.4 discusses the notion that various galaxies contain various proportions of the OME and of each DME. People might say that, to the extent, stuff in the core of such a galaxy rotates in the same (circular) direction, each ensemble's contribution to the overall 4G4& somewhat analog to the magnetic field adds to the overall somewhat analog field. People might say that, to the extent, objects away from the core of such a galaxy rotate in the same (circular) direction (meaning, same direction as the core and other objects), each ensemble's contribution to the overall 4G4& somewhat analog to the magnetic field adds to the overall effect.

Section 4.8 The spacecraft flyby anomaly

Section 4.8 discusses the spacecraft flyby anomaly. People might say that, depending on models people use, this work may point to overlooked phenomena that would help resolve the anomaly.

~ ~ ~

This subsection introduces the spacecraft flyby anomaly.

People say that the observed energies of spacecraft that return to near the earth (and use the earth for velocity boosts) may differ from anticipated values. Generally, people design the trajectory for such a spacecraft so that, at some time after leaving the vicinity of the earth, the spacecraft nears earth and so that the earth's gravity can redirect the trajectory.

~ ~ ~

This subsection discusses possible explanations of the flyby anomaly.

People might say that Table 4.8.1 pertains.

Table 4.8.1 Possible explanations for the spacecraft flyby anomaly

1. Possibly, people do not consider effects of the 4G4& somewhat analog to electromagnetism's magnetic field. (See Table 2.15.10.)
    1.1. The following might produce such effects.
        1.1.1. Rotation of the sun.
        1.1.2. Motions of planets.
        1.1.3. Rotation of planets.
2. Possibly, people do not consider effects of asteroid-like objects (or other objects) composed primarily of dark matter.
3. Other.

People might say that Table 4.8.2 suggests possible opportunities for research.

Table 4.8.2 A research opportunity related to the spacecraft flyby anomaly

1. To what extent does work in this monograph correlate with an explanation for the spacecraft flyby anomaly?

Section 4.9 Quasars

Section 4.9 discusses the formation of quasars and quasar jets. People might say that at least one of the r^{-4} and r^{-8} spatial dependence G-family-mediated interactions produces quasars.

~ ~ ~

This subsection provides information about quasars.

Quasars feature jets of material that move away from very massive black holes. Possibly, each quasar features two jets that move in opposite directions from the relevant black hole.

People correlate each observed quasar with a black hole. For each of many black holes, observations do not detect quasar-like activity.

~ ~ ~

This subsection suggests phenomena that produce quasars.

People might say that Table 4.9.1 pertains.

Table 4.9.1 Chronology and phenomenology of a quasar

1. While and after a black hole first forms, the density of stuff increases based on the influence of 4G4&.
2. Eventually, the density may become large enough that repulsion based on at least one of the r^{-4} and r^{-8} spatial dependence G-family-mediated interactions dominates 4G4& attraction. (See Table 2.8.3 and Table 4.2.5.)
    - 2.1. Collapse slows.
    - 2.2. For at least some of the stuff in the black hole, collapse may reverse.
        - 2.2.1. A quasar may develop.
3. Possibly, quasar jets follow field lines correlating with the gravitational somewhat analog to electromagnetism's magnetic field. (See Table 2.15.10.)

Section 4.10 Cosmic microwave background cooling

Section 4.10 discusses the cooling (or redshift) over time of CMB (cosmic microwave background radiation). People might say that models for CMB cooling can correlate with Doppler shifts.

~ ~ ~

This subsection discusses the cooling (or redshift) over time of CMB (cosmic microwave background radiation).

After about 380,000 years (or, $10^{5.6}$ years (See Section 4.3.)) after the big bang, a temperature associated with (say, the peak of) the energy distribution of CMB has decreased.

People might say that, in traditional physics, the universe has expanded and, that, as the universe expands, the wavelengths of CMB photons decrease. Our work does not necessarily correlate with an expansion of space-time.

~ ~ ~

This subsection discusses the concept of co-moving reference frames.

Consider two observers who measure, from the same region of the universe and at the same time, properties of CMB. Assume that one observer determines that the pattern of wavelengths of CMB does not vary much with respect to the direction from which the CMB comes. Assume that the second observer moves adequately fast relative to the first observer that the second observer detects a blueshift for light coming generally from some direction and a redshift for light coming generally from the opposite direction. (Here, for example, if the second observer is moving directly toward the first observer, the observed direction of most blueshift correlates with the direction toward the first observer.) People would say that the motion of the first observer correlates with a coordinate system based on co-moving coordinates. People might say that the first observer moves in way that represents an average motion of some aspects (including CMB) of nature. People would say that the motion of the second observer does not correlate with a co-moving reference frame.

~ ~ ~

This subsection discusses an explanation (that does not depend on assuming that space-time expands) for the cooling of CMB.

As a thought experiment, imagine an emitter of a CMB MM1MS1-photon. Select an emitter that is (or, more precisely, after the CMB has fully formed will be) approximately at rest in a co-moving reference frame. Image that the emitter remains essentially at rest in that reference frame. Imagine an object that - after the MM1MS1-photon is emitted - will receive the MM1MS1-photon. Imagine that the object will be, at the time of receipt, approximately at rest in a co-moving reference frame.

Imagine that an observer that is at rest in the emitter's reference frame determines an E for the MM1MS1-photon (or, perhaps more correctly, that the observer measures an E for a similar MM1MS1-photon).

At the time of MM1MS1-photon emission, the eventual-receiver object lies some non-zero distance (in the emitter's reference frame) from the emitter. In the emitter's reference frame, the object is already moving away from the emitter, based on effects of 4G2468&. Such motion continues over time.

When the object absorbs the MM1MS1-photon, the object (in effect) detects a MM1MS1-photon E (in the object's reference frame) that is smaller than the E detected by the original observer.

People might say that this effect correlates with CMB cooling. People might use the term Doppler shift to characterize this effect. This CMB-related Doppler shift differs from local Doppler shifts that people discuss regarding motions of objects within (say) a galaxy or galactic cluster. Also, this CMB-related Doppler shift differs from Doppler shifts that people correlate just with motions with respect to a co-moving reference frame.

Section 4.11 Quark-based plasmas

Section 4.11 discusses concepts possibly related to quark-based plasmas. People might say that we suggest possible bases for phases for which traditional research would not include the bases or the phases.

~ ~ ~

This subsection discusses traditional notions about quark-based plasmas.

People term a period from about 10^{-12} second to about 10^{-6} second after the big bang the quark epoch. People think that, by the end of the epoch, quarks and gluons formed hadrons. A term for a subsequent period is hadron epoch.

People attempt experiments to create stuff that might be similar to stuff correlating with the quark epoch.

People use the term phase to discuss states of stuff. For example, phases for water include gaseous (such as water molecules {not water droplets} in the atmosphere), liquid (such as water in a lake), and solid (such as ice).

People discuss phases that quarks and gluons might exhibit under circumstances correlating with various combinations of density and temperature. People might say that some of these phases would exhibit 1Q+2Y properties that do not correlate with COMPAR concepts.

~ ~ ~

This subsection suggests some phases that might correlate with quark-based plasmas.

Perhaps, people might consider more (than people traditionally consider) possibilities for phases (or, states of matter) than people might correlate with traditional physics.

Suppose 2O bosons exist. 2O bosons have non-zero mass. People might say that work above suggests that individual bosons would have limited ranges. Effects of 2O bosons would be most apparent above some density (number per unit volume) of quarks. It would take some energy to produce 2O bosons (possibly, in minimum units of pairs or triplets of such bosons). In nature, effects of 2O would be most apparent at above some temperature.

Table 4.11.1 pertains. (See, also, Table 2.13.40 and Table 2.13.41.)

Table 4.11.1 Possible research topics regarding quark-based plasmas and related physics

1. To what extent can states of matter exhibit primarily 1Q+2O properties?
   1.1. Characterize phases for such matter.
2. To what extent can states of matter exhibit primarily mixtures of 1Q+2O properties and 1Q+2Y properties?
   2.1. Characterize phases for such matter.
3. Consider possible similar phenomena regarding ...
   3.1. 1R+2YO.
   3.2. 3QIRD+4YO.
   3.3. Mixtures of 1QR+2YO and 3QIRD+4YO.

Chapter 5 From MM1MS1 models to traditional models

Chapter 5 (From MM1MS1 models to traditional models) starts with models this monograph features, discusses extents to which people might correlate or integrate our models with traditional physics models, and discusses extents people might use our models to extend or improve on traditional physics models.

~ ~ ~

People might say that results in Chapter 5 (From MM1MS1 models to traditional models) might correlate with or provide insight about the following aspects of physics. Additions to the Standard Model. Details regarding aspects of the cosmology timeline. Additions to the cosmology timeline.

Section 5.1 The Standard Model

Section 5.1 discusses blending aspects of work in this monograph with aspects of the Standard Model. We show an approximate value for the weak mixing angle. People might say that this result fills a gap within the current scope of the Standard Model. We note the possibility that a lack of inclusion of gravity in the early twenty-first century Standard Model may correlate with notions that non-zero mass must pertain for at least one neutrino flavor. We discuss processes for blending MM1MS1 models and results and Standard Model techniques and results. We list candidate particles people might add to the Standard Model set of elementary particles and composite particles. Candidates correlate with the MM1 meta-model and, therefore, pertain each of ENS48 models, ENS6 models, and ENS1 models. We suggest steps for harmonizing and extending the Standard Model and MM1MS1 models.

~ ~ ~

This subsection describes aspects of the Standard Model.

The particle-physics Standard Model features attempts to catalog elementary particles, composite particles, and other subatomic particles and features attempts to describe interactions among such particles. Much development of the Standard Model took place during the second half of the twentieth century. Regarding elementary-particle predictions, successes include quarks and the Higgs boson. People found these

T.J. Buckholtz, *Models for Physics of the Very Small and Very Large*,
Atlantis Studies in Mathematics for Engineering and Science 14,
DOI 10.2991/978-94-6239-166-6_5

particles in nature. Regarding interactions, successes include math-based models correlating with electromagnetism, the weak interaction, the strong interaction, and relationships among these three interactions. Successes include calculating anomalous magnetic dipole moments for the electron and the muon. People say that the early twenty-first century Standard Model likely does not include all elementary particles and does not include all fundamental interactions. If nature exhibits quantum gravity, one deficiency features the graviton and the gravitational interaction. Other opportunities for predicting or cataloging particles and interactions exist.

Other gaps and opportunities exist. For example, as far as we know, the early twenty-first century Standard Model does not predict (in retrospect) the weak mixing angle.

~ ~ ~

This subsection notes that work above provides a possible basis for estimating the weak mixing angle.

In the Standard Model, people define the weak mixing angle θ_W by the formula $\cos(\theta_W)$ = m'(W boson) / m'(Z boson). Here, m' denotes rest mass. Possibly, traditional theories do not predict this ratio. This monograph includes models correlating with the approximation m'(W boson) / m'(Z boson) $\approx (7/9)^{1/2}$. And, this monograph provides for possible contributions that may change the estimated mass for the W boson from being within 3 standard deviations of observations to being closer to observations. (See Table 3.7.4 and Table 3.7.6.)

~ ~ ~

This subsection provides some thoughts about mass, gravity, and the Standard Model.

People say that non-zero-mass elementary particles have non-zero mass because of interactions with an entity people call the Higgs field. People say that the existence of neutrino oscillations (or, flavor mixing) implies that at least one flavor of neutrino must have non-zero rest mass. People say that people have as yet not found a way to include non-zero mass neutrinos in the Standard Model.

Perhaps, until theory adequately includes quantum gravity, people might consider that some such statements about mass are premature. Perhaps, people would say that theory that adequately dovetails with one but not both of mass-correlated energy and gravity falls short regarding completeness and/or consistency.

Perhaps, this monograph provides some steps forward toward resolving some such issues. Coupling gravity (as a force) to energy-momentum (not rest mass) may help. Another concept that might help features aspects of our models that couple oscillator P0, the Higgs boson, and the correlation that (for ground-state INTERN LADDER solutions) N(P0) = 0 correlates with m' ≠ 0.

~ ~ ~

This subsection provides thoughts about relationships between work in this monograph and work regarding the Standard Model.

Table 5.1.1 shows some possible paths for improving models regarding elementary particles, composite particles, and so forth. People might say that, for any such path, new results from experiments, observations, and inferences about the universe can be key to making progress.

Table 5.1.1 Possible paths for improving models regarding physics related to elementary particles

1. MM1MS1 centric.
    - 1.1. Process.
        - 1.1.1. Start from MM1MS1 models.
        - 1.1.2. Add models that correlate with Standard Model results that MM1MS1 models do not yet produce.
        - 1.1.3. And so forth.
    - 1.2. Notes.
        - 1.2.1. People might say that people could gain confidence in this approach to the extent ...
            - 1.2.1.1. People verify possible predictions people can make based on MM1MS1 models.
            - 1.2.1.2. People replicate some Standard Model successes (especially ones for which the fine-structure constant plays a role in results), perhaps based on ...
                - 1.2.1.2.1. An expanded set of candidate elementary particles (especially particles correlating with G-family solutions).
                - 1.2.1.2.2. A simplified perturbation-like theory (that may avoid traditional seeming needs for renormalization) that correlates with at least one of the following.
                    - 1.2.1.2.2.1. The theory involves the expanded set of particles.
                    - 1.2.1.2.2.2. The theory successfully estimates anomalous magnetic dipole moments, kaon CP-violation, and neutral B meson flavor oscillation.
                - 1.2.1.2.3. Perhaps, estimates of some interaction strengths (especially for G-family elementary particles).

1.2.1.2.3.1. Here, a hope would be that a few estimates lead to results that correlate with many experiments.

1.2.1.3. People show how to redevelop, based on MM1MS1 concepts, aspects of the Standard Model that correlate with ...

1.2.1.3.1. Action-based or Lagrangian expressions that some statements of Standard Model physics feature.

1.2.1.3.2. The physics of composite particles and other objects that feature more than one elementary particle.

2. Standard Model centric.
    2.1. Process.
        2.1.1. Start from the Standard Model.
        2.1.2. Add models that correlate with MM1MS1 results that the Standard Model has yet to produce.
        2.1.3. And so forth.
3. Hybrid.
    3.1. Process.
        3.1.1. Work back and forth between MM1MS1 models and the Standard Model.
        3.1.2. At each step, try to ...
            3.1.2.1. Take use aspects from one set of models to extend the other set of models.
4. Other.

Below, we offer concepts that may help people merge work in this monograph and results from Standard Model work. (See the third item in Table 5.1.1.)

~ ~ ~

This subsection discusses candidates for particles that, if found, people might add to the Standard Model.

Table 5.1.2 pertains. Here, $n \geq 2$ pertains.

Table 5.1.2 Standard-Model and possible beyond-the-Standard-Model particle subfamilies

1. The Standard Model (as of 2015) includes ...
    1.1. Elementary particles correlating with the subfamilies 1C, 1N, 1Q, 0H, 2W, 2G2&, and 2Y.
    1.2. Some composites of the form n × 1Q+2Y with n = 2 or n = 3.
    1.3. Possibly, some composites of the form n × 1Q+2Y with n = 4 or n = 5.

   1.4. Other non-elementary particles.
2. Possible OME beyond-the-Standard-Model particles include ...
   2.1. G-family elementary particles other than 2G2&. (See Table 3.3.4.)
   2.2. Elementary particles correlating with the subfamilies 0O, 1R, 2O, 3N, 3QIRD, and 4YO.
   2.3. Composites of the forms ...
      2.3.1. n × 1Q+2Y, with n ≥ 6.
      2.3.2. n × 0O.
      2.3.3. n × 1Q+2O.
      2.3.4. n × 1R+2YO.
      2.3.5. n × 3QIRD+4YO.
3. Possible DME and DEE beyond-the-Standard-Model additions include ...
   3.1. Assuming ENS48 pertains, ...
      3.1.1. 47 other instances of elementary particles (other than non-2G2& G-family elementary particles) and composite particles.
      3.1.2. 23 other instances of G-family elementary particles with spans of 2.
      3.1.3. 7 other instances of G-family elementary particles with spans of 6.
   3.2. Assuming ENS6 pertains, ...
      3.2.1. 5 other instances of elementary particles (other than non-2G2& G-family elementary particles) and composite particles.
      3.2.2. 2 other instances of G-family elementary particles with spans of 2.
   3.3. Assuming ENS1 pertains, ...
      3.3.1. None (at least, that MM1MS1 models would currently correlate with).

~ ~ ~

This subsection discusses concepts that people might use regarding harmonizing and extending the Standard Model and MM1MS1 models regarding interactions involving spin-1 and spin-0 elementary bosons this monograph discusses.

Table 5.1.3 discusses the acronym CTCW.

Table 5.1.3 CTCW

1. The acronym CTCW ...
   1.1. Abbreviates the phrase current that correlates with.
   1.2. Regarding the Standard Model, ...
      1.2.1. Correlates with a Standard Model current.
   1.3. Regarding MM1MS1 models, ...
      1.3.1. Correlates with a MM1MS1 analog of current regarding MM1MS1 phenomena.

1.3.1.1. For example, CTCW2G2& correlates with a current that correlates with charge and MM1MS1-photonics.

Table 5.1.4 discusses steps to harmonize and extend the Standard Model and MM1MS1 models regarding interactions involving spin-1 and spin-0 elementary bosons this monograph discusses. Different steps and/or other sequences of steps may pertain.

Table 5.1.4 Steps to harmonize and extend the Standard Model and MM1MS1 models regarding interactions involving spin-1 and spin-0 elementary bosons this monograph discusses

1. Harmonize, in each of MM1MS1 and the Standard Model, models regarding electromagnetism.
    - 1.1. To the extent appropriate, ignore aspects pertaining to elementary particle magnetic dipole moments.
2. Harmonize, regarding elementary particle magnetic dipole moments.
    - 2.1. Determine the extent MM1MS1 concepts related to 2G24& enhance Standard Model models.
3. Explore the extent to which the combination of adding 2G24& to the Standard Model and assuming some ENS48 models (such as ENS48" models) or some ENS6 models (other than ENS6' models) pertain for MM1MS1 correlates with opportunities to measure interactions that span more than one ensemble.
    - 3.1. See Table 5.1.5.
4. Add to the Standard Model other G-family spin-1 elementary particles.
    - 4.1. Those particles include (and, for %68even models, are) 2G68& and 2G468&.
    - 4.2. Perhaps, explore the extent to which the combination of adding 2G68& and/or 2G468& to the Standard Model and assuming ENS48 models or ENS6 models pertain for MM1MS1 correlates with opportunities to measure interactions that span more than one ensemble.
5. Harmonize, in each of MM1MS1 and the Standard Model, models regarding weak-interaction bosons and regarding the Higgs boson.
    - 5.1. Perhaps, this involves little more than matching aspects of models.
6. Correlate gluon aspects of the Standard Model with MM1MS1 models regarding the 2Y-subfamily.
    - 6.1. Perhaps, this involves little more than matching aspects of models.
    - 6.2. Perhaps, find that this work ...
        - 6.2.1. Adds to the Standard Model aspects regarding (1;3)< symmetries and other (a`;b`)< symmetries.

7. Regarding the weak-interaction bosons, consider how to characterize a progression of models that starts with MM1MS1 representations in (in essence) a quantum analog to energy-momentum space and ends with Standard Model vertex factors for weak-interaction bosons.
8. Add to the Standard Model phenomena correlating with the 2O- and 0O-subfamilies.
   8.1. Perhaps, find that this work ...
      8.1.1. Uses vertex factors that have similarity to Standard Model vertex factors for the W- and H-families.
         8.1.1.1. Perhaps, the vertex factors for 2O correlate with the γ^μ (1/2) (1 – γ^5) expression that people correlate with 2W vertex factors.
      8.1.2. Adds understanding regarding (1;3)< symmetries and other (a`;b`)< symmetries.
9. Summarize lessons-learned and anticipate reusing such lessons-learned regarding adding (to the Standard Model) interactions involving spin-2 elementary bosons.

Table 5.1.5 discusses relationships between 2G2&-related models and 2G24&-related models and photons.

Table 5.1.5 Relationships between 2G2&-related models and 2G24&-related models and photons (ENS48 models and ENS6 models)

1. To the extent models for 2G24& (including considerations for oscillator swap symmetries) adequately represent photon coupling (or lack of photon coupling) to DME elementary-particle magnetic dipole moments, ...
   1.1. Models for photons might correlate with models combining 2G2& and 2G24&.
2. If nature does not exhibit OME photon coupling to DME elementary-particle magnetic dipole moments, ...
   2.1. Perhaps, people can develop models that correlate photons with 2G2& and 2G24&.
3. If nature exhibits OME photon coupling to DME elementary-particle magnetic dipole moments, ...
   3.1. Perhaps, people can update photon-related aspects of the Standard Model accordingly.
4. In any event, ...
   4.1. Experiments to find or rule out (to some degree of confidence) 2G24&-coupling and/or photon-coupling to DME elementary-particle magnetic dipole moments might provide useful results.

~ ~ ~

This subsection discusses concepts that people might use regarding harmonizing and extending the Standard Model and MM1MS1 models regarding interactions involving spin-2 elementary bosons this monograph discusses.

Table 5.1.6 discusses steps to harmonize and extend the Standard Model and MM1MS1 models regarding interactions involving spin-2 elementary bosons this monograph discusses. Different steps and/or other sequences of steps may pertain.

Table 5.1.6 Steps to harmonize and extend the Standard Model and MM1MS1 models regarding interactions involving spin-2 elementary bosons this monograph discusses

1. Try adding (to the Standard Model) currents and interactions correlating with the 4O-subfamily.
   1.1. Possibly, the 4O-subfamily does not pertain to nature. Even so, ...
      1.1.1. One might learn something about modeling spin-2 elementary bosons.
      1.1.2. One might learn about particle properties and/or about modeling particle properties.
         1.1.2.1. People might want to consider the property (if any) that at least 4O4 and 4O3 transfer.
            1.1.2.1.1. This topic parallels the concept that the property that 2O2 and 2O1 transfer is charge.
            1.1.2.1.2. For the 2O-subfamily, transference of charge parallels transference of charge by 2W-subfamily particles.
               1.1.2.1.2.1. MM1MS1 models correlate with nonexistence of a 4W-subfamily.
2. Perhaps, regarding photons (for which S = 1), consider how to characterize a progression of models that starts with MM1MS1 representations in (in essence) a quantum analog to energy-momentum space and ends with Standard-Model vertex factors for photons.
3. Develop candidate vertex factors for use in describing interactions between 4G4& and charged leptons.
   3.1. Perhaps, regarding gravitons, develop and use a progression similar to the progression (from models correlating with a MM1MS1 quantum energy-momentum space to Standard-Model vertex factors) pertaining to photons.
   3.2. Perhaps, consider using ...
      3.2.1. Spinors with more than 4-components
      3.2.2. (Square) gamma-like matrices with numbers of elements suitable for the spinors.

4. Add (to the Standard Model) a current and interactions correlating with 4G4&.
    4.1. Perhaps, find that this work (when combined with work in Table 5.1.4 and Table 5.1.5) ...
        4.1.1. Provides a basis for modeling a Standard Model particle correlating with 4G268& and a Standard Model particle correlating with 4G2468&.
5. Add (to the Standard Model) phenomena correlating with 4G268& and with 4G2468&.
6. Try adding (to the Standard Model) phenomena correlating with the 4Y-subfamily.
    6.1. (Even to the extent the 4Y-subfamily does not pertain to nature, one might learn something about modeling spin-2 elementary bosons.)

~ ~ ~

This subsection discusses interactions between gravity and zero-mass elementary particles for which $\sigma = +1$.

In MM1MS1 models, 4G4& bosons interact with energy-momentum 4-vector MM1MS1-currents and not with m' (or, rest mass).

A graviton interacts with a MM1MS1-photon via the cloud of virtual particles that correlates with the energy (and magnitude of the energy-momentum 4-vector) of the MM1MS1-photon.

Regarding augmenting the Standard Model, this suggests gravitons interact with a photon via the cloud of virtual particles that correlates with the photon. Regarding MM1MS1 models and regarding the Standard Model, people may also want to consider concepts related to photons interacting with a graviton via the cloud of virtual particles that accompanies the graviton.

Similar remarks pertain to interactions between gravity and MM1MS1-neutrinos and between gravity on 3N-subfamily elementary particles.

~ ~ ~

This subsection discusses concepts that people might use regarding adding (to the Standard Model) 3N-subfamily elementary particles and that people might use regarding modeling neutrinos.

Table 5.1.7 discusses a plan for adding (to the Standard Model) 3N-subfamily fermions this monograph discusses and for understanding models related to neutrinos.

Table 5.1.7 Possible steps for adding (to the Standard Model) 3N-subfamily fermions this monograph discusses and for understanding models related to neutrinos

1. Conduct experiments (and/or possibly make observations) sufficient to determine the extent to which neutrinos have non-zero rest masses.
   - 1.1. Do not necessarily correlate observations of neutrino oscillations with a necessity that at least one flavor of neutrino must have non-zero rest mass.
   - 1.2. Possibly, consider that, perhaps, experiments have yet to measure neutrino speeds (in a near vacuum) significantly less than c.
2. To the extent experiments (and/or observations) show that all neutrinos have zero rest mass, ...
   - 2.1. Integrate work regarding MM1MS1-neutrinos into the Standard Model.
     - 2.1.1. For example, possibly assume that base states for neutrinos correlate with Dirac-fermion behavior.
3. To the extent experiments (and/or observations) show that neutrinos have non-zero rest mass, ...
   - 3.1. Determine a progression of models that starts from an appropriate (for the Standard Model) model for neutrinos and ends with models for MM1MS1-neutrinos.
4. Summarize lessons-learned and anticipate reusing such lessons-learned regarding adding (to the Standard Model) aspects related to the 3N-subfamily.
5. Try adding (to the Standard Model) fields and particles correlating with the 3N-subfamily.
   - 5.1. (Even to the extent the 3N-subfamily does not pertain to nature, one might learn something about modeling.)

~ ~ ~

This subsection discusses relative rest masses of some particles that are generation-1 elementary fermions for which $m' \neq 0$ or that feature generation-1 elementary fermions for which $m' \neq 0$.

Baryons have larger rest masses than do pions, at least in part because baryons have more capacity (than do pions) to interact with gravitons. The extra capacity features more relevant non-zero-mass fermions and more ability to create aspects of clouds of virtual particles. Pions have more rest mass than do electrons, at least in part because pions have more capacity (than do electrons) to interact with gravitons. The extra capacity features (two, not one) non-zero-mass fermions and more ability to create aspects of clouds of virtual particles.

People might say that (for some particles that are or that feature generation-1 elementary fermions for which m' ≠ 0) the above thinking correlates with a hierarchy of rest masses.

~ ~ ~

This subsection discusses other possible additions to the Standard Model. Table 5.1.8 pertains.

Table 5.1.8 Possible steps for adding (to the Standard Model) other aspects correlating with work in this monograph

1. Consider work, in various places in this monograph, regarding rest masses.
2. Try adding (to the Standard Model) models sufficient to correlate with rest masses for elementary particles.
   2.1. At least, try adding enough to correlate with ratios of rest masses for W- and H-family bosons.
3. Consider possibilities for adding (to the Standard Model) models (based on MM1MS1 models for elementary-particle rest masses) for composite particles.
4. Try adding (to the Standard Model), as appropriate, as yet not added aspects regarding 1R- and 3QIRD-subfamilies of possible elementary particles.
5. To the extent ENS48 models or ENS6 models pertain to nature, explore further implications of and/or correlations regarding physics-relevant SU(17) symmetries, symmetries related to subgroups of SU(17), SU(7) symmetries, and symmetries related to subgroups of SU(7).

Section 5.2 The cosmology timeline

Section 5.2 discusses additions that might pertain for narratives about cosmology. We position, in cosmological timelines, various known and possible phenomena. We note mechanisms (related to the G-family of elementary particles) correlating with the rate of expansion of the universe. We discuss a mechanism (correlating with spin-3/2 elementary fermions and spin-2 elementary bosons) that could correlate with an inflationary epoch. We note a mechanism (correlating with 2O-subfamily bosons) that could lead to matter/antimatter imbalance. We note mechanisms (correlating with electromagnetism) that could lead to varying proportions of matter and dark matter in objects such as galaxies. We discuss mechanisms (correlating clustering and anti-clustering of clumps of perhaps galaxy size with G-family bosons fostering the clustering and anti-clustering) that could lead to subtle patterns in CMB. We discuss mechanisms (correlating with G-family bosons) that could correlate with formation of quasars.

~ ~ ~

This subsection discusses features that people might want to add to narratives regarding the evolution of the universe.

Table 5.2.1 pertains.

Table 5.2.1 . Possible additions to narratives about the evolution of the universe

1. During the big bang and times thereafter, ...
   1.1. Interactions govern the so-called rate of expansion of the universe.
   1.2. The interaction that correlates with the G-family boson 4G2468& drives the big bang.
   1.3. Stuff clumps.
   1.4. Smaller clumps transit from r^{-8} repulsion (caused by 4G2468&) to r^{-6} attraction (caused by other G-family bosons) to r^{-4} repulsion (caused by yet other G-family bosons) to r^{-2} gravitational attraction (with electromagnetic attraction and/or repulsion being possible) sooner than do larger clumps.
2. To the extent the concept of an inflationary epoch pertains, ...
   2.1. Transitions, such as from 4O-physics to 4Y-physics and/or from 4O-physics or 4Y-physics to matter-antimatter annihilations, take place during that epoch.
3. To the extent the ratio of charged matter to charged antimatter changes, ...
   3.1. Non-zero-mass elementary bosons catalyze the change.
      3.1.1. 2O charged bosons convert antimatter quarks to matter quarks.
      3.1.2. 2W charged bosons convert antimatter charged leptons to antimatter neutrinos.
      3.1.3. 2W charged bosons convert matter neutrinos to matter charged leptons.
   3.2. The change takes place before 2O-subfamily bosons lose influence.
   3.3. 2O-subfamily bosons lose influence before formation of nucleons.
4. To the extent galaxies that contain both ordinary matter and dark matter pertain and that ENS48 models or ENS6 models pertain, ...
   4.1. Electromagnetic-like interactions play key roles (within each ensemble) in clumping leading to galaxy formation and galaxy-clustering.
   4.2. At least some such clumping occurs after the formation of nucleons.

4.3. At least some such clumping occurs before and/or during significant formation of clumps that evolve into galaxies.

5. To the extent some subtle patterns in CMB (or, cosmic microwave background radiation) exist and that ENS48 models or ENS6 models pertain, ...

   5.1. Interactions mediated by G-family bosons lead to clustering and anti-clustering (of clumps of galactic size and of perhaps somewhat smaller sizes and of perhaps somewhat bigger sizes) that eventually lead (indirectly) to such patterns.

6. Regarding quasars (and known large black holes), ...

   6.1. Quasars form based on, in effect, competition between attractive G-family-mediated interactions (with spatial dependences of r^{-2} and/or r^{-6}) and repulsive G-family-mediated interactions (with spatial dependences of r^{-4} and/or r^{-8}).

   6.2. Known quasars correlate with black holes.

   6.3. Possibly, known quasars correlate with repulsive G-family-mediated interactions with spatial dependences of r^{-4}.

   6.4. Possibly, a gravitational somewhat analog to electromagnetism's magnetic field helps focus jets of stuff that quasars eject.

   6.5. Quasars form after black holes form.

Chapter 6 From MM1MS1 models to traditional theories

Chapter 6 (From MM1MS1 models to traditional theories) starts with models this monograph features, discusses extents to which people might correlate or integrate our models with traditional physics theories, and discusses extents people might use our models to extend or improve on traditional physics theories.

~ ~ ~

People might say that results in Chapter 6 (From MM1MS1 models to traditional theories) might correlate with or provide insight about the following aspects of physics. Similarities and differences regarding concepts that pertain regarding MM1MS1 models, traditional quantum physics, and classical physics. The Dirac equation. Relationships among quadratic operators, linear operators, and conservation laws. Relationships between models correlating with conservation laws and models correlating with internal properties of elementary particles. For the hydrogen atom, energy levels, fine-structure splitting, hyperfine splitting, and the Lamb shift. Possible alternatives to some traditional-physics correlations between some aspects of quantum chromodynamics and some applications of concepts correlating with special relativity. Facets of, uses of, and limits regarding models based on general relativity.

Section 6.1 Classical-physics models, QMUSPR models, and MM1MS1 models

Section 6.1 discusses and contrasts some aspects of classical-physics models, QMUSPR models, and MM1MS1 models. We discuss, for elementary particles, representations for spin and concepts related to helicity, chirality, and/or handedness. We discuss magnetic dipole moments not associated with moving charges. We discuss properties that correlate with G-family physics. We discuss roles of $\hbar$. We discuss an operator-centric interpretation of relationships between work in this monograph and aspects of traditional physics. We discuss a possible conceptual derivation of the Dirac equation. We discuss relationships among quadratic operators, linear operators, and conservation laws. We discuss relationships between models correlating with

T.J. Buckholtz, *Models for Physics of the Very Small and Very Large*,
Atlantis Studies in Mathematics for Engineering and Science 14,
DOI 10.2991/978-94-6239-166-6_6

conservation laws and models correlating with internal properties of elementary particles.

~ ~ ~

This subsection provides perspective about contrasting new and traditional concepts.

When contrasting new and traditional concepts, people may start with the traditional ones and then discuss the new ones. Doing so establishes context.

But, if the new concepts improve on the traditional ones, an old-to-new order can anchor discussion in untoward ways.

We prefer, in this section, to assume that other work in this monograph provides adequate context. We assume that starting from concepts we develop and then discussing traditional concepts can be appropriate.

~ ~ ~

This subsection provides information about properties of particles.

People catalog elementary, composite, and other particles based on properties of those particles. This monograph discusses various properties, including spin, rest mass, and charge. This monograph does not discuss, for example, composite-particle properties such as isospin and strangeness.

~ ~ ~

This subsection compares some aspects of MM1MS1 models, QMUSPR models, and classical-physics models.

Table 6.1.1 pertains. (See Table 2.1.7 and Table 2.1.8 regarding terminology for classes of physics theories and models.) Table 6.1.1 presents some concepts that readers should take as conceptual but not literal. Work in this monograph does not depend on statements in this table. In this table, we do not attempt to be thorough regarding concepts, regarding aspects, or regarding wording. To the extent, QMUSPR models or classical-physics models make use of a concept in a way this monograph does not, people may have opportunities to extend work beyond coverage in this monograph and/or to explore other means to unify work in this monograph with traditional models. People may want to explore, for each of some aspects the table lists, the relative ease (across the three classes of theories and models) of expressing the aspect. This monograph does not feature representations for wave functions for zero-mass particles. The acronym MDM abbreviates the term magnetic dipole moment.

Table 6.1.1 Comparisons of some aspects of MM1MS1 models, QMUSPR models, and classical-physics models

1. Spin (S) for elementary particles.
   1.1. In MM1MS1 models, ...

- 1.1.1. 2S is an integer that, for elementary particles, correlates with ...
  - 1.1.1.1. For other than G-family particles, a number (#'P) of QP-like oscillators relevant to solutions.
  - 1.1.1.2. For G-family particles, net polarization.
  - 1.1.1.3. For particles for which m' ≠ 0, the S relevant to solutions for wave functions.
- 1.2. In QMUSPR models, ...
  - 1.2.1. S, for elementary particles, correlates with ...
    - 1.2.1.1. Contributions toward total angular momentum.
    - 1.2.1.2. The S relevant to solutions for wave functions.
- 1.3. In classical-physics models, ...
  - 1.3.1. Perhaps, S and a vector parallel to direction of spin are aspects people can use to add spin-related angular momentum to angular momentum that correlates with an object's physical rotation.

2. Helicity, chirality, and/or handedness for elementary particles.
   - 2.1. In MM1MS1 models, ...
     - 2.1.1. For at least elementary particles correlating with the 1C- and 1N-subfamilies, we think that at least some of helicity, chirality, and/or handedness correlate with the one harmonic oscillator (P2R or P2L) that pertains for a relevant LADDER solution.
       - 2.1.1.1. People might say that these representations correlate with energy-momentum space.
       - 2.1.1.2. People might say that these representations avoid possible problems arising from considering whether a particle moves toward or away from an observer.
   - 2.2. In QMUSPR models, ...
     - 2.2.1. Explanations of these concepts correlate with models based on space-time coordinates.
3. 4-vectors.
   - 3.1. In MM1MS1 models, for elementary particles, ...
     - 3.1.1. The energy-momentum 4-vector ...
       - 3.1.1.1. Correlates with $\sigma = +1$ and SPATIM-related conservation laws.
       - 3.1.1.2. Correlates with elementary-fermion non-zero rest masses that interrelate via generally exponential relationships.
       - 3.1.1.3. Correlates with elementary-boson rest masses that interrelate generally via linear relationships involving squares of masses.

3.1.2. The electric-charge-current 4-vector ...
3.1.2.1. Correlates with quantum-property-transfer-related conservation laws.
3.1.2.2. Correlates with elementary-particle rest charges that interrelate via linear relationships.
3.2. In classical-physics models, for special relativity, ...
3.2.1. People might say that the energy-momentum 4-vector and the electric-charge-current 4-vector belong to the same set of 4-vectors.
4. Magnetic dipole moment (not including anomalous magnetic dipole moment).
4.1. For other than elementary particles and coherent magnetic dipole moment states, ...
4.1.1. For each of MM1MS1 models, QMUSPR models, and classical-physics models, ...
4.1.1.1. Magnetic dipole moment of an object correlates with the motions of charges inside that object.
4.2. For elementary particles, ...
4.2.1. In MM1MS1 models, ...
4.2.1.1. MM1MS1-MDM correlates with G-family aspects pertaining to the P2L-and-P2R and P4L-and-P4R oscillator pairs.
4.2.2. In QMUSPR models, ...
4.2.2.1. People might say that the size of magnetic dipole moment for charged leptons correlates with use of models based on the Dirac equation.

Regarding helicity, chirality, and/or handedness for elementary particles, Table 6.1.2 provides symbols that dovetail with remarks in Table 6.1.1. Here, for example, 2W2L' refers to a fermion's losing a unit of negative charge by absorbing a W^+ boson. (Compare with the second row of Table 3.5.2 and the second row of Table 3.5.3 for a particle that can absorb a W^+. Compare with Table 3.6.4 for notation regarding erasing a unit of color charge.)

Table 6.1.2 Notation for emission or absorption of charged W-family elementary particles

Particle	Symbol	Symbol related to emission	Symbol related to absorption
W^+ boson	2W1	2W2L	2W2L'
W^- boson	2W2	2W2R	2W2R'

~ ~ ~

This subsection lists, for elementary particles, properties related to G-family bosons.

Table 6.1.3 provides acronyms for properties correlating with %68even G-family bosons. Here, we also provide a column labelled s/3v. Interactions correlating with type-3v carry information about a magnetic-dipole-like state or a spin-like state (or, to use classical-physics terminology, a 3-vector). For MM1MS1-MDM, people might (from a perspective of classical physics) correlate the 3-vector with the term axis of spin. (See Table 3.3.15.) Interactions correlating with type-s do not carry information about a magnetic-dipole-like state or a spin-like state. Here, we sort the entries so as to group spin-1 G-family bosons before spin-2 G-family bosons. Here, we sort, within each of the two sets, items in order of increasing n"(%).

Table 6.1.3 Properties correlating with %68even G-family bosons

Property	Term for QE-like aspects	Term for QP-like aspects	s/3v	SDI	Particle	S
PR2G2&	Electric charge	Electric current	s	r^{-2}	2G2&	1
PR2G24&	MM1MS1-MDM		3v	r^{-4}	2G24&	1
PR2G68&	Spin		3v	r^{-4}	2G68&	1
PR2G468&			3v	r^{-6}	2G468&	1
PR4G4&	Energy	Momentum	s	r^{-2}	4G4&	2
PR4G268&			3v	r^{-6}	4G268&	2
PR4G2468&			3v	r^{-8}	4G2468&	2

~ ~ ~

This subsection provides thoughts regarding particle properties related to G-family bosons.

Table 6.1.4 pertains.

Table 6.1.4 Thoughts regarding properties correlating with G-family bosons

1. Of the properties Table 6.1.3 lists, ...
   1.1. The following properties seem not to appear in classical-physics models or in QMUSPR models ...
      1.1.1. PR2G468&.
      1.1.2. PR4G268&.
      1.1.3. PR4G2468&.
   1.2. People might want to explore the extents to which to consider that the following properties correlate with (perhaps new types of) anomalous (or, other term) magnetic dipole moments or spin moments.
      1.2.1. PR2G468&.

1.2.2. PR4G268&.
1.2.3. PR4G2468&.

2. People might explore, regarding QMUSPR models (and, perhaps, regarding classical-physics models) ...
   2.1. The extents to which to keep separate and/or to combine effects of the spin-1 particles Table 6.1.3 lists.
   2.2. The extents to which to keep separate and/or to combine effects of the spin-2 particles Table 6.1.3 lists.

~ ~ ~

This subsection notes a touchpoint, regarding uncertainty, between work in this monograph and QMUSPR models.

Work in this monograph correlates the physics number ħ with particle properties such as spin.

QMUSPR models correlate ħ with particle properties such as spin.

QMUSPR models also correlate ħ with uncertainty. Uncertainty pertains to models that feature linear combinations of two or more base states. Some such models have roots in classical-physics models.

Much work in this monograph tends not much to discuss linear combinations of base states.

Some work in this monograph features quantum operators and quantum wave functions. For example, work in Section 2.10 uses such constructs.

People might say that some work in Section 2.10 pertains to concepts somewhat related to traditional quantum-physics notions of uncertainty.

People might say that work in Section 2.10 adequately injects, into work in this monograph, roles for ħ related to traditional quantum-physics roles for ħ related to uncertainty.

~ ~ ~

This subsection correlates aspects of work in this monograph and aspects of traditional physics.

Much traditional physics correlates with aspects of interactions, within a context people might call a system, between subsystems. A subsystem can be as conceptually small as (at most) an elementary particle. A subsystem can be as conceptually large as (at least) the universe except for an elementary particle.

Table 6.1.5 discusses, at least conceptually, relationships between an elementary particle and the rest of the universe.

Table 6.1.5 Conceptual relationships between a particle and the universe and relationships between work in this monograph and aspects of traditional physics

1. We define an abbreviation.

$$\text{Eg"E abbreviates } E^2 - c^2P^2 \tag{6.1}$$

1.1. Here, each of E and P is an operator.
1.2. Here, g" is a metric.

2. We assume the following pertains.

$$Eg''E = m^2c^4 \tag{6.2}$$

2.1. Here, m has dimensions of mass.

3. We divide a system (for which E, g", and m pertain) into two subsystems. For example, the following pertains.

$$E = E\{1\} + E\{2\} \tag{6.3}$$

4. We assume the following pertain.

$$E\{1\}g''E\{1\} = (m\{1\})^2c^4 \tag{6.4}$$

$$E\{2\}g''E\{2\} = (m\{2\})^2c^4 \tag{6.5}$$

$$E\{1\}g''E\{2\} + E\{2\}g''E\{1\} = m\{1\}m\{2\}c^4 + m\{2\}m\{1\}c^4 \tag{6.6}$$

4.1. People might say that the following pertain.
4.1.1. We derive the equations by stating terms correlating with $(E\{1\} + E\{2\})g''(E\{1\} + E\{2\}) = \ldots$.
4.1.2. We assume minimal, but not no, interactions between the two subsystems.

5. We assume that ...
5.1. Subsystem 1 correlates with a free-ranging (or, $\sigma = +1$) elementary particle in a FRERAN context.
5.1.1. People might say that examples can include (at least) particles correlating with the 1C- and 1N-subfamilies.
5.2. Subsystem 2 correlates with the DIU universe, except for subsystem 1.

6. Regarding the equation ...

$$E\{1\}g''E\{2\} + E\{2\}g''E\{1\} = m\{1\}m\{2\}c^4 + m\{2\}m\{1\}c^4 \tag{6.7}$$

6.1. ... The following pertain.
6.1.1. We consider that the interaction between subsystem 2 and subsystem 1 correlates with a point (regarding space-time coordinates).
6.1.2. We assume we can approximate aspects related to subsystem 2 by the following expression.

$$m\{2\}c^2 \tag{6.8}$$

6.1.3. Here, $m\{2\}c^2$ does not necessarily correlate with an energy of a specific object.

7. We assume that the effective $m\{2\}c^2$ is finite.
7.1. Even if the universe is infinite, the DIU (at a specific point, with respect to a set of space-time coordinates) is finite.
7.2. The speed of light is finite.
7.3. Any part of the universe beyond the DIU has (at a specific point, with respect to a set of space-time coordinates) yet to influence subsystem 1.

8. The following pertain.

$$E\{1\} + E\{1\} = m\{1\}c^2 + m\{1\}c^2 \quad (6.9)$$

$$E\{1\} = m\{1\}c^2 \quad (6.10)$$

   8.1. Here, (6.10) dovetails with (6.1).
      8.1.1. E{1} is a sum of four terms, with one term correlating with QE-like aspects and three terms correlating with QP-like aspects.

9. People might say that ...
   9.1. E{1}g"E{1} = $(m\{1\})^2c^4$ ...
      9.1.1. Correlates with internal properties of subsystem 1.
      9.1.2. For subsystem 1 being an elementary particle, ...
         9.1.2.1. Correlates with INTERN-related work in this monograph.
         9.1.2.2. Correlates with FERTRA-related work in this monograph.
      9.1.3. Features work for which units of energy squared pertain.
   9.2. E{1} = m{1}c^2 ...
      9.2.1. Correlates with kinematics of subsystem 1.
      9.2.2. For subsystem 1 being an elementary particle, ...
         9.2.2.1. Correlates with ENVIRO-related work in this monograph.
         9.2.2.2. Correlates with much traditional physics.
      9.2.3. Seems to feature work for which units of energy pertain. (See Table 2.5.2.)
      9.2.4. Can (and perhaps should) be considered to feature work for which units of energy squared pertain, based on its derivation via the following equation. (See Table 2.5.2.)

$$m\{2\}c^2 E\{1\} = m\{2\}c^2 m\{1\}c^2 \quad (6.11)$$

10. People might say that, for spin-1/2 elementary fermions, (6.10) correlates (at least conceptually) with the Dirac equation.
   10.1. Regarding the left side of (6.10), each of the four operators would include a gamma matrix.

Table 6.1.6 pertains. Possibly, the table points to opportunities for research.

Table 6.1.6 Linear operators, conservation laws, SPATIM aspects, quadratic operators, elementary-particle properties, INTERN aspects, and aspects related to #b' and #b"

1. People might say that work in Table 6.1.5 provides perspective for integrating and for contrasting work in this monograph and much of traditional physics.

2. Some aspects (such as approximate squares of masses for elementary bosons) of MM1MS1 models correlate (assuming, in Table 6.1.5, subsystem 1 correlates with an elementary particle and subsystem 2 correlates with the DIU universe) with operators such as (6.4) shows.
    2.1. People might say that some such aspects correlate with ...
        2.1.1. Quadratic operators.
        2.1.2. INTERN aspects.
3. Some conservation laws correlate with operators such as the left side of (6.10) shows.
    3.1. People might say that some such conservation laws correlate with ...
        3.1.1. Linear operators.
        3.1.2. SPATIM ENVIRO aspects.
    3.2. People might say that such ostensibly linear-operator concepts also correlate with quadratic-operator concepts, per, for example, (6.7).
4. People might say that angular momentum provides an interesting case.
    4.1. People might say that the classical-physics conservation law correlates (at least somewhat) with a linear operator.
    4.2. People might say that MM1MS1 aspects related (at least) to elementary-particle spin correlate with a quadratic operator. (See (2.104) in Table 2.12.9.)
5. People might say that invariance regarding RESENE provides an interesting case. (See Table 2.12.9.)
6. People might say that, regarding the example (in Table 6.1.5) of subsystem 1 being an elementary particle and subsystem 2 being the DIU universe, ...
    6.1. For the elementary particle, ...
        6.1.1. #b' correlates with ...
            6.1.1.1. (6.4) and (6.10).
            6.1.1.2. Some aspects of (6.6) and (6.8).
    6.2. For the DIU, ...
        6.2.1. #b" correlates with ...
            6.2.1.1. (6.5).
            6.2.1.2. Some aspects of (6.6) and (6.8).
7. People might say that possible research could feature contrasting and trying to unify concepts in this table.
8. People might say that possible research could feature attempts to interpret quadratic operators (such as each component of the sum that Eg"E represents, as per (6.2) in Table 6.1.5) as a construct combining a field-centric operator and a particle-centric operator.
9. People might say that possible research could feature attempting to develop an analog, for phenomena correlating with $\sigma = -1$, to Table 6.1.7 (which correlates with phenomena for $\sigma = +1$).

9.1. For example, try to determine (in more breadth and/or depth than this monograph presents) invariances, symmetries and related (mathematics) groups, operator-based equations, and so forth that might pertain.

Regarding some models for phenomena for which $\sigma = +1$, Table 6.1.7 correlates some invariances, symmetries, conservation laws, and facets of elementary-particle internal properties. (Regarding the aspect column, see, for example, Table 2.12.1. Regarding the two rightmost columns, see Section 2.12.) People might say that Table 6.1.7 shows aspects of relationships among quadratic operators, linear operators, and conservation laws.

Table 6.1.7 Some aspects, regarding some traditional models and MM1MS1 models for phenomena for which $\sigma = +1$, of invariances, symmetries, conservation laws, and elementary-particle internal properties

	Aspect	MM1MS1 ENVIRO	MM1MS1 INTERN
1.	Focus	Conservation laws	Internal properties
2.	Can measurements vary by observer?	Yes	No
3.	(1×1; ..., ..., ...)>	E [scalar] (energy)	N (number of excitations)
4.	(...; 1×3, ..., ...)>	P [vector] (momentum)	$\{ N(Pj) \mid 0 \le j \le \#P \}$ (specific particle)
5.	(...; ..., 1×3, ...)>	J = L + S [vector] (angular momentum)	S [scalar] [or, for S being an operator, $<(2S)^2> - <(2S)>^2$] (spin)
6.	(...; ..., ..., 1×3)>	v [vector] (relative velocity)	generation [for fermions]; RESENE (rest energy)
7.	Principle for modeling	Conceptually, transfers of each of Æ_{QE} and Æ_{QP} to (or from) an elementary boson from (or to) ENVIRO and/or INTERN properties of an elementary fermion	ŒE = 0
8.	Sub-focus	QMUSPR	QMPRPR

	Aspect	MM1MS1 ENVIRO	MM1MS1 INTERN
9.	Theme for units centric to some MM1MS1 modeling	(energy squared) / energy	energy squared

Section 6.2 Atomic physics

Section 6.2 discusses models possibly pertaining to objects similar to hydrogen atoms and possibly pertaining to other aspects of atomic and molecular physics. We speculate regarding correlating four spin-1 G-family bosons, respectively, with energy levels related to the simplest electromagnetic interactions between two particles comprising a hydrogen-like atom, fine-structure splitting, hyperfine splitting, and the Lamb shift. We speculate about ATOMOL symmetries. We speculate about possible correlations between ATOMOL symmetries and the group SO(4). We point toward modeling techniques (based on G-family bosons) that might generate results similar to results (involving the presence of the fine-structure constant) that people develop via traditional approaches. People might say that work in this section is speculative.

~ ~ ~

This subsection provides perspective about this section.

People might say that work in this section is speculative.

Above, this monograph shows possibilities that MM1MS1 models lead to non-traditional symmetries regarding COMPAR models (or aspects of QCD physics) or to possibly new useful ways to interpret traditionally known symmetries regarding COMPAR models (or aspects of QCD physics).

People might say that models for harmonic oscillators and models for atoms similar to (and including) hydrogen atoms represent two categories of physics models for which people can relatively easily derive detailed widely-useful results analytically regarding nature. (The previous sentence does not pertain to computer-based simulations.)

This section discusses possibilities that aspects of MM1MS1-like models pertain to atoms and molecules. People might say that work in this section lies outside the scope of MM1MS1 models.

Molecular physics tends, from a modeling standpoint, to be complex. People might say that the entirety of an approximate analytic exploration regarding molecules and atoms would be difficult.

People might say that an exploration of two-body atomic-like physics can be appropriate.

This section features discussion pertaining to atom-like objects that include exactly 2 fermions, each having spin-1/2 and each having a magnitude of charge equal to the magnitude of charge for the electron. Examples of such objects include hydrogen atoms for which one fermion is an electron and for which the nucleus is a proton, bound states of a proton and a non-electron lepton (such as a muon), and positronium (which consists of an electron and a positron).

For such an object, we use the term H1-atom.

~ ~ ~

This subsection discusses possibilities for models, for the H1-atom, based on math for isotropic harmonic oscillators.

People might say that Table 6.1.5, Table 6.1.6, and Table 6.1.7 correlate with uses, for internal properties, of quadratic operators and of models correlating with physics units of energy squared.

Traditional models of H1-atoms include a potential energy that is proportional to r^{-1}, for which r is a radial coordinate correlating with 3 dimensions.

Table 6.2.1 pertains. The table provides possible opportunities for research. Below, we speculate about roles for (a`;b`)< symmetries and/or for (a`;b`)> symmetries. (See Table 6.2.3.)

Table 6.2.1 A possibility for modeling H1-atoms

1. Regarding H1-atoms (and possibly other systems), ...
    - 1.1. A term proportional to the square of the potential energy would be proportional to r^{-2}.
    - 1.2. People might say that, regarding (2.39) specifically and Table 2.5.6 generally, one can add a term correlating with (the square of) the potential to the term Ωr^{-2}.
    - 1.3. People might say that, regarding (2.38), a model for an H1-atom should not feature significant effects of the term $(\xi'/2)\,\eta^{-2}\,r^2$.
    - 1.4. We deemphasize the term $(\xi'/2)\,\eta^{-2}\,r^2$ by positing that we can consider the limit $(\xi'/2)\,\eta^{-2} \rightarrow 0$.
    - 1.5. For this work, we assume $(\xi'/2)\,\eta^2$ remains constant.
    - 1.6. With that assumption, $(\xi'/2)\,\eta^2\,\Omega$ is a constant.
    - 1.7. People might say that we can set that constant to include a linear combination of terms, with each term proportional to one of the following.
        - 1.7.1. A factor (for example, $S(S + 1)$) correlating with spin.
        - 1.7.2. A factor (for example, $L(L + 1)$) correlating with orbital angular momentum.

- 1.7.3. A factor (for example, J(J + 1)) correlating with total angular momentum.
- 1.7.4. r^2 times the square of a potential (for which the potential is proportional to r^{-1} and for which the potential has units of energy).

2. To what extent can such a process lead to useful models for ...
   - 2.1. H1-atoms?
   - 2.2. Other systems?
3. Assuming such a process can lead to useful models for H1-atoms (or other systems), to what extent do ...
   - 3.1. Symmetries similar to (a`;b`)< correlate with such models?
   - 3.2. Symmetries similar to (a`;b`)> correlate with such models?

~ ~ ~

This subsection discusses possibilities for ATOMOL models for the H1-atom. Table 6.2.2 pertains. This table contains speculation.

Table 6.2.2 Concepts that might underlie ATOMOL models for the H1-atom

1. Assume that the H1-atom contains ...
   - 1.1. 2 elementary fermions.
     - 1.1.1. Assume that absolute value of the charge of each fermion is $|q_e|$.
     - 1.1.2. Assume that S = 1/2 pertains for each fermion.
   - 1.2. Elementary bosons that include at least 2G2&.
2. Assume that one can treat as an input to MM1MS1-like models the ground-state energy of the H1-atom.
   - 2.1. Let EGS denote the negative of the ground-state energy.
   - 2.2. EGS depends on the masses of the fermions.
3. Regarding EGS and fermion masses, ...
   - 3.1. EGS depends on the masses (as well as charges) of the two fermions.
     - 3.1.1. Traditional models tend to ...
       - 3.1.1.1. Include, for each fermion, a term correlating with linear momentum for the fermion.
         - 3.1.1.1.1. Include, in such a term, a factor correlating with mass of the fermion.
       - 3.1.1.2. Restate the model in terms of ...
         - 3.1.1.2.1. A center-of-mass component.
         - 3.1.1.2.2. Other components.
       - 3.1.1.3. Then, compute energy levels.

- 3.2. Beyond EGS, as far as we know, traditional work does not include factors correlating with mass.

4. Regarding 4G4&, ...
   - 4.1. 4G4& correlates with the oscillator pair E0-and-P0.
5. Regarding 2G2&, ...
   - 5.1. Note that the following formula provides energy levels for the H1-atom.

$$-\text{EGS}\ (1/n^2) \tag{6.12}$$

     - 5.1.1. Here, ...
       - 5.1.1.1. n is an integer.
       - 5.1.1.2. $1 \leq n < \infty$.
       - 5.1.1.3. People use for n the term principal quantum number.
   - 5.2. Note that, in the limit of large mass asymmetry between the two fermions, the following formula pertains.

$$\text{EGS} \approx m'c^2\alpha^2 / 2 \tag{6.13}$$

     - 5.2.1. Here, m' equals the mass of the lower-mass fermion.
   - 5.3. People might say that ...
     - 5.3.1. This spectrum of energies correlates with effects of 2G2&.
6. Regarding 2G24&, ...
   - 6.1. Note that the following formula provides an approximate factor that multiplies the energy levels stated in the previous item (in this table) for the H1-atom.

$$1 + (\alpha^2 / n^2)\ ((n/(j + 1/2)) - 3/4) \tag{6.14}$$

     - 6.1.1. Here, people use for j the term total angular momentum.
     - 6.1.2. People might say that ...
       - 6.1.2.1. These fine-structure effects correlate with 2G24&.
7. Regarding 2G468&, ...
   - 7.1. In traditional physics, people correlate some hyperfine splitting effects with magnetic dipole moments.
   - 7.2. People might say that ...
     - 7.2.1. Hyperfine splitting effects correlate with 2G468&.
8. Regarding 2G68&, ...
   - 8.1. In traditional physics, people correlate the Lamb shift with interactions with the vacuum.
   - 8.2. People might say that ...
     - 8.2.1. Lamb shift effects correlate with 2G68& and interactions with virtual particles.

~ ~ ~

This subsection discusses possibilities for (a`;b`)< ATOMOL symmetries pertaining to H1-atoms.

Table 6.2.3 pertains. After the first item in the table, each item in the table constitutes speculation. We do not speculate about the relevance of a symmetry that the table correlates with the symbol a`. Assuming a symmetry that correlates with a` is relevant, we do not speculate about values for a`.

Table 6.2.3 Speculation about concepts for ATOMOL symmetries for H1-atoms

1. People say that symmetries that correlate with the group SO(4) correlate with models for objects such as H1-atoms.
2. People might say that ...
    - 2.1. Each of the following, when its field is paired with the field for a spin-1/2 fermion, correlates with a different instance of an ATOMOL (a`;1)<-like symmetry ...
        - 2.1.1. 2G2&.
        - 2.1.2. 2G24&.
        - 2.1.3. 2G468&.
        - 2.1.4. 2G68&.
    - 2.2. Based on the 4 instances (and considering only spin-1 G-family bosons), ...
        - 2.2.1. ATOMOL (a`;4)< symmetries pertain to each combination correlating with one fermion and all of the 4 bosons.
    - 2.3. Based on the H1-atom having 2 such fermions, ...
        - 2.3.1. ATOMOL (a`;4)> symmetries pertain to the H1-atom.
3. People might say that ...
    - 3.1. ATOMOL (a`;4)> symmetries correlate with SO(4).
    - 3.2. For ATOMOL symmetries, a` might be one of ...
        - 3.2.1. 0.
        - 3.2.2. Irrelevant.

~ ~ ~

This subsection notes possible extensions for work above in this section. Table 6.2.4 pertains.

Table 6.2.4 A possible basis for estimating quantities for which people find that the fine-structure constant provides contributions

1. Regarding the G-family member 2G24&, ...
    - 1.1. A formula of the form ...

$$(|q_e|^2/(\,(4\pi\varepsilon_0)r^2\,))\ (R^2/r^2)\ \text{(correlation of two spin-like states)} \qquad (6.15)$$

1.1.1. ... Pertains to an interaction between two elementary particles for which each elementary particle has charge $|q_e|$, with ...
  1.1.1.1. R being independent of the masses of the two interacting charged particles.
1.2. Possibly, for the hydrogen atom (specifically and not necessarily for H1-atoms generally), ...
  1.2.1. R^2 ...
    1.2.1.1. Correlates with a function of $<r^2>$ for the atom and/or of $(<r^4>)^{1/2}$ for the atom.
    1.2.1.2. Correlates with notions that ...
      1.2.1.2.1. The fine-structure constant (α) equals $(|q_e|^2/(4\pi\varepsilon_0))$ / $(\hbar c)$.
      1.2.1.2.2. The fine-structure constant pertains to known data about the fine structure of the spectrum of photons emitted by hydrogen atoms.
2. Possibly, similar considerations pertain for other spin-1 G-family particles for which the sub-list % has at least 2 members.
3. People might say that this work provides a basis for an alternative (to traditional models) approach to making quantitative estimates regarding phenomena that people correlate with the fine-structure constant. (See Section 5.1.)
  3.1. To the extent such an alternative approach pertains, ...
    3.1.1. Perhaps, such an approach correlates with simpler (than traditional) ways to estimate anomalous magnetic moments, (for example, for the electron, muon, and tauon). (See Table 8.1.1.)

~ ~ ~

This subsection notes possible extensions for work above in this section. Table 6.2.5 pertains.

Table 6.2.5 Possible generalizations for and/or from applications of possible ATOMOL (a`;4)< symmetries and possible ATOMOL (a`;4)> symmetries

1. Consider boson (as well as fermion) component particles or objects.
2. Consider spins other than 1/2.
3. Consider non-zero charges with magnitudes other than $|q_e|$.
4. Consider charge-neutral atoms that have (an atomic nucleus and ...) more than one lepton.
5. Consider ions.
6. Consider molecules that include more than one atom.
7. Consider interactions between molecules.

Section 6.3 Special relativity

Section 6.3 discusses concepts which with people correlate the term special relativity and notes a possibility that people may find that work based on other concepts can prove beneficial compared to traditional work. We list concepts people correlate with the term special relativity. People might say that, for some aspects related to QCD (or, quantum chromodynamics), work based on MM1MS1 models might provide (compared to work based on traditional models) more robust results and/or results based on simpler methods.

~ ~ ~

This subsection provides perspective about special relativity.

Special relativity unifies various classical-physics concepts regarding the motions of objects and light. Special relativity provides bounds on candidate developments in quantum-physics theories and models.

~ ~ ~

This subsection notes concepts people correlate with the term special relativity and discusses compatibility between special relativity and this monograph's models.

Table 6.3.1 notes some concepts that differ, but for which people might find overlaps.

Table 6.3.1 Concepts with which people correlate the term special relativity

1. Theory modeling ...
    - 1.1. Observations of motion of objects, including motion influenced by electromagnetic fields.
    - 1.2. Aspects of electromagnetism.
2. Mathematical symmetries people associate with theory to which the previous item alludes.
3. Instances of those symmetries (to which the immediately previous item alludes) within math that people associate with quantum mechanical models of nature.

People might say that, regarding physics that this monograph correlates with $\sigma = +1$ FRERAN SPATIM, (1;3)> symmetries correlate with the second item in Table 6.3.1. Work above shows models for all $\sigma = +1$ elementary particles and all composite particles correlate with the third item in the table. People might say that these correlations correlate with #b' = 3.

~ ~ ~

This subsection discusses correlations between some aspects of phenomena related to quarks and gluons and special relativity.

People might say that, regarding models that this monograph correlates with $\sigma = -1$, (1;3)> symmetries need not fully pertain.

As far as we know, no observations of quarks or gluons as free particles have been made. When discussing QCD (or, quantum chromodynamics) and the strong interaction, people use the term asymptotic freedom in conjunction with an inferred long-range spatial dependence of $\sim r^0$ for the interaction.

People might say that, such results are not incompatible with COMPAR notions that, for 1Q particles, $\sigma = -1$ pertains and that (0;1)< symmetries pertain. Such results are not incompatible with notions that, for 2Y-related particles, $\sigma = -1$ pertains and that (1;2)< symmetries pertain. For 1Q+2Y, (1;3)< symmetries pertain. For $n \geq 2$ and $n \times$ 1Q+2Y, σ equals +1 and (1;3)> symmetries pertain. Free-ranging particles correlate with (1;3)> symmetries.

~ ~ ~

This subsection notes that some models pertaining to phenomena related to quarks and gluons display aspects that correlate with special relativity.

People might say that much of traditional particle-physics methodology, including methods related to Feynman diagrams, display aspects that correlate with the second and third items in Table 6.3.1. People might say that, for example, the notions correlating with propagators, correlate with the second and third items in Table 6.3.1. People might say that such work includes work pertaining to quarks and gluons. People might say that people have gained much by using these methods. People might say that math related to these methods can be complicated or delicate.

People might say that modeling the previous paragraph discusses correlates with FRERAN modeling techniques.

~ ~ ~

This subsection points to possible opportunities for research regarding theory and/or models.

Table 6.3.2 pertains.

Table 6.3.2 Possible research topics regarding models that correlate COMPAR models

1. To what extents (regarding particles and/or fields) does each of the following operator equations pertain for COMPAR models for which $\sigma = -1$?
$$E^2 - c^2P^2 = \sigma(m')^2c^4 \qquad (6.16)$$
$$E^2 - c^2P^2 = (m')^2c^4 \qquad (6.17)$$
2. What operator-based equations and/or other mathematics correlate with each of the various COMPAR (a`;b`)< symmetries?
   - 2.1. To what extent might people benefit by using such symmetries, equations, or other mathematics?
   - 2.2. To what extent might techniques based on COMPAR models lead to ...
     - 2.2.1. Results that are (compared to traditional results) more robust?
     - 2.2.2. Methods that are (compared to traditional methods) simpler?
3. What operator-based equations and/or other mathematics correlate with the following expression?
$$(1;3)< + (1;3)< = (1;3)> \qquad (6.18)$$

Section 6.4 General relativity

Section 6.4 discusses relationships between aspects of curvature inherent in uses of space-time coordinates and various models, including models this monograph discusses and including models based on general relativity. We discuss the extent to which the universe may exhibit large-scale flatness. We discuss the topic of #b" for G-family bosons. We discuss the concept that ENS48 models may not be compatible with notions that elementary particles and small mass objects traverse geodesics in a space-time having non-zero curvature. People might say that some work in this section is speculative.

~ ~ ~

This subsection discusses aspects of general relativity.

People use general relativity to model astrophysical and cosmological aspects of nature. People interpret cosmology-scale data based on models based on general relativity. We discuss elsewhere in this monograph various applications of general relativity. (See Section 4.6, Section 4.8, and Section 4.9.) Other applications exist.

People might say that coordinate systems used with general relativity correlate with mathematical notions of non-zero curvature. Such curvature would be independent of choices of coordinate systems. People assume that such mathematical non-zero curvature characterizes aspects of space-time.

People assume that trajectories of low rest energy objects correlate with geodesics of space-time.

People have had difficulties in trying to develop theories that unify quantum models with general relativity.

~ ~ ~

This subsection provides perspective we think is relevant to this section.

Table 6.4.1 provides perspective about discussions, in this monograph, related to general relativity.

Table 6.4.1 Perspective about discussions, in this monograph, related to general relativity

1. People might say that this monograph might not need to include a section regarding general relativity.
    - 1.1. Perhaps, Table 6.1.5 points to adequate separation between general relativity models for large things (such as the universe) and relevant models for small things (such as elementary particles).
    - 1.2. Perhaps, Table 6.1.5 and Table 6.1.6 point to adequate decoupling between models related to #b" and models related to "b'.
    - 1.3. Perhaps, we offer a basis for modeling (via techniques other than ones people correlate with general relativity) phenomena people model via general relativity. (See, for example, Section 4.6.)
2. Nevertheless, we think it appropriate to try to address possible correlations and/or lack of correlation between models based on general relativity and MM1MS1 models.
3. People might say that some aspects of this section exhibit speculation.
4. We discuss possibilities regarding large-scale curvature (or flatness) for the universe.
5. We discuss the possibility that ENS48 models are not compatible with concepts of geodesic motion that correlate with applications of general relativity. (See Table 6.4.6.)
6. We discuss attempting to merge at least four sets of concepts - general relativity, quantum gravity, models pertaining to gravitons, and aspects regarding ENS48 models, ENS6 models, or ENS1 models.
7. We discuss QE-like coordinates and QP-like coordinates possibly related to a general-relativistic approach to modeling.

8. People might say that aspects of this section leave untouched work in other sections regarding quantum gravitation. (See, for example, Section 3.3.)

~ ~ ~

This subsection provides perspective about the topic of curvature correlating with coordinate systems and the topic of curvature of space-time.

Applications of special relativity use coordinate systems and metrics for which people say that there is no curvature. For linear coordinates, people call an often used metric the Minkowski metric.

People correlate mathematical expressions with the possible existence of space-time and with possible geometric interpretations of space-time.

For applications of general relativity, the Minkowski metric does not apply (except, perhaps, approximately in some cases). Applications of general relativity use coordinate systems and metrics that people say correlate with the notion that curvature of space-time is non-zero.

As interpreted via models based on general relativity, observations may not be incompatible with an interpretation that, on large scales, space-time has zero curvature. (See Table 6.4.2 and Table 6.4.3.)

People might also consider the extent to which various models correlate notions of the rate of expansion of the universe with notions of curvature of space-time.

~ ~ ~

This subsection discusses types of tests that might have possibilities to indicate that the universe is not flat.

Table 6.4.2 lists possibilities that would indicate against the universe being flat.

Table 6.4.2 Types of results that might indicate against the universe being flat

1. Observations provide data not compatible with flatness.
2. A combination, of a theory and calculations based on that theory, that both ...
   - 2.1. Correlates with data.
   - 2.2. Depends on assumptions correlating with the universe not being flat.
3. Other, to be determined.

~ ~ ~

This subsection reviews data regarding large-scale curvature of the universe.

Table 6.4.3 pertains. Ω_0 is not related to the symbol Ω this monograph uses. People might say that this data correlates with interpretations made based on traditional theory and models. People might say that this data does not correlate with the first item in Table 6.4.2.

Table 6.4.3 Observations pertaining to large-scale curvature of the universe

1. People discuss the extent to which to consider that, on large scales, the universe exhibits ...
   1.1. Positive curvature (or, is spherical or has $\Omega_0 > 1$).
   1.2. No curvature (or, is flat or has $\Omega_0 = 1$).
   1.3. Negative curvature (or, is hyperbolic or has $\Omega_0 < 1$).
2. Observations indicate that $\Omega_0 \approx 1$ may pertain. (See the Ω_{tot} item in Table 8.1.1. Here, $\Omega_0 = \Omega_{tot}$. See, also, NASA (2014).)

~ ~ ~

This subsection suggests that interactions may have always correlated with near or actual large-scale flatness.

Table 6.4.4 pertains. (See Section 4.2.)

Table 6.4.4 Cosmology that may help indicate that flat coordinates can pertain regarding adequately large-scale phenomena

1. The stuff that currently comprises the largest objects people can observe ...
   1.1. Currently exhibits behavior governed by era FE3 (or, r^{-4}) interactions.
   1.2. Previously was governed by era FE2 (or, r^{-6}) interactions.
   1.3. And yet earlier, was governed by era FE1 (or, r^{-8}) interactions.
2. To the extent era FE3 behavior correlates with flatness, it may be likely that behavior during eras FE1 and FE2 correlated with flatness.

The G-family boson that dominates era FE1 behavior is 4G2468& and is a spin-2 boson. Era FE2 behavior would be dominated by at least one of the 4G268& boson and the 2G468& boson. 4G268& has S = 2. 2G468& has S = 1.

People might say that such notions may support concepts that the existence of spin-2 G-family bosons need not necessarily correlate with non-zero curvature of the universe. People might consider extending this thinking to pertain to era FE4 behavior and, thus, to pertain to gravity.

~ ~ ~

This subsection notes that $\Omega_0 = 1$ need not necessarily correlate with a concept of a critical density.

People associate the notion of a critical density of the universe with $\Omega_0 = 1$. People might say that, at least to the extent (1;3)> symmetries are global symmetries for all G-family elementary particles other than 4G4& and 2G2& (or, #b" = 3 for all G-family elementary particles other than 4G4& and 2G2&), work in this monograph suggests that $\Omega_0 = 1$ and that Ω_0 need not depend on the density of the universe.

~ ~ ~

This subsection discusses assumptions we make regarding values of #b".
Table 6.4.5 pertains regarding the remainder of this section.

Table 6.4.5 Assumptions regarding values of #b"

1. Values of #b" for other than G-family elementary particles are not necessarily relevant.
   1.1. Individual particles are local.
2. For each G-family elementary particle other than 4G4& the next equation pertains.

   #b" = 3 (6.19)

   2.1. Here, we assume that #b" = 3 for each G-family boson for we correlate with eras FE1, FE2, or FE3.
   2.2. Here, we assume that #b" = 3 for 2G2&.
      2.2.1. Regarding large-scale effects, we correlate 2G2& with era FE4.
      2.2.2. People might say that precision estimates, related to the fine-structure constant, based on traditional theories might support this assumption.
      2.2.3. People might say that traditional treatments, related to photonics, based on general relativity do not contradict this assumption.
3. Thus, we assume that the following equation pertains for, at most, models and phenomena correlating with 4G4&.

   #b" = 2 (6.20)

~ ~ ~

This subsection discusses possible correlations or lack of correlations between applications of general relativity and each of ENS48 models, ENS6 models, and ENS1 models.

Applications of general relativity correlate concepts regarding mass-energy, gravity, and curvature of space-time. As far as we know, applications of general relativity depend on uses (or implied uses) of space-time coordinates.

Table 6.4.6 pertains.

Table 6.4.6 Possible correlations or lack of correlations between applications of general relativity and each of ENS48 models, ENS6 models, and ENS1 models

1. Regarding ENS48 models, ...
   1.1. The models feature eight instances of gravity, with each instance correlating with different stuff.
   1.2. Consider a thought experiment.

- 1.2.1. Suppose that a model accommodates each of the following.
  - 1.2.1.1. Eight gravitationally independent instances of stuff and gravity.
  - 1.2.1.2. Linkages (based on general relativity) between space-time coordinates and gravity.
  - 1.2.1.3. Linkages (based on general relativity) between non-zero space-time curvature and gravity.
  - 1.2.1.4. Traditional interpretations that motions of zero-mass and small-mass objects follow geodesics.
- 1.2.2. Consider two scenarios.
  - 1.2.2.1. In the first scenario, the universe features a clump of stuff from the OME and the trajectory of the clump in space-time.
  - 1.2.2.2. In the second scenario, the universe features a similar clump of stuff made of one DEE and the same trajectory.
- 1.2.3. Presumably, geodesics for the second scenario match geodesics for the first scenario.
- 1.2.4. Consider the path and characteristics of an OME photon that travels near the clump.
  - 1.2.4.1. In the first scenario, gravity (associated with the OME clump) should affect the trajectory of the photon and should provide for blueshifting and/or redshifting of the photon.
  - 1.2.4.2. In the second scenario, gravity (associated with the DEE clump) should not affect the trajectory of the photon and should not provide for blueshifting and/or redshifting of the photon.

1.3. People might say that physics models cannot accommodate the entirety of the following list.

- 1.3.1. Eight gravitationally independent instances of stuff and gravity.
- 1.3.2. Linkages (based on general relativity) between space-time coordinates and gravity.
- 1.3.3. Linkages (based on general relativity) between non-zero space-time curvature and gravity.
- 1.3.4. Traditional interpretations that motions of zero-mass and small-mass objects follow geodesics.

1.4. People might say that, for ENS48 models, ...

- 1.4.1. Models (based on general relativity and geodesic motion of small-mass objects) may not be able to link non-zero space-time curvature and gravity.

2. People might say that, for ENS48" models, the following pertain regarding attempting to correlate space-time coordinates with properties of stuff.
    2.1. For a model for which space-time coordinates correlate with non-zero curvature, ...
        2.1.1. To the extent curvature correlates with trajectories, ...
            2.1.1.1. Any interaction that can correlate with non-zero curvature must link all ensembles.
            2.1.1.2. The only interaction that can correlate with non-zero curvature is 4G2468&. (Note that we are discussing ENS48" models and not necessarily other ENS48 models.)
                2.1.1.2.1. All stuff is relevant.
                2.1.1.2.2. For 4G2468&, #b" = 2.
            2.1.1.3. For 4G4&, ...
                2.1.1.3.1. #b" = 3.
    2.2. For models for which, for 4G4&, #b" = 3, ...
        2.2.1. Possibly, people can reproduce useful results people obtain via general relativity ...
            2.2.1.1. By using models not based on general relativity.
3. People might say that, for ENS48 models, for applying traditional general-relativistic modeling (for which #b" = 2 for 4G4&), ...
    3.1. Small-mass objects follow geodesics ...
        3.1.1. Related to a model that ...
            3.1.1.1. Correlates coordinate-related curvature with one instance of 4G4&.
            3.1.1.2. Possibly compensates (for example, via the cosmological constant) regarding other instances of 4G4&.
            3.1.1.3. Possibly deemphasizes effects of G-family bosons other than 4G4& and 2G2&.
        3.1.2. Not fully related to a model correlating with concepts of curvature for an overall space-time.
    3.2. For small-mass objects, conservation of RESENE does not necessarily pertain.
3. Regarding ENS6 models and ENS1 models, ...
    3.1. The models feature one instance of gravity.
    3.2. People might say that, based on this concept alone, ...
        3.2.1. Models (based on general relativity and geodesic motion of small-mass objects) may be able to link non-zero space-time curvature and gravity.

~ ~ ~

This subsection discusses the notion that MM1MS1 models may correlate with some aspects of attempting to address non-zero curvature.

Table 6.4.7 provides an analog to Table 3.2.6 and adds information from Table 3.2.7. People might say that, here, regarding models that correlate with solutions 022G2& and 244G4& and that can correlate with a Minkowski-like (or. flat with one fold) metric, a model requires no more than 3 QE-like space-time coordinates and requires no more than 5 QP-like space-time coordinates.

Table 6.4.7 QE-like coordinates and QP-like coordinates possibly related to a general-relativistic approach

Term	Energy-momentum-space coordinates	244G4&-specific space-time coordinates	Observer-specific space-time coordinates
QE-like	e2, e1, e0	t2, t1, t0	t
QP-like	p0, p1, p2, p3, p4	x0, x1, x2, x3, x4	x, y, z

Table 6.4.8 provides possible opportunities for research.

Table 6.4.8 Possibly useful questions, correlating with general relativity regarding representations for energy-momentum space and for curvature people might correlate with space-time

1. To what extent might models need only 3, not 5, QP-like coordinates of the form pj?
   1.1. Here, we assume 0 ≤ j ≤ 2 pertains.
      1.1.1. Possibly, Table 6.4.9 pertains.
   1.2. People might say that p0, p1, and p2 suffice for each G-family elementary boson other than 4G4&.
      1.2.1. People might say that, for example, for 2G24&, more than 3 QP-like axes would correlate with a calculation of S = 1 not necessarily working. (See Table 3.3.5 and remarks before that table.)
   1.3. Perhaps, having just three such coordinates correlates with having exactly 3 QP-like observer-specific space-time coordinates.
2. To what extent might models need only 1, not 3, QE-like coordinates of the form ej?
   2.1. Here, we assume j = 0 pertains.
   2.2. Perhaps, having just one such coordinate correlates with having exactly 1 QE-like observer-specific space-time coordinate.
3. To what extent would (in a scenario involving just e0, p0, p1, and p2, with possibly mutually orthogonal p0, p1, and p2) off-diagonal elements in an energy-momentum space metric correlate with off-diagonal elements in a stress-energy tensor in a model based on general relativity?

Table 6.4.9 augments Table 6.4.8.

Table 6.4.9 Correlations between some spin-2 G-family effects and energy-momentum-space coordinates

1. The two P4j oscillators need not correlate (ultimately) with additional QP-like space-time coordinates.
   1.1. Models deal with circular polarization.
   1.2. People can choose coordinates such that ...
      1.2.1. The coordinate p1 associates with oscillators P4L and P4R via $(244G4L + 244G4R)/2^{1/2}$.
         1.2.1.1. See Table 3.2.5 and Table 3.2.9.
      1.2.2. The coordinate p2 associates with oscillators P4L and P4R via $(244G4L - 244G4R)/(\{-i\}2^{1/2})$.
         1.2.2.1. See Table 3.2.5 and Table 3.2.9.

~ ~ ~

This subsection discusses thoughts about uses of and interpretation of general relativity.

Above, we offer possibilities that people can reproduce results (that people obtain via models based on general relativity) without using general-relativity math-based models. Here, we provide other possibly useful concepts.

People might say that, assuming flatness pertains to the universe and to some useful models of the universe, general relativity provides, for some phenomena, a useful alternative model. General relativity models feature a prominent role for a stress-energy tensor. That role generally implies needs to use coordinate systems for which non-zero curvature pertains.

In other areas of physics, people derive meaningful results based on stress-energy concepts related to solids and liquids. Solids and liquids feature collective and statistical phenomena. Independent of the extent to which people can (and do) actually supplant general-relativistic models via use of models that correlate with zero-curvature, people might want to explore correlations between general-relativistic models and the statistical-physics train of thought we suggest in Section 2.12.

~ ~ ~

Reference 17 NASA (2014)

NASA, http://map.gsfc.nasa.gov/universe/uni_shape.html (2014)

Chapter 7 From the MM1 meta-model to perspective

Chapter 7 (From the MM1 meta-model to perspective) discusses the extent to which models in this monograph might provide components for broader understanding of nature. We show results from applying the models. We suggest possible opportunities for research regarding and applications of models. We suggest possible applications regarding results from models.

~ ~ ~

People might say that results in Chapter 7 (From the MM1 meta-model to perspective) might correlate with or provide insight about the following aspects of physics. Various aspects of physics, assuming specific choices of models that may close gaps between known data and traditional theory. Various aspects of physics, assuming a specific choice of models that may correlate with traditional theory via which people interpret data. A table of families of known and possible elementary particles and of some properties correlating with those particles. Possible opportunities for research, centric to aspects of the MM1 meta-model and MM1MS1 models that correlate with the MM1 meta-model. General opportunities, regarding modeling, for research and applications. General opportunities, regarding physics, for research and applications.

Section 7.1 The MM1 meta-model and ENS48, ENS6, and ENS1 models

Section 7.1 compares aspects of ENS48 models, ENS6 models, and ENS1 models. People might say that ENS48 models and ENS6 models possibly correlate better with some data about dark matter than do ENS1 models. People might say that ENS6 models and ENS1 models possibly correlate better with some aspects of general relativity that do ENS48 models.

~ ~ ~

This subsection provides perspective.

The MM1 meta-model features one set of solutions that may correlate with known and candidate elementary particles. Across MM1MS1 models, roles of and needs for an individual particle can vary. (See, for example, Section 4.3.)

T.J. Buckholtz, *Models for Physics of the Very Small and Very Large*,
Atlantis Studies in Mathematics for Engineering and Science 14,
DOI 10.2991/978-94-6239-166-6_7

This monograph discusses three differing sets of models for elementary particles. We term one set ENS48 models. We term one set ENS6 models. We term the other set ENS1 models. A difference between the three sets correlates with interpretations of the instance-related symmetry we call INSSYM7.

People might say that each of some of the differences between ENS48 models, ENS6 models, and ENS1 models correlates (for the purposes of work in this monograph) with some aspects of dark matter and dark-energy stuff. People might say that some such differences do not necessarily (for the purposes of work in this monograph) correlate with each of some aspects of elementary-particle physics. People might say that some such differences do not necessarily (for the purposes of work in this monograph) correlate with each of some aspects of cosmology and astrophysics. (See Table 4.1.1.)

~ ~ ~

This subsection compares aspects of ENS48 models, ENS6 models, and ENS1 models.

Possibly, any of the three sets of models suffices to correlate with known physics phenomena. Some differences pertain. Table 7.1.1 pertains.

Table 7.1.1 Some similarities and differences regarding ENS48 models, ENS6 models, and ENS1 models

1. Regarding possible elementary particles, ...
    1.1. One catalog of possible elementary particles pertains to ENS48 models, to ENS6 models, and to ENS1 models.
        1.1.1. The catalog correlates with the MM1 meta-model. (See, for example, Table 2.4.1.)
2. Regarding the composition of dark matter, ...
    2.1. ENS48 and ENS6 models may not necessarily require the existence of 3N or 3QIRD (and 4YO) particles.
    2.2. ENS1 models likely correlate with existence of at least some of the possible examples of SEDMS that Table 4.3.14 lists.
3. Regarding the composition of dark-energy stuff, ...
    3.1. ENS48 models may not necessarily require the existence of 3N or 3QIRD (and 4YO) particles.
    3.2. ENS6 models and ENS1 models likely correlate with existence of at least some of the possible examples of SEDES that Table 4.3.13 lists.
4. Possibly, if ENS48 models to pertain to nature, the only required elementary particles with $S > 1$ are G-family particles.
5. Regarding inferred ratios of density of DMS to density of OMS, ...
    5.1. ENS48 models and ENS6 models provide an estimate for the ratios.
    5.2. ENS1 models do not provide estimates for the ratios.
6. Regarding inferred ratios of density of DES to density of OMDMS, ...

6.1. ENS48 models provide an estimate for an upper bound (over time) on the ratios.
   6.1.1. Such an upper bound might not be approached within a few billion years from now.
6.2. ENS6 models and ENS1 models do not provide estimates for an upper bound (over time) on the ratios.

~ ~ ~

This subsection speculates about comparisons between ENS48 models, ENS6 models, and ENS1 models.

Table 7.1.2 pertains.

Table 7.1.2 Possible perspective regarding applications of ENS48 models, ENS6 models, and ENS1 models

1. People might say that current physics data is not incompatible with ENS48 models, ENS6 models, or ENS1 models.
2. People might say that ENS48 models and ENS6 models correlate with results that are (compared to ENS1 models) ...
   2.1. More numerous.
   2.2. More definitive.
3. People might say that current interpretations of physics data are not sufficient to determine ...
   3.1. The overall usefulness of ENS48 models, ENS6 models, and ENS1 models.
   3.2. The relative usefulness of ENS48 models, ENS6 models, and ENS1 models.
4. People might say that ENS6 models and ENS1 models possibly correlate better with some aspects of general relativity that do ENS48 models.

Section 7.2 Bases for tables of elementary particles

Section 7.2 discusses characteristics regarding the periodic table for elements and characteristics that this monograph discusses regarding elementary particles. We show possible bases for analogs, for elementary particles, to the periodic table for elements.

~ ~ ~

This subsection provides perspective about this section.

This section provides perspective regarding possibly desired attributes for a table for elementary particles.

~ ~ ~

This subsection notes features of the periodic table for elements.

Table 7.2.1 pertains.

Table 7.2.1 Characteristics of the periodic table for elements

1. The periodic table for elements provides a 2-dimensional array in which ...
    1.1. To the extent one knows little about the components of atoms, ...
        1.1.1. One dimension correlates with chemical interactions.
        1.1.2. One dimension correlates with atomic weight.
    1.2. To the extent one knows enough about the components of atoms, ...
        1.2.1. Atomic number (number of protons in nuclei, or, equivalently, number of electrons in electrically charge-neutral atoms) provides a basis for structuring the table.
        1.2.2. The quantum mechanics of electron orbitals (for example, those orbitals that underlie properties of chemical reactions) provides a basis for structuring the table.
        1.2.3. Measured atomic weights correlate with ...
            1.2.3.1. Physically possible isotopes (and nuclear physics).
            1.2.3.2. Circumstances under which atomic nuclei are created.
            1.2.3.3. Effects of radioactive properties of atomic nuclei.
2. After the time such periodic tables were first proposed (There were and remain alternative arrays for the same data.) and used, ...
    2.1. Some positions in the tables correlated with possibilities for discovering (in nature) or creating new elements.
    2.2. Some positions in the table correlated with possibilities for discovering (in nature) or creating new isotopes for known elements.

~ ~ ~

This subsection notes features of possible arrays for displaying known elementary particles and possible elementary particles.

Table 7.2.2 pertains.

Table 7.2.2 Characteristics that tables might feature regarding known and possible elementary particles

1. Each of the following provides a possible index for a table of known and possible elementary particles.
    1.1. σ.
        1.1.1. Is the value −1 or +1?
    1.2. S.
        1.2.1. Is the value 0, 1/2, 1, 3/2, or 2?
    1.3. m'.
        1.3.1. Is the value zero or non-zero?
        1.3.2. For non-zero values, what are the magnitudes or relative magnitudes?
    1.4. Families and subfamilies of particles.
        1.4.1. Which families and/or subfamilies should one list?
        1.4.2. Based on what order, array, or other paradigm should one structure a display of families, subfamilies, or particles?
            1.4.2.1. For example, to what extent might one distinguish particles for which #'E = 'Ø (that is, leptons) from particles for which #'E is a nonnegative integer?
    1.5. Number of instances.
        1.5.1. Assuming some particles exhibit more than one instance (as in ENS48 models and ENS6 models), ...
            1.5.1.1. For a given particle, how many instances exist?
            1.5.1.2. Or, for a given number of instances, which particles pertain?
    1.6. SPATIM symmetries.
        1.6.1. For example, ...
            1.6.1.1. Group some elementary particles based on COMPAR symmetries.
    1.7. Interactions.
        1.7.1. For example, for elementary fermions, ...
            1.7.1.1. With which W- and 0-family bosons does each elementary fermion interact?
    1.8. Experimental status, for specific particles.
        1.8.1. Known?
        1.8.2. Being sought?
        1.8.3. Planned for being sought?
        1.8.4. Other status?
    1.9. Perceived needs to find (or rule out) specific particles.
        1.9.1. For example, which theories depend on which particles?
    1.10. Directness of correlations between what people consider to be particles and MM1MS1 solutions.
        1.10.1. For example, ...

1.10.1.1. For solutions correlating with particles (other than gluons) presently in the Standard Model, there are one-to-one correlations between solutions and particles.

1.10.1.2. For 2Y solutions, pairs of solutions (with one element in a pair representing erasing a color charge and with one element in the pair representing painting a color charge) correlate with components of gluons.

1.10.1.3. For 4Y solutions, something similar (to the statement about 2Y solutions) pertains regarding spin-2 analogs of gluons.

1.11. Characteristics particular to some types of particles.

1.11.1. For elementary fermions, ...

1.11.1.1. Number of generations.

1.11.1.2. Number of color charges.

1.11.2. For elementary bosons, ...

1.11.2.1. Range or spatial dependence of interactions associated with the particle.

1.11.2.2. The extents of conservation or lack of conservation, in interactions, of fermion generation.

1.11.3. For G-family elementary bosons, ...

1.11.3.1. Type of coupling to fermion generation.

1.11.3.2. Oscillator swap symmetries. (See Table 2.12.10.)

1.12. Directness of correlations between #P and 2S.

1.12.1. For all fermion elementary particles, 2S = #P − 1.

1.12.2. For all boson elementary particles except G-family particles, 2S = #P.

1.12.3. For G-family elementary particles 2S does not necessarily correlate directly with #P.

1.13. The extent to which instances of the expression D + 2ν pertain in models correlating with particles.

1.14. Other. (See, for example, Table 7.3.4.)

Section 7.3 Tables showing elementary particles and related concepts

Section 7.3 discusses a table of subfamilies of elementary particles and related instances and symmetries. We show a table of all known and some possible elementary particles. We include, in the table, various properties of the particles. We note some intricacies that people might find useful regarding constructing other tables of elementary particles.

~ ~ ~

This subsection provides perspective about this section.

This section features a table for known and possible subfamilies of elementary particles. (See Table 7.3.2.) People might develop other tables.

People might say that Table 7.3.1 exemplifies, regarding Table 7.3.2, notions that pertain to the MM1-meta model (and, therefore, to all MM1MS1 models) and notions that pertain to some but not all MM1MS1 models.

Table 7.3.1 Some contrasts between aspects of the MM1 meta-model and aspects of MM1MS1 models

1. People might say that the rows in Table 7.3.2 correlate with the MM1 meta-model.
    1.1. We note that the table omits rows that would correlate with G-family possible %68odd particles.
2. People might say that each of the following columns in Table 7.3.2 correlates with the MM1 meta-model.
    2.1. The σ column.
    2.2. The S column.
    2.3. The m' column.
3. People might say that aspects of entities such as each of the following (in Table 7.3.2) correlate with MM1MS1 models.
    3.1. The GEN column.
    3.2. The Cg column.
    3.3. The ENS48" INST column.
    3.4. The COMPAR column.

~ ~ ~

This subsection provides a table of known and possible subfamilies of elementary particles.

Table 7.3.2 pertains. We limit the table to show, for the G-family, just the %68even set of particles. The table shows particles for the H- and G-families. The table shows subfamilies for the other families. The 0O-subfamily includes just one particle. Regarding known examples, the table abbreviates charged leptons as ch leptons. The table abbreviates Higgs boson as Higgs b. The table correlates MM1MS1-neutrino with neutrino. The table correlates MM1MS1-photon with photon. The table abbreviates number of generations as GEN. The concept of generations (or, for MM1MS1-neutrinos, flavors) pertains for elementary fermions only. The table assumes #GEN(3/2) = 15. (See Table 2.6.4.) The table abbreviates spatial dependence of interaction as SDI. The table shows spatial dependences of interactions for G-family particles only. For the G-family, the table abbreviates couples to generations as cg. The cg column pertains for interactions between two elementary fermions. Here, a 0 denotes that the interaction does not differentiate by generation of elementary fermion. A 1 denotes that the interaction differentiates by generation for spin-1/2 elementary fermions and does not differentiate by generation for spin-3/2 elementary fermions. A 2 denotes that the interaction differentiates by generation for spin-3/2 elementary fermions and does not differentiate by generation for spin-1/2 elementary fermions. (See Table 2.13.12, Table 3.3.3, and Table 3.3.4.) Table 7.3.2 abbreviates number of solutions that correlate with particles as sol. Here, we use the traditional-physics convention of counting 1 for each particle-and-antiparticle pair. (For the G-family, we use traditional-physics counting. We count 1 for each mode-and-anti-mode pair. See Table 3.1.3.) The table abbreviates particles per instance as PPI. Again, we use the convention of counting 1 for each particle-and-antiparticle pair. The table abbreviates instances (of subfamilies or particles) correlating with ENS48" models as ENS48" INST. For other ENS48 models, the number of instances for 2G24& and/or 4G2468& might differ from results the table shows. For ENS6 models, the number of instances (for a subfamily or particle) would be 6 for each non-G-family row and would not exceed 6 for each G-family row. For ENS1 models, the number of instances would, for each subfamily or particle, be 1. For QIRD- and YO-subfamilies, the COMPAR column shows COMPAR symmetries (assuming a` = 0 pertains for the QIRD-subfamilies).

Table 7.3.2 Subfamilies of possible elementary particles (assuming the %68odd set of possible G-family elementary particles does not pertain) and characteristics for those subfamilies

σ	S	m'	Subfamily or particle	Known examples	GEN	SDI	Cg	Sol	PPI	ENS48" INST	COMPAR
+1	1/2	≠ 0	1C	ch leptons	3			3	3	48	
+1	1/2	0	1N	neutrinos	3			3	3	48	
+1	3/2	0	3N		15			15	15	48	
+1	0	≠ 0	0H0	Higgs b.				1	1	48	
+1	1	≠ 0	2W	W, Z				2	2	48	
+1	1	0	2G2&	photon		r^{-2}	0	1	1	48	

σ	S	m'	Subfamily or particle	Known examples	GEN	SDI	Cg	Sol	PPI	ENS48" INST	COMPAR
+1	2	0	4G4&			r^{-2}	1	1	1	8	
+1	1	0	2G24&			r^{-4}	0	1	1	24	
+1	1	0	2G68&			r^{-4}	2	1	1	24	
+1	2	0	4G268&			r^{-6}	1	1	1	8	
+1	1	0	2G468&			r^{-6}	1	1	1	8	
+1	2	0	4G2468&			r^{-8}	0	1	1	1	
−1	0	≠ 0	0O0					1	1	48	
−1	1/2	≠ 0	1Q	quarks	3			6	6	48	(1;3)<
−1	3/2	≠ 0	3Q		15			30	30	48	(0;2)<
−1	1	0	2Y	gluons				3	8	48	(1;2)<
−1	2	0	4Y					10	24	48	(1;1)<
−1	1	≠ 0	2O					2	2	48	(1;2)<
−1	2	≠ 0	4O					3	3	48	(1;1)<
−1	1/2	0	1R		3			6	6	48	(0;1)<
−1	3/2	≠ 0	3I		15			30	30	48	(0;2)<
−1	3/2	0	3R		15			30	30	48	(0;2)<
−1	3/2	0	3D		15			30	30	48	(0;2)<

Table 7.3.3 summarizes (as of 2015) known elementary particles and possible (per MM1MS1 models) elementary particles.

Table 7.3.3 Known and possible elementary particles (assuming the %68odd set of possible G-family elementary particles does not pertain)

1. Each elementary particle correlating with a subfamily or particle for which Table 7.3.2 lists a known example has been discovered.
   1.1. The above statement ...
      1.1.1. Pertains for the ordinary-matter instance of the particle.
      1.1.2. Does not pertain for possible non-ordinary-matter instances of the particle.
         1.1.2.1. As of 2015, no non-ordinary-matter instances had been found.
2. Each candidate elementary particle correlating with a subfamily or particle for which Table 7.3.2 does not list a known example has not been discovered.

~ ~ ~

This subsection discusses some intricacies that might prove useful for catalyzing further work or for developing concepts for other tables of known and possible elementary particles.

MM1MS1 models underlying Section 2.2 are somewhat straightforward. Table 2.1.1 lists some types of physics observations that yield discreet results. The list seems tractable. But, Chapter 2 and Chapter 3 are intricate. And, in general, nature seems complicated.

Table 7.3.4 lists aspects of intricacy regarding MM1MS1 models.

Table 7.3.4 Aspects of intricacy regarding MM1MS1 models

1. SPATIM symmetries.
    1.1. The W- and O-families vary from other families, based on ...
        1.1.1. The W-family's lacking one QP-like reflection symmetry.
            1.1.1.1. This correlates with traditional physics.
        1.1.2. The possibilities that ...
            1.1.2.1. The 2O-subfamily lacks one QE-like reflection symmetry.
            1.1.2.2. The 4O-subfamily lacks one QE-like reflection symmetry.
2. COMPAR symmetries.
    2.1. All $\sigma = +1$ solutions that correlate with particles have (1;3)> SPATIM symmetries, while $\sigma = -1$ solutions that correlate with particles ...
        2.1.1. Have (a`;b`)< COMPAR symmetries.
        2.1.2. Do not (except for 0O0) have (1;3)< COMPAR symmetries.
3. G- and Y-family solutions.
    3.1. For all but the G- and Y-families, a solution that might correlate with an elementary particle would correlate with exactly one of ...
        3.1.1. One matter elementary particle.
        3.1.2. One antimatter elementary particle.
        3.1.3. One elementary particle that is its own antiparticle.
        3.1.4. Zero elementary particles.
    3.2. For the G-family, modes pertain.
    3.3. For the G-family, for models that feature the SOMMUL set, each of some particles correlates with more than one solution.
    3.4. Regarding the Y-family, each elementary particle (such as a gluon) does not correlate directly with an individual solution.
4. Generations.
    4.1. The concept of generations pertains only to elementary fermions.
5. Instances, in ENS48 models.
    5.1. The only family with particles having other than 48 instances is the G-family.
6. Instances, in ENS6 models.
    6.1. The only family with particles having other than 6 instances is the G-family.

Section 7.4 Necessity and sufficiency of some particles and models

Section 7.4 discusses the extent to which particles and MM1MS1 models this monograph discusses are necessary and/or sufficient to match known data about nature and/or to match some people's conjectures about nature. We list types of known data and conjectured phenomena. We provide, for each listed item, comments based on work in this monograph. We discuss sets of elementary particles that may be necessary and/or sufficient to correlate with all known data.

~ ~ ~

This subsection provides perspective about this section.

Regarding a theory or model, people might want estimates regarding various aspects. How much of the theory or model likely pertains? How well does the theory or model likely pertain?

This section discusses, for some aspects of nature, some such questions.

~ ~ ~

This subsection discusses the extent to which particles and models this monograph discusses are necessary and/or sufficient to match known data about nature and/or to match some people's conjectures about nature.

Table 7.4.1 pertains.

Table 7.4.1 Aspects of and conjectures about nature, and discussion about models this monograph discusses

0.	Aspect of or conjecture about nature. (The symbol * denotes conjecture.) 0.1. Work in this monograph and/or other concepts.
1.	Known elementary particles (the Higgs boson, charged leptons, neutrinos, quarks, weak-interaction bosons, gluons, and the photon) exist. 1.1. The 0H0 particle; 1C-, 1N-, and 1Q-subfamilies particles; W-family particles; 2Y-related particles; and the 2G2&-and-(single-ensemble aspects of ...)-2G24& combination particle (or, photon) correlate with the known elementary particles.
2.	* People anticipate that other elementary particles exist. 2.1. This monograph discusses possible not-yet-found elementary particles.
3.	Composite particles exist. 3.1. The n × 1Q+2Y constructs correlate with known (n = 2 and n = 3) particles.

0. Aspect of or conjecture about nature. (The symbol * denotes conjecture.)
    - 0.1. Work in this monograph and/or other concepts.

- 3.2. The n × 1Q+2Y constructs correlate with n = 4 and n = 5 particles.
    - 3.2.1. People may be on the verge of declaring that people have detected tetraquarks and pentaquarks.

4. Some elementary particles have non-zero magnetic dipole moments.
    - 4.1. 2G24& correlates with and transmits information about known elementary-particle magnetic dipole moments.
5. The fine-structure constant pertains to various aspect of physics.
    - 5.1. (See Table 3.12.1.)
6. Gravity exists.
    - 6.1. 4G4& particles mediate the gravitational interaction.
7. The universe expands and the rate of expansion changes.
    - 7.1. The observed rate of expansion pertains to ordinary matter. (* The observed rate of expansion pertains to dark matter.)
    - 7.2. Interactions that correlate with some G-family particles propelled the big bang and govern the rate of expansion of the universe.
8. Dark-energy stuff exists.
    - 8.1. Each of ENS48 models, ENS6 models, and ENS1 models correlates with candidate particles that could be dark-energy particles. (See Table 4.3.13.)
    - 8.2. ENS48 models correlate with existence of eight similar units, one of which encompasses ordinary matter and dark matter. The other seven units are dark-energy stuff.
9. Dark matter exists.
    - 9.1. Each of ENS48 models, ENS6 models, and ENS1 models correlates with particles that could be dark-matter particles. (See Table 4.3.14.)
    - 9.2. Each of ENS48 models and ENS6 models correlates with existence of six similar units, one of which encompasses ordinary matter. The other five units are dark matter.
10. The spatial distribution of cosmic microwave background (CMB) radiation exhibits granularity.
    - 10.1. Some aspects of the granularity may result for large-scale clustering and/or anti-clustering of ordinary matter, of dark matter, and/or of dark-energy stuff. (ENS48 models)
    - 10.2. Some aspects of the granularity may result for large-scale clustering and/or anti-clustering of ordinary matter and/or of dark matter. (ENS6 models)
11. Objects that include ordinary matter and dark matter exist.
    - 11.1. The gravity that pertains to ordinary matter pertains to dark matter.
    - 11.2. G-family-mediated interactions provide possibilities for, before galaxy formation, the clustering of ordinary matter and the clustering of dark matter. (ENS48 models and ENS6 models)

0. Aspect of or conjecture about nature. (The symbol * denotes conjecture.)
   0.1. Work in this monograph and/or other concepts.

   11.3. G-family-mediated interactions provide possibilities for, before and during galaxy formation, clustering and anti-clustering of ordinary matter and/or various of five units of dark matter. (ENS48 models and ENS6 models)
12. * The ratio the amount of charged matter to the amount of charged antimatter may have changed.
   12.1. People do not know - for the very early universe - a numeric ratio of the amount of charged matter to the amount of charged antimatter.
   12.2. People sometimes assume that the ratio was approximately 1 : 1.
   12.3. Assuming the ratio changed over time, ...
      12.3.1. Lasing of 2O-subfamily bosons converted antimatter quarks to matter quarks.
      12.3.2. Concurrently, effects of W-family bosons converted antimatter charged leptons to antimatter MM1MS1-neutrinos and converted matter MM1MS1-neutrinos to matter charged leptons.
13. * Neutrinos may be Dirac fermions or neutrinos may be Majorana fermions.
   13.1. Physics seems unsettled on the topic of the extent to which neutrinos are Dirac fermions (meaning that antimatter neutrinos are particles distinct from matter neutrinos) or Majorana fermions (meaning that each neutrino is its own antiparticle).
   13.2. Models in this monograph correlate with MM1MS1-neutrino interaction vertices for which the property of Dirac fermion pertains and the property of Majorana fermion does not pertain.
14. Neutrino oscillations (or, flavor mixing) occur.
   14.1. Traditional physics theory seems to indicate that neutrino oscillations correlate with at least one flavor of neutrino having non-zero rest mass.
   14.2. Models in this monograph ...
      14.2.1. Correlate with various possible mechanisms for creating MM1MS1-neutrino oscillations. (See Table 3.11.1.)
      14.2.2. Do not require that any flavor of MM1MS1-neutrino has non-zero mass.
15. Galaxies that feature ordinary matter do not shed matter quickly (except perhaps, for example, during collisions with other galaxies).
   15.1. We offer a mechanism (other than effects of gravity correlating with dark matter) that traditional models may have overlooked regarding binding material within a galaxy.
   15.2. The mechanism features attraction associated with parallel motions of stuff.

0. Aspect of or conjecture about nature. (The symbol * denotes conjecture.)
    0.1. Work in this monograph and/or other concepts.

        15.2.1. The effect arises from a gravitational somewhat analog to electromagnetism's magnetic field. (See Table 2.15.10.)
16. Quasars exist and correlate with black holes. Most black holes do not correlate with quasars.
    16.1. Under some circumstances, effects of at least one of 2G24&, 2G68&, and 4G2468& can lead to a black hole's expelling matter.
17. * Sterile neutrinos may exist.
    17.1. 3N-subfamily elementary particles may correlate with some notions of sterile neutrinos and may exist.
18. * Primordial black holes may exist.
    18.1. Clusters of composite particles may have formed primordial black holes.
19. * WIMPs (weakly interacting massive particles) may exist.
    19.1. WIMPs may feature composite particles for which the composites include 0-family particles.
20. * Axions may exist.
    20.1. Possibly, 2O particles correlate with people's concepts for axions.
    20.2. Possibly some G-family particles correlate with people's concepts for axions.
21. The hydrogen atom exhibits spectral lines correlating with various phenomena.
    21.1. Possibly, various G-family particles correlate with aspects of the spectral lines. (This is a speculative possible result. See Section 6.2.)

Table 7.4.2 discusses sets of elementary particles needed to match data about nature.

Table 7.4.2 Conjectures about a smallest set of elementary particles sufficient to correlate with data known to pertain to nature or thought likely to pertain to nature (%68even)

1. Based on present knowledge about elementary particles, ...
    1.1. These elementary-particle subfamilies and these elementary particles are necessary.
        1.1.1. 1C, 1N, 0H, 2W, 2G2&, 4G4&, 1Q, and 2Y-related particles.
    1.2. To correlate work in this monograph with magnetic dipole moments associated with elementary particles, this candidate elementary particle may be necessary.
        1.2.1. 2G24&.
2. To correlate work in this monograph with gravity, this candidate particle needs to exist.
    2.1. 4G4&.

3. To correlate work in this monograph with known data about the rate of expansion of the universe, these candidate G-family particles need to exist.
   3.1. At least one of 2G24& and 2G68&.
   3.2. At least one of 4G268& and 2G468&.
   3.3. 4G2468&.
4. To correlate work in this monograph with known data and the theoretic concept that the ratio of the amount of charged matter to the amount of charged antimatter was, early in the evolution of the universe, closer to one than it is today, these proposed 2O-subfamily particles need to exist.
   4.1. 2O2 and 2O1.
5. To correlate work in this monograph with known data about neutrino oscillations, at least one of these proposed elementary particles needs to exist.
   5.1. 2G68& and 2O0.
6. To the extent ENS48 models (including ENS48") pertain, ...
   6.1.1. With some assumptions, possibly the particles mentioned above in this table suffice.
7. To the extent ENS6 models pertain (and ENS48 models ENS1 models do not pertain), some of these subfamilies of particles likely pertain. (See Table 4.3.13.)
   7.1. 3N.
   7.2. 1R.
   7.3. 4Y and at least one of 3R or 3D.
8. To the extent ENS1 models pertain (and ENS48 models ENS6 models do not pertain), some of these subfamilies of particles likely pertain. (See Table 4.3.13 and Table 4.3.14.)
   8.1. 3N.
   8.2. 1R.
   8.3. 4Y and at least one of 3R or 3D.
   8.4. 4Y and at least one of 3Q or 3I.

Section 7.5 Possible MM1MS1-correlated opportunities for research

Section 7.5 discusses possible opportunities for further research, based on MM1MS1 considerations.

~ ~ ~

This subsection suggests context for considering further exploration of topics this monograph addresses.

This monograph discusses the MM1 meta-model and possibly useful MM1MS1 models for various aspects of elementary-particle, astrophysical, and cosmological physics. Opportunities exist to test, verify, modify, refute, or extend aspects of this monograph. Opportunities exist to blend work this monograph shows and work from traditional physics.

~ ~ ~

This subsection suggests topics for further exploration.

People might want to explore math-related topics and/or modeling-related topics to which Table 7.5.1 alludes.

Table 7.5.1 Possible math-related topics and/or modeling-related topics for further exploration

1. To what extent can people develop a broader range of solutions for a broader interpretation of the term multi-dimensional harmonic oscillator math?
    1.1. Possibly, broader ranges of values (possibly including complex numbers) can pertain for parameters such as D*, Ω, ν, S, and D.
2. To what extent can people develop practical applications of harmonic oscillator math beyond traditional applications and applications this monograph discusses?
3. To what extent can people develop useful math for a broader range of pairs of multi-dimensional harmonic oscillators?
    3.1. Possibly, one or more of the following pertain.
        3.1.1. Each of the sets of QE-like and QP-like oscillators has an even-integer number of members.
        3.1.2. Less restrictive concepts (than concepts this work correlates with the term isotropic) pertain.
        3.1.3. Applications for which Œ can be non-zero pertain.
4. To what other topics can people apply math pertaining to isotropic pairs of isotropic quantum harmonic oscillators?
5. To what extent can people develop useful math for a broader range of sets of multi-dimensional harmonic oscillators?
    5.1. Possibly, more than just the two sets (for which this work uses, respectively, the terms QE-like and QP-like) of oscillator indices pertain.
    5.2. Possibly, less restrictive concepts (than concepts this work correlates with isotropic) pertain.

6. To what extent can people develop simpler bases for and/or descriptions of work in this monograph?
7. To what extent might people benefit from exploring or using the following generalization? (This generalization correlates with Table 2.5.6 and Table 2.5.8. Specifically, work in Table 2.5.6 and Table 2.5.8 correlates with setting k = 2.)

$$\xi\, r^{(k-2)}\, \eta^{-(k-2)}\, \Psi(r) = (\xi'/2)\,(-\eta^2\, \nabla^2 + \eta^{-2(k-1)}\, r^{2(k-1)})\, \Psi(r) \quad (7.1)$$

$$\Psi(r) \propto r^{\nu} \exp(-r^k / (k\eta^k)) \quad (7.2)$$

   7.1. The following algebraic equations pertain.

$$\xi = (D + 2\nu + k - 2)\, (\xi'/2) \quad (7.3)$$

$$\Omega = \nu(\nu + D - 2) \quad (7.4)$$

   7.2. The parameter η does not appear in the two equations.
   7.3. Each of r, η, and k appears to the left of the equal sign in the first equation above. (See (7.1).)
8. To what extent might people benefit from exploring or using the following generalization?

$$\Omega = \pm\, S(S + D - 2) \pm L(L + D - 2) \pm J(J + D - 2) \quad (7.5)$$

   8.1. Here, each of 2S, L, and 2J can be (at least) a nonnegative integer.
   8.2. This generalization can pertain to Ω in, say, either of ...
      8.2.1. (2.39) in Table 2.5.6.
      8.2.2. (7.4) regarding an analog to (2.39) regarding (7.1) in Table 7.5.1.
   8.3. See also, Table 6.2.1.
9. To what extent might people benefit from using various operator-based equations (such as $E\{1\} = m\{1\}c^2$) that Table 6.1.5 shows?

To the extent work in this monograph correlates with nature or with possibly useful models, people might want to explore topics to which Table 7.5.2 alludes. (People might say that overlaps between Table 7.5.2 and Table 7.5.3 exist.)

Table 7.5.2 Possible physics topics and/or physic modeling topics for further exploration

1. To what extent - and how - can people verify or refute possible predictions with which work in this monograph correlates?
2. To what extent can people use work in this monograph to suggest possibly useful new observational approaches to infer or experiments to detect gravitons or other possible elementary particles?
3. To what extent does work in this monograph provide insight regarding nuclear physics, nuclear forces, or neutron stars?
   3.1. For example, to what extent would including 0-family phenomena affect models for such phenomena?

4. To what extent does work in this monograph provide insight regarding QCD (or, quantum chromodynamics) and other physics for which models correlating with $\sigma = -1$ can pertain?
   4.1. For example, to what extent might people correlate various (a`;b`)< symmetries with phenomena, existing theory, and/or new models?
   4.2. For example, to what extent might people benefit by exploring the notion that, for phenomena for which models correlate with $\sigma = -1$, in DIFEQU equations for which Œ = 0, the QP-like term involving Ω correlates with QE-like mathematical aspects and/or QE-like phenomena?
5. To what extent can people extend work in this monograph to correlate with decay rates for elementary particles?
6. To what extent do decay rates (for elementary particles, compound particles, and nuclei) depend on influences of environmental factors such as gravity?
7. To what extent can people use results based on work in this monograph to substitute quantum-based approximations for classical-physics-based approximations in simulations (for example, simulations for which efforts span quantum-physics/classical-physics boundaries)?
8. To what extent can work based on this monograph shed light on the relative appropriateness, for various aspects of physics, of using models based on work in this monograph and models based on various traditional models?
9. To what extent might people benefit by developing, regarding interactions between subsystems, models based on methodology that Table 6.1.5 shows?
10. To what extent can people derive insight about dark matter by exploring interactions intermediated by 2G24& particles?
11. To what extent can people derive insight about dark matter by exploring interactions intermediated by 2G68& particles?
12. To what extent can people derive insight about the 0-family by exploring possible interactions between 2O0 bosons and neutrinos?
13. To what extent can people use directly, hone, or extend work in this monograph to predict masses for 0-family elementary bosons?
14. To what extent can people extend work in this monograph to predict masses for non-zero-mass spin-3/2 elementary fermions?
15. To what extent can people derive insight from this monograph about the topic of elementary-particle magnetic dipole moments?
16. To what extent can people derive insight from this monograph about the topic of magnetic monopoles?
17. To what extent might people gain insight regarding the concept of wave-particle duality, based on work in this monograph?

- 17.1. For example, people might consider implications of wave-like solutions (for which η^2 can be any nonnegative number) and particle-like solutions (for which $\eta^2 \to 0$).
- 17.2. For example, people might consider work regarding interaction vertices and models for sizes of interaction vertices.

18. To what extent can people merge techniques and models in this monograph with traditional physics models and techniques?
19. To what extent can people use results in this monograph to explore possible desirability for emphasizing or deemphasizing traditional physics models or techniques?
    - 19.1. For example, to what extent is supersymmetry incompatible with Table 7.3.2?
    - 19.2. For example, to what extent is string theory incompatible with (2.65) in Table 2.10.4 or with the limit $\eta^2 \to 0$ per (2.47) in Table 2.6.1?
20. To what extent can people use models based on isotropic pairs of isotropic quantum harmonic oscillators to recreate models for traditional action-based physics?
21. To what extent can people use models based on isotropic pairs of isotropic quantum harmonic oscillators to explore limitations regarding using traditional action-based models in physics?
22. To what extent can people describe the population of ...
    - 22.1. MM1MS1-photon states (and other G-family-particle states) for an all-quantum description of the binding of a proton and electron to form a hydrogen atom?
    - 22.2. MM1MS1-photon states (and other G-family-particle states) for an all-quantum description of the interaction between two charged objects, each having zero spin and zero angular momentum?
23. To what extents should people consider $\hbar/2$ to be a basis (paralleling $|q_e|/3$ or m'(Z boson)/3) for one particle property or for more than one particle property?
24. To what extents should people consider $\hbar$ to be a basis (perhaps paralleling $1/(4\pi\varepsilon_0)$ or G_N) for a possible property of something other than particles?
25. To what extent should people consider properties with which G-family bosons interact to be correlated?
    - 25.1. For example, ...
        - 25.1.1. We show $(m'(W)/m'(Z))^2 = \cos^2(\theta_W) \approx 7/9$.
        - 25.1.2. Thus, $\sin^2(\theta_W) \approx 2/9$.
        - 25.1.3. To what extent might people consider that the $(D + 2\nu)_j = 2$ term that contributes to $(m'(W))^2$ correlates aspects of charge with aspects of mass?

25.2. For example, to what extent might a characterization of strength of 2G24& correlate with characterizations of the strengths of 2G2& and 4G4&?

26. To what extent can people use work based on this monograph to reduce the number of physics constants that people perceive as being independent?
27. To what extent, for ENS48 models, do the following items correlate?
    27.1. The SU(7)-based instance-related symmetry and the number, 48, of generators for this group.
    27.2. The Standard Model SU(3)×SU(2)×U(1) symmetry (for the which component symmetries have, respectively 8, 3, and 2 generators) and the product, 48, of the number of generators for the three groups.
    27.3. Telescoping (or, nesting), for spans of G-family-mediated interactions. (See Table 2.13.17.)
28. To what extent, for at least G-family physics and ENS48 models or ENS6 models, would math correlating with subgroups of SU(7) correlate with physics-relevant symmetries?
    28.1. For example, ...
        28.1.1. Consider ...
            28.1.1.1. SU(7) ⊃ SU(5) × SU(2) × U(1).
            28.1.1.2. SU(5) ⊃ SU(3) × SU(2) × U(1).
        28.1.2. To what extent might such relationships correlate with telescoping (or, nesting), for spans of G-family-mediated interactions? (See Table 2.13.17.)
        28.1.3. To what extent might each SU(2) correlate with an SU(2) pertaining to the Poincare group?
        28.1.4. To what extent might each U(1) correlate with an approximate symmetry related to CPT-related symmetries?
        28.1.5. To what extent might aspects of the following not-fully-mathematically-appropriate notion pertain? (Here, perhaps, 'SU(1)' correlates with no symmetry.)
            28.1.5.1. "SU(3) ⊃ 'SU(1)' × SU(2) × U(1)."
29. To what extent does the exponent 6 in expression β^6 (for the relative strengths per channel of electromagnetism and gravity) correlate (at least mathematically) with the ratio 6 (for ENS48 models and ENS6 models) of number of ensembles linked by one instance of gravitons and number of ensembles linked by one instance of MM1MS1-photons? (See, for example, Table 3.13.4.)
30. To what extent does work in this monograph provide insight regarding problems for which people associate the terms vacuum zero-point energy, mass of the quantum vacuum, vacuum catastrophe, and/or the cosmological constant?

30.1. For example, to what extent might people benefit from considering (at least in regard to vacuum zero-point energy and/or mass of the quantum vacuum) Œ = 0?

31. To what extent does work in this monograph provide insight that people can use regarding theories that might feature or hypothesize properties for or structure of space-time?

32. To what extent can work in this monograph provide insight regarding the topic of arrow of time?

    32.1. For example, explore the extent to which the following correlate.

        32.1.1. Arrow-of-time concepts.

        32.1.2. Asymmetries between $\sigma = +1$ physics and $\sigma = -1$ physics.

        32.1.3. Effects of the interaction correlating with the 4G2468& boson.

        32.1.4. Possible use of QE-like radial coordinates correlating with existence of the big bang or correlating with other phenomena or models.

        32.1.5. The notion that some 4W solutions might correlate ...

            32.1.5.1. With math that could correlate with imaginary masses (See Table 3.8.2.) and therefore ...

            32.1.5.2. Possibly with math for imaginary energy and therefore ...

            32.1.5.3. Possibly with math that includes an imaginary component for a time coordinate and therefore ...

            32.1.5.4. Possibly with a decay-centric branch and a non-decay-centric branch in the math, depending on the sign of the imaginary component for the time coordinate.

        32.1.6. Entropy.

33. To what extent does the existence of the 0H0 particle correlate with the existence of elementary particles with non-zero mass?

34. To what extent can people usefully deploy concepts paralleling (a`;b`)< symmetries (or, paralleling $\sigma = -1$) to enhance understanding of bound-state systems (such as an atom and its electrons) or partly-bound, partly-free systems (such as (valence) electrons below and (conduction) electrons above a band-gap in a semiconductor)?

35. To what extent might people benefit by exploring possible physics-relevance of correlations between the following?

    35.1. Correlations between SO(3,1) and FRERAN (1;3)> symmetries.

    35.2. Correlations between SO(4) and ATOMOL (a`;4)> symmetries.

36. Including and beyond traditional conservation laws and some math-based models this monograph discusses, to what extent do math-based models precisely describe nature? (See Section 2.1.)

37. To what extent can people develop simpler bases for and/or descriptions of work in this monograph?

To the extent work in this monograph correlates with nature or with possibly useful models, people might want to explore topics we suggest elsewhere. Table 7.5.3 points to some such topics. (People might say that overlaps between Table 7.5.2 and Table 7.5.3 exist.) Each item in Table 7.5.3 provides the number of a table and the name of that table.

Table 7.5.3 Pointers to some possible physics topics and/or physics modeling topics for further exploration

1. Table 2.9.6.
   1.1. Possible lines of inquiry related to the concept of R_0 for elementary particles
2. Table 2.12.7.
   2.1. Possible correlation between measurement of a property of an elementary fermion by an elementary boson and conservation of that property
3. Table 2.13.35.
   3.1. Speculation regarding relationships among $\sigma = -1$, $\sigma = +1$, and abilities of QIRD-family elementary particles to interact with YO-family elementary particles
4. Table 2.13.43.
   4.1. Possible steps for research regarding possibilities that #b' =2
5. Table 2.14.5.
   5.1. Possible opportunities for research regarding, for elementary fermions, relationships among spin, generations, color charge, charge, and SPATIM symmetries
6. Many of the topics to which Table 2.15.7 alludes.
   6.1. Some topics correlating with the MM1 meta-model and with choices available when developing a MM1MS1 model
7. Table 3.3.14.
   7.1. Some possibilities for detecting dark matter and/or cross-ensemble interactions meditated by 2G24& (some ENS48 models and some ENS6 models)
8. Table 3.5.4.
   8.1. Aspects regarding an interaction between an extant elementary fermion and a WO-family boson
9. Table 3.5.6.
   9.1. Possible opportunities for research regarding neutrinos and the possible 2O0 boson
10. Table 3.9.6.
   10.1. Possible opportunities for research, based on m'(tauon) and G_N

11. Table 3.9.11.
    11.1. A possible hint toward additional significance related to d" and some applications of trigonometric functions
12. Table 3.10.2.
    12.1. Possible relationships between R_0 (and similar lengths) and some strengths of interactions
13. Table 3.13.3.
    13.1. A possible link between masses of elementary bosons and the strength of electromagnetism
14. Table 3.13.4.
    14.1. Possible ratios of G-family vertex strengths, beyond for just 2G2&, 4G4&, and interactions involving ordinary matter
15. Table 3.13.5.
    15.1. A possible link between β and a number (or between various masses and interaction strengths)
16. Table 4.2.6.
    16.1. An example of possible opportunities for research based (at least in part) on observations of large-scale phenomena
17. Table 4.3.11.
    17.1. Observations to possibly attempt regarding the distributions of ordinary matter, dark matter, and dark-energy stuff
18. Table 4.3.12.
    18.1. Experiments or observations possibly to attempt regarding the nature of dark matter and/or the nature of dark-energy stuff
19. Table 4.5.1.
    19.1. Possibility for experiments or inference regarding balance between matter neutrinos and antimatter neutrinos
20. Table 4.6.2.
    20.1. Possible research opportunities related to general relativity and alternative models
21. Table 4.8.2.
    21.1. A research opportunity related to the spacecraft flyby anomaly
22. Table 4.9.1.
    22.1. Chronology and phenomenology of a quasar
23. Table 4.11.1.
    23.1. Possible research topics regarding quark-based plasmas and related physics
24. Table 5.1.1.
    24.1. Possible paths for improving models regarding physics related to elementary particles
25. Table 5.1.4.

    25.1. Steps to harmonize and extend the Standard Model and MM1MS1 models regarding interactions involving spin-1 and spin-0 elementary bosons this monograph discusses
26. Table 5.1.5.
    26.1. Relationships between 2G2&-related models and 2G24&-related models and photons (ENS48 models and ENS6 models)
27. Table 5.1.6.
    27.1. Steps to harmonize and extend the Standard Model and MM1MS1 models regarding interactions involving spin-2 elementary bosons this monograph discusses
28. Table 5.1.7.
    28.1. Possible steps for adding (to the Standard Model) 3N-subfamily fermions this monograph discusses and for understanding models related to neutrinos
29. Table 5.1.8.
    29.1. Possible steps for adding (to the Standard Model) other aspects correlating with work in this monograph
30. Table 5.2.1.
    30.1. Possible additions to narratives about the evolution of the universe
31. Table 6.1.4.
    31.1. Thoughts regarding properties correlating with G-family bosons
32. Table 6.1.6.
    32.1. Linear operators, conservation laws, SPATIM aspects, quadratic operators, elementary-particle properties, INTERN aspects
33. Table 6.2.1.
    33.1. A possibility for modeling H1-atoms
34. Table 6.2.3.
    34.1. Speculation about concepts for ATOMOL symmetries for H1-atoms
35. Table 6.2.4.
    35.1. A possible basis for estimating quantities for which people find that the fine-structure constant provides contributions
36. Table 6.2.5.
    36.1. Possible generalizations for and/or from applications of possible ATOMOL (a`;4)< symmetries and possible ATOMOL (a`;4)> symmetries
37. Table 6.3.2.
    37.1. Possible research topics regarding models that correlate COMPAR models
38. Table 6.4.8.
    38.1. Possibly useful questions, correlating with general relativity regarding representations for energy-momentum space and for curvature people might correlate with space-time
39. Table 7.5.2.

39.1. Possible physics topics and/or physic modeling topics for further exploration

Section 7.6 Possible general opportunities

Section 7.6 discusses advances in physics modeling. We list some previous advances. We list possible opportunities for forthcoming advances. We list generic possible steps forward.

~ ~ ~

This subsection lists some previous and ongoing advances in physics and in models for physics.

Table 7.6.1 pertains.

Table 7.6.1 Some advances in physics and in models for physics

1. Perceived motions of heavenly bodies.
   - 1.1. Before Copernicus, people used epicycle math to model perceived motions of heavenly bodies. Copernicus started a change. Kepler completed it. Ellipses apply better than do epicycles. Newton explained elliptical motion.
2. Catalog of chemical elements or of atoms.
   - 2.1. Before Mendeleev, people proposed ways to catalog known elements. Mendeleev listed elements in a two-dimensional table. People correlated gaps in the table with unfound elements. People found or made such elements.
     - 2.1.1. More specifically, Mendeleev's catalog was one of various proposed catalogs. People recognize the organizing principles Mendeleev's catalog has. People find those principles useful. People found ways to better depict a catalog of elements. Today, a standard depiction prevails. Other depictions exist.
   - 2.2. Today, cataloging principles ...
     - 2.2.1. Correlate with knowledge of components of atoms.
     - 2.2.2. Permit cataloging of isotopes.
3. Unification of electricity and magnetism.
   - 3.1. For part of the nineteenth century, people considered electricity and magnetism to be separate topics. Maxwell developed a unified treatment. The topic of electromagnetism arose.
4. Models for quantum phenomena.

- 4.1. Starting in the early twentieth century, quantum mechanics provides useful models for small things.
5. Relativistic models for motion and forces.
    - 5.1. Starting in the early twentieth century, Einstein's special relativity and general relativity provide useful models for some small things and some large things.
6. Improved narratives regarding quantum models.
    - 6.1. Starting in the second half of the twentieth century, people deployed new useful techniques and notation (for example, the use of Feynman diagrams) regarding aspects of quantum modeling.
7. Fostering of practical applications.
    - 7.1. Feynman encouraged pursuit of small-scale electromechanical devices.

~ ~ ~

This subsection lists some advances people might derive from paralleling or continuing work in this monograph.

Table 7.6.2 pertains. Names in parentheses allude to possible precedents. (See Table 7.6.1.)

Table 7.6.2 Some advances people might derive from paralleling or continuing work in this monograph

1. Simplify models for phenomena people already model. (Copernicus, Kepler).
    - 1.1. Phenomena might include ...
        - 1.1.1. Phenomena people discuss via the Standard Model.
        - 1.1.2. Possibly, phenomena within atomic nuclei.
            - 1.1.2.1. For example, the nuclear shell model.
        - 1.1.3. Possibly, phenomena within neutron stars.
        - 1.1.4. Possibly, atomic and molecular phenomena.
        - 1.1.5. Possibly, phenomena involving non-bound electrons in metals, semiconductors, or superconductors.
2. Provide theories of how to explain models of what. (Newton)
    - 2.1. Provide (especially, regarding elementary particles) paradigms, theories, or models for how nature works (and not just for what people measure or infer).
3. Develop a table of known and possible elementary particles. (Mendeleev)
    - 3.1. Include in such a table ...
        - 3.1.1. Ordinary-matter particles.
        - 3.1.2. Particles other than ordinary-matter particles.
4. Unify models and theories across various scales of size. (Maxwell)

   4.1. For example, unify elementary-particle models and cosmology models.
5. Provide useful modeling techniques. (Einstein, Feynman)
   5.1. Develop models that are more fundamental than extant models.
   5.2. Develop models and/or narratives that enhance utility of modeling techniques and/or models.
6. Promote underexplored types of practical applications. (Feynman)
   6.1. Point to possible applications based on possible or newly discovered elementary particles.

~ ~ ~

This subsection lists some possible steps forward.
Table 7.6.3 pertains.

Table 7.6.3 Possible steps forward

1. Start from one or more meta-models (perhaps similar to the one this monograph discusses).
2. Work on and improve ...
   2.1. Meta-models.
   2.2. Models.
   2.3. Experimentation, observation, and inference.
   2.4. Data.
   2.5. Narratives.

Chapter 8 Appendices

Chapter 8 (Appendices) shows some data and math this monograph uses as inputs.

Section 8.1 Some physics numbers

Section 8.1 shows some physics numbers this monograph uses.

Table 8.1.1 shows results of experiments and observations. (K.A. Olive (2014) provides these numbers.) For some items, this monograph supplies a symbol. For items (for which the symbol is a) pertaining to magnetic dipole moment anomalies, $g_S = 2$.

Table 8.1.1 Some physics numbers

Symbol	Units	Number	Description
c	m s^{-1}	2.99792458×10^{8}	speed of light in a vacuum
$\hbar$	J s	$1.054571726(47)\times10^{-34}$	Planck constant, reduced
$\hbar$	MeV s	$6.58211928(15)\times10^{-22}$	Planck constant, reduced
q_e	C	$-1.602176565(35)\times10^{-19}$	charge of an electron
m_e	kg	$9.10938291(40)\times10^{-31}$	mass of an electron
m_e	MeV/c^2	0.510998928(11)	mass of an electron
ε_0	F m^{-1}	$8.854187817\times10^{-12}$	permittivity of free space
α	(no units)	$7.2973525698(24)\times10^{-3}$	fine-structure constant
α^{-1}	(no units)	137.035999074(44)	
G_N	m^3 kg^{-1} s^{-2}	$6.67384(80)\times10^{-11}$	gravitational constant
$\Sigma(m')_\nu$	eV/c^2	<0.23	sum of neutrino masses
Ω_Λ		$0.685^{+0.017}_{-0.016}$	dark energy density of the CDM Universe
Ω_b		0.0499(22)	baryon density of the Universe
Ω_{cdm}		0.265(11)	cold dark matter density of the universe
$\Omega_m = \Omega_{cdm} + \Omega_b$		$0.315^{+0.016}_{-0.017}$	pressureless matter density of the Universe
$\Omega_{tot} = \Omega_m +$		$0.96^{+0.4}_{-0.5}$	curvature
$\ldots + \Omega_\Lambda$		1.000(7)	
m'(Higgs)	GeV/c^2	125.7(4)	mass of a Higgs boson

T.J. Buckholtz, *Models for Physics of the Very Small and Very Large*,
Atlantis Studies in Mathematics for Engineering and Science 14,
DOI 10.2991/978-94-6239-166-6_8

Symbol	Units	Number	Description
m'(Z)	GeV/c^2	91.1876(21)	mass of a Z boson
m'(W)	GeV/c^2	80.385(15)	mass of a W boson
m'(tauon)	MeV/c^2	$1.77682(16)\times10^3$	mass of a tauon
m'(muon)	MeV/c^2	105.6583715 ± 0.0000035	mass of a muon
a(electron)	(no units)	$(1159.65218076 \pm 0.00000027)\times10^{-6}$	electron magnetic dipole moment anomaly $(g - g_S)/g_S$
a(muon)	(no units)	$(11659209\pm6)\times10^{-10}$	muon magnetic dipole moment anomaly $(g - g_S)/g_S$
a(tauon)	(no units)	> -0.052 $< +0.013$ (each with a confidence level of 95%)	tauon magnetic dipole moment anomaly $(g - g_S)/g_S$

~ ~ ~

Reference 18 K.A. Olive (2014)
K.A. Olive et al. (Particle Data Group), *Chin. Phys. C*, 38, 090001 (2014). (http://pdg.lbl.gov/2014/html/authors_2014.html)

Section 8.2 Numbers of generators for groups SU(j)

Section 8.2 notes relationships between numbers of generators for groups SU(j) for various j.

The number of generators associated with each mathematical group SU(j) (for which j is an integer, with $j \geq 2$) is $j^2 - 1$.

Table 8.2.1 shows numbers of generators for some values of j. The term multiple of k refers to multiple of k number of generators. The table lists integer multiples.

Table 8.2.1 Ratios of numbers of generators for various SU(j)

j	Number of generators	Multiple of 8	Multiple of 24	Multiple of 48	Multiple of 288
3	8	1	-	-	-
5	24	3	1	-	-
7	48	6	2	1	-
9	80	10	-	-	-

j	Number of generators	Multiple of 8	Multiple of 24	Multiple of 48	Multiple of 288
11	120	15	5	-	-
13	168	21	7	-	-
15	224	28	-	-	-
17	288	36	12	6	1
289	83520	10440	3480	1740	290

Table 8.2.2 shows other relationships between numbers of generators.

Table 8.2.2 Other arithmetic relationships between numbers of generators for various SU(j)

1. (1 + 2/1) (generators for SU(3)) = (3/1) 8 = 24 = generators for SU(5).
2. (1 + 1/1) (generators for SU(5)) = (2/1) 24 = 48 = generators for SU(7).
3. (1 + 2/3) (generators for SU(7)) = (5/3) 48 = 80 = generators for SU(9).
4. (1 + 1/2) (generators for SU(9)) = (3/2) 80 = 120 = generators for SU(11).
5. (1 + 2/5) (generators for SU(11)) = (7/5) 120 = 168 = generators for SU(13).
6. (1 + 1/3) (generators for SU(13)) = (4/3) 168 = 224 = generators for SU(15).
7. (1 + 2/7) (generators for SU(15)) = (9/7) 224 = 288 = generators for SU(17).
8. Regarding the factors (1 + l) in items above, each of the next items pertains to a series.
   - 8.1. l = 2/1, 2/3, 2/5, and 2/7 pertain respectively for the first, third, fifth, and seventh relationships above.
   - 8.2. l = 1/1, 1/2, and 1/3 pertain respectively for the second, fourth, and sixth relationships above.

Chapter 9 Compendia

Chapter 9 (Compendia) provides a list of acronyms this monograph uses and provides lists for use in finding items in this monograph.

Section 9.1 Acronyms

Section 9.1 lists acronyms this monograph uses.

1. ATOMOL - atoms and molecules
2. CHN - channels
3. CMB - cosmic microwave background (radiation)
4. COMPAR - composite particles
5. CORMAT - core relevant mathematical solutions
6. CORPHY - core solutions relevant to elementary-particle physics
7. CTCW - current that correlates with
8. DEE - dark-energy ensemble or dark-energy ensembles
9. DES - dark-energy stuff
10. DIFEQU - (usually) radial-coordinate (generally) multidimensional partial differential equation
11. DIU - directly inferable universe
12. DME - dark-matter ensemble(s)
13. DME-1 - the one DME that interacts with the OME via instances of G-family bosons that have spans of 2
14. DME-4 - four DME for which no G-family boson for which the span is 2 intermediates interactions between those ensembles and the OME
15. DMS - dark-matter stuff
16. EACUNI - each G-family elementary particle correlates with a unique G-family solution
17. EGS - negative of the ground-state energy
18. ENS1 - models for which #ENS = 1
19. ENS48 - models for which #ENS = 48
20. ENS6 - models for which #ENS = 6
21. ENVIRO - environmental
22. FERTRA - fermion transformational
23. FRERAN - free-ranging

T.J. Buckholtz, *Models for Physics of the Very Small and Very Large*,
Atlantis Studies in Mathematics for Engineering and Science 14,
DOI 10.2991/978-94-6239-166-6_9

24. G#XY - symbols combining (in the following order) the letter G, a nonnegative integer, a type of ensemble (OME, DME, or DEE) stated without the last E, and a type of ensemble (OME, DME, or DEE) stated without the last E
25. GEN(3/2)≠15 - models correlating with other than 15 generations for spin-3/2 elementary fermions
26. GEN(3/2)15 - models correlating with 15 generations for spin-3/2 elementary fermions
27. H1-atom - atom-like objects that include exactly 2 fermions, each having spin-1/2 and each having a magnitude of charge equal to the magnitude of charge for the electron
28. INSSYM - instance-related symmetry
29. INSSYM7 - symmetry or symmetries correlating with INSSYM
30. INST - number of instances of a field or particle
31. INTERN - internal
32. LADDER - ladder operators or ladder-operator based
33. MDM - magnetic dipole moment
34. MM1 - meta-model 1
35. MM1MS1 - meta-model 1, model set 1
36. OMDME - ordinary-matter plus dark-matter ensemble(s)
37. OMDMS - the combination of OMS and DMS and possibly some instances of G-family bosons
38. OME - ordinary-matter ensemble
39. OMS - ordinary-matter stuff
40. QCD - quantum chromodynamics
41. QED - quantum electrodynamics
42. QMPRPR - quantum model that provides elementary-particle properties
43. QMUSPR - quantum model that uses assumed elementary-particle properties
44. RESENE - rest energy
45. SDI - spatial dependence of interactions
46. SEDES - single-ensemble dark-energy stuff
47. SEDMS - single-ensemble dark-matter stuff
48. SOMMUL - some G-family particles correlate with multiple G-family solutions (or, more than one G-family solution)
49. SPAN - number of ensembles an instance of a particle spans
50. SPATIM - space-time coordinate (symmetries)
51. WIMP - weakly interacting massive particle

Section 9.2 Summaries of sections and names of tables

Section 9.2 lists summaries of sections and names of tables.

Chapter 1 Overview

Section 1.1 Some perspective

Section 1.1 discusses context for this monograph. We identify needs for models to predict aspects of nature. We review a past scenario in which modeling improved physics. We draw parallels between that scenario and today's situation. We discuss an opportunity to improve physics. We suggest an agenda for capturing that opportunity.

Section 1.2 This monograph

Section 1.2 discusses relationships between this monograph and the agenda the previous section suggests. People might say that work in this monograph provides impetus to tackle the agenda. We think this monograph provides examples of trying to take steps the agenda features. We note that models we show produce candidates for physics predictions. We note that some of candidates may prove useful and that some may prove to be wrong. We suggest next steps people might want to take. We summarize, from a modeling perspective, chapters in this monograph. We summarize, from a physics perspective, types of insight and examples of predictions that following the agenda might produce.

Chapter 2 From data to the MM1 meta-model and MM1MS1 models

Section 2.1 Math-based models for quantum phenomena

Section 2.1 discusses modeling. We discuss differences between inputs to models and outputs from models. We discuss the desirability to extend modeling practices to include models that produce as outputs information that traditional models use as inputs but do not produce as outputs. We point to some physics data that generally accepted physics models treat as inputs and generally do not produce as outputs. We discuss goals for the MM1 meta-model and for MM1MS1 models. We note that trying to develop models that use quantum harmonic oscillator math may be useful. We introduce terminology regarding classes of physics theories, classes of models this monograph shows, and some physics concepts.

Table 2.1.1 Types of physics observations that yield discreet numbers

Table 2.1.2 A discreet number (#ENS) for which much traditional physics assumes one value and for which MM1MS1 models explore more than one value

Table 2.1.3 Notation of the form #Z

Table 2.1.4 Categories of some discreet physics numbers

Section 2.2 LADDER models for isotropic quantum harmonic oscillators

Section 2.2 discusses LADDER models for isotropic pairs of isotropic quantum harmonic oscillators. We discuss aspects regarding solutions to equations. We discuss discreet-math models pertaining to 1-dimensional harmonic oscillators. We discuss discreet-math models pertaining to multi-dimensional isotropic quantum harmonic oscillators. We define a constraint, Œ = 0, that each MM1MS1 LADDER model features. We define, in the context of LADDER models, the term solution. We show notation for labeling columns in tables that show aspects of LADDER models. People might say that we show a ground-state solution that traditional physics models may have underutilized.

Section 2.3 INTERN applications of LADDER models

Section 2.3 discusses relevant solutions within models for isotropic pairs of isotropic quantum harmonic oscillators. We feature subsets of the solutions. We focus on INTERN LADDER subsets for which the solutions might correlate with data about elementary particles. We provide names for families of solutions. We characterize solutions pertaining to various families. People might say that we show or point to solutions that correlate with all known and some possible elementary particles. People might say that those solutions correlate with properties that include at least spin and whether rest mass is zero or non-zero.

Table 2.3.14 CORPHY ground-state INTERN LADDER solutions for which $\sigma = -1$

Section 2.4 One MM1 meta-model and various MM1MS1 models

Section 2.4 discusses relationships between the MM1 meta-model and MM1MS1 models.

Table 2.4.1 Relationships between the MM1 meta-model and MM1MS1 models

Section 2.5 DIFEQU models for isotropic quantum harmonic oscillators

Section 2.5 discusses DIFEQU models for isotropic quantum harmonic oscillators. We discuss continuous-math models pertaining to lone harmonic oscillators. We discuss continuous-math models pertaining to multi-dimensional isotropic quantum harmonic oscillators. We discuss projecting representations into smaller numbers of dimensions than originally pertain to the representations. We discuss, in the context of DIFEQU models, the notion of solution. We note that the topic of normalization has physics-relevance. We show two types of solutions that normalize. People might say that we show solutions that traditional physics models may have underutilized. We discuss ranges of parameters that pertain for each of traditional solutions and non-traditional solutions. People might say that solutions we characterize correlate with known elementary particles and with possible elementary particles. People might say that we show integers that we later correlate with masses of elementary bosons.

Table 2.5.1 Notation and math pertaining to a lone quantum harmonic oscillator (DIFEQU solutions)

Table 2.5.2 Differences in units for ξ (and ξ') between some aspects of traditional physics and some aspects of work in this monograph

Table 2.5.3 Notation and math pertaining to a DIFEQU solution for a multi-dimensional isotropic quantum harmonic oscillator

Table 2.5.4 Process for creating and preparing to use a partly linear-coordinate, partly radial-coordinate representation for a DIFEQU isotropic quantum harmonic oscillator

Table 2.5.5 Possible applications for models correlating with D* equaling each of 3, 2, and 1

Table 2.5.6 Math and notation for radial-coordinate solutions for QP-like isotropic quantum harmonic oscillators

Table 2.5.7 Solutions, based on Hermite polynomials, for quantum harmonic oscillators for which $D = 1$

Table 2.5.8 Some radial-coordinate solutions for isotropic quantum harmonic oscillator math

Table 2.5.9 Normalization for solutions for which $D + 2\nu > 0$

Table 2.5.10 Normalization for solutions for which $D + 2\nu = 0$

Table 2.5.11 Definitions, regarding a solution and a number of dimensions, of inside, edge, and outside (DIFEQU solutions)

Table 2.5.12 Some ([Physics:] traditional) solutions (DIFEQU solutions)

Table 2.5.13 Some ([Physics:] non-traditional) solutions, including an example of a non-traditional ground state for 3-dimensional isotropic quantum harmonic oscillators (DIFEQU solutions)

Table 2.5.14 Parameters for solutions (based on radial coordinates) that can be normalized (DIFEQU solutions)

Table 2.5.15 The number σ

Table 2.5.16 Relationships between some parameters, for some solutions for which D* = 3 (DIFEQU solutions)

Table 2.5.17 Relationships between some parameters, for some solutions for which D* = 2 (DIFEQU solutions)

Table 2.5.18 Values of S that could arise for the choice D* = 1

Section 2.6 Applications of DIFEQU models

Section 2.6 discusses, in the context of DIFEQU models, solutions. We focus on DIFEQU subsets for which the solutions might correlate with data about elementary particles. We correlate some subsets with the physics terms fermion and boson. We correlate some subsets with the physics terms fields and particles. We discuss aspects of projecting mathematical representations into smaller than the original numbers of dimensions. We show factors contributing to numbers of physics-relevant solutions. We list, for some DIFEQU subsets, numbers of relevant solutions. People might say that we discuss possible time-like uses of radial coordinates.

Table 2.6.1 Correlations among ν, η, particles, and fields, for solutions based on DIFEQU representations (for D* = 3 or D* = 2)

Table 2.6.2 Correlations between particles and dimensions, for boson solutions based on radial coordinates (for D* = 3)

Table 2.6.3 Definition of min(j, k)

Table 2.6.4 Factors contributing toward maximum numbers of solutions (for D* = 3 and ν = −1/2 and for D* = 3 and ν = −3/2)

Table 2.6.5 Numbers of candidate solutions, for some solutions based on radial coordinates (for D* = 3)

Table 2.6.6 Perspective regarding possible QE-like uses of DIFEQU models

Section 2.7 FERTRA and other models linking LADDER and DIFEQU models

Section 2.7 discusses correlations between LADDER models and DIFEQU models. We correlate LADDER solutions and DIFEQU solutions for the WHO-families of solutions. We discuss overlaps and gaps in the applicability of LADDER and DIFEQU models. We discuss the concept of composite particles. We show FERTRA LADDER solutions for the CN-families and for the QIRD-families.

Table 2.7.1 Ground states for D* = 3, ν = −1, and S ≤ 2

Table 2.7.2 Concepts pertaining to FERTRA LADDER solutions

Table 2.7.3 Some FERTRA LADDER solutions for the CN-families and the QIRD-families

Table 2.7.4 Correlations between m' and N(P0)

Table 2.7.5 Correlations between solutions this monograph discusses and m', plus extrapolations from DIFEQU results for some solutions to similar results for other LADDER solutions

Section 2.8 Generation, color charge, and INTERN models for bosons

Section 2.8 discusses the extents to which elementary bosons impact, for elementary fermions, the properties of generation and color charge. We consider INTERN models correlating with G-family solutions. We discuss conservation of fermion generation for some interactions involving elementary fermions and G-family bosons. We consider INTERN models correlating with WHO-family solutions. We discuss conservation of fermion generation for some interactions involving elementary fermions and WHO-family bosons. We consider INTERN models correlating with Y-family solutions. We note a lack of conservation of fermion color charge for Y-family-related interactions involving fermions for which $\sigma = -1$.

Table 2.8.1 G-family ground-state INTERN LADDER solutions

Table 2.8.2 G-family ground-state INTERN LADDER solutions, showing QE-like possible interaction-related symmetries

Table 2.8.3 Interaction-related phenomena, G-family bosons that intermediate the interactions, and numbers of "G symmetries

Table 2.8.4 Assumptions about generation-based variation of elementary-fermion properties, regarding G-family interactions

Table 2.8.5 Values of #"G for WHO-family elementary bosons

Table 2.8.6 Aspects regarding generation-based variation of elementary-fermion properties, regarding interactions with WHO-family bosons

Table 2.8.7 Y-family ground-state INTERN LADDER solutions

Table 2.8.8 Reinterpretation of Y-family ground-state INTERN LADDER solutions

Table 2.8.9 Lack of conservation of color charge

Section 2.9 Schwarzschild radius, Planck length, and R_0

Section 2.9 discusses a series of lengths. The series features the Schwarzschild radius, the Planck length, and a length we call R_0. People might say that we show physics-relevance for R_0.

Table 2.9.1 Types of patterns this section discusses

Table 2.9.2 A series of lengths that includes the Planck length and a length derived by applying the formula for the Schwarzschild radius to the mass of an electron

Table 2.9.3 A series of lengths relevant to electrons and positrons

Table 2.9.4 Concepts indicating possible significance for lengths R_0

Table 2.9.5 Some values of R_0, for various particles

Table 2.9.6 Possible lines of inquiry related to the concept of R_0 for elementary particles

Section 2.10 Models related to vertices and to particle sizes, masses, and ranges

Section 2.10 discusses models related to sizes, masses, and ranges of elementary particles. We discuss clouds of virtual particles. We discuss MM1MS1-currents related to energy and momentum. We discuss interactions between gravity and MM1MS1-currents related to energy and momentum. We discuss aspects of models for interaction vertices (or, for vertices in Feynman diagrams). People might say that our work correlates with notions that, in some models, elementary particles have sizes of zero. We point to models and integers related to actual or approximate masses of elementary bosons. People might say that we show models that correlate with the spatial ranges of elementary bosons.

Section 2.11 SPATIM symmetries

Section 2.11 discusses symmetries that correlate with the Poincare group and with special relativity. We provide perspective regarding possible uses of symmetries. We introduce SPATIM models correlating with symmetries that can correlate with solutions. We discuss rules for how, for a field that correlates with a combination of a fermion (elementary particle) field and a boson (elementary particle) field, to compute symmetries for the combined field based on the symmetries of the two component fields.

Section 2.12 Invariances, symmetries, and conservation laws

Section 2.12 discusses the extent to which invariances, symmetries, and conservation laws pertain. We discuss relationships among concepts of invariances, symmetries, and conservation laws. We provide perspective regarding invariances, symmetries, and conservation laws that correlate with SPATIM-correlated phenomena. We show examples of MM1MS1-related symmetries that traditional physics models may not feature. One such symmetry correlates with conservation of fermion generation for some interactions. We suggest categories for conservation laws. We suggest a quantum-related basis for symmetries correlating with conservation of angular momentum. People might say that we provide new perspective regarding approximate conservation laws people associate with the terms C-symmetry, P-symmetry, and T-symmetry. We compare MM1MS1 models and traditional interpretations regarding conservation of charge.

Section 2.13 FRERAN and COMPAR applications of LADDER models

Section 2.13 discusses LADDER models that correlate with internal properties of elementary particles, motions of particles, and interactions between elementary particles. We define INSSYM7, an instance-related symmetry. We provide perspective about transforming INTERN LADDER solutions into SPATIM LADDER solutions. We transform INTERN LADDER solutions so as to include SPATIM symmetries and INSSYM7 symmetry. People might say that, for some interpretations, INSSYM7 symmetry correlates with the possibility the universe includes multiple ensembles, with each ensemble including a set of elementary particles that includes at least the Standard Model set of elementary particles. People might say that one ensemble correlates with ordinary matter, five ensembles correlate with dark matter, some ensembles possibly correlate with dark-energy stuff, and some elementary particles do not correlate with an ensemble. People might say that we correlate G-family solutions with various symmetries, with various force laws, and with the concept of the span on an instance of a force. People might say that COMPAR-related symmetries and solutions correlate with phenomena correlating with QCD (or, quantum chromodynamics) and with aspects of composite particles. We discuss possibilities for states of matter (other than as found in hadrons) that involve quarks.

Table 2.13.1 Two concepts possibly correlating with INSSYM7

Table 2.13.2 Steps for constructing a FRERAN SPATIM LADDER solution from one INTERN LADDER solution

Table 2.13.3 $\sigma = +1$ non-G-family ground-state FRERAN SPATIM LADDER solutions

Table 2.13.4 A second depiction of $\sigma = +1$ non-G-family ground-state FRERAN SPATIM LADDER solutions

Table 2.13.5 A third depiction of $\sigma = +1$ non-G-family ground-state FRERAN SPATIM LADDER solutions

Table 2.13.6 Instance-related symmetry and FRERAN SPATIM symmetries for $\sigma = +1$ non-G-family ground-state ENVIRO LADDER solutions

Table 2.13.7 Correlation of aspects of FRERAN modeling with aspects of phenomena involving elementary particles for which $\sigma = -1$

Table 2.13.8 $\sigma = -1$ boson ground-state FRERAN SPATIM LADDER solutions

Table 2.13.9 $\sigma = -1$ fermion ground-state FRERAN SPATIM LADDER solutions

Table 2.13.10 Instance-related symmetry and FRERAN SPATIM symmetries for $\sigma = -1$ ground-state FRERAN SPATIM LADDER solutions

Table 2.13.11 Two ways to model channels related to G-family elementary particles

Table 2.13.12 Channels and (bases for) FRERAN SPATIM symmetries for G-family ground-state LADDER solutions (assuming EACUNI models and that #b' = 3 pertain for each solution or subfamily)

Table 2.13.13 Math related to possible interpretations of INSSYM7

Table 2.13.14 Possible correlation between observations and instance-related symmetries

Table 2.13.15 Definitions of OME, DME, DEE, and hypothetical definitions of sets of instances of non-2G2& G-family particles (ENS48 models)

Table 2.13.16 Discussion regarding a possible cataloging method for ensembles and other particles, plus definitions of DME-1, DME-4, and OMDME (ENS48 models)

Table 2.13.17 Principles and characteristics of assumptions, regarding interactions mediated by G-family bosons, of telescoping or nesting (ENS48 models)

Table 2.13.18 Possible aspects correlating with deemphasizing the notion that non-G-family elementary bosons mediate interactions that span ensembles

Table 2.13.19 Possible values of #E` for G-family bosons for which #E = 0 (ENS1 models, ENS6 models, and ENS48 models)

Table 2.13.20 Values of #E` for G-family solutions for which #E = 0 (ENS48" models)

Table 2.13.21 Values of #E` for G-family solutions for which #E = 0 (ENS6' models)

Section 2.14 EXTINT LADDER models for fermion generations and color charge

Section 2.14 discusses EXTINT LADDER solutions that correlate with numbers of generations for elementary fermions and with numbers of color charges for elementary fermions existing in COMPAR environments. People might say that MM1MS1 models correlate with 5 color charges for elementary fermions for which $\sigma = -1$ and $S = 3/2$. People might say that MM1MS1 models provide, for spin-1/2 elementary fermions for which σ = +1, some correlation between numbers of generations and numbers of charges.

Table 2.14.1 Possible concepts regarding color charge

Table 2.14.2 A depiction of σ = −1 fermion ground-state EXTINT LADDER solutions

Table 2.14.3 Ground-state fermion INTERN LADDER solutions for σ = +1

Table 2.14.4 Ground-state fermion EXTINT LADDER solutions for σ = +1, correlating with generations

Table 2.14.5 Possible opportunities for research regarding, for elementary fermions, relationships among spin, generations, color charge, charge, and SPATIM symmetries

Section 2.15 The MM1 meta-model and various MM1MS1 models

Section 2.15 discusses using the MM1 meta-model. We define the MM1 meta-model by featuring a list of topics that correlate with choices people can make when using the MM1 meta-model to define models. We provide a process for using the MM1 meta-model to select and work with models. We list some key parameters (such as #ENS) that people can set in order to specify a model. We list some characteristics that models share. We discuss some aspects of ENS48 models, ENS6 models, and ENS1 models. We describe models we discuss in subsequent chapters.

Table 2.15.1 Topics that correlate with limitations this monograph places on the MM1 meta-model

Table 2.15.2 Steps for using the MM1 meta-model and using a MM1MS1 model

Table 2.15.3 Some topics correlating with the MM1 meta-model and possibly with all MM1MS1 models

Table 2.15.4 A topic for which people might not be able to make choices when using the MM1 meta-model to develop one or more types of MM1MS1 models

Table 2.15.5 Steps for specifying a model (an example)

Table 2.15.6 Possible further steps for specifying a model (an example)

Table 2.15.7 Some topics correlating with the MM1 meta-model and with choices available when developing a MM1MS1 model

Table 2.15.8 Assumptions regarding lack of physics-relevance for some solutions

Table 2.15.9 Possible lack of physics-relevance for some solutions, plus definitions of the terms %68even, %68both, and %68odd

Table 2.15.10 Characteristics shared by many models that correlate with the MM1 meta-model

Table 2.15.11 #ENS-related models we discuss in this monograph

Chapter 3 From the MM1 meta-model to particles and properties

Section 3.1 Introduction

Section 3.1 discusses some results that span the previous chapter and this chapter. We list families of known and possible elementary particles. We discuss notation for solutions and notation for elementary particles. We note that the catalog of known and candidate elementary particles correlates with the MM1 meta-model and, therefore, with all MM1MS1 models.

Table 3.1.1 Families of elementary particles, plus some composite particles

Table 3.1.2 Notes about the possible existence of suggested yet-to-be-found elementary particles

Table 3.1.3 Notes about correlations between and notation for solutions and elementary particles

Table 3.1.4 Correlations between #ENS and aspects of elementary-particle physics

Section 3.2 Aspects of models for MM1MS1-photons and gravitons

Section 3.2 discusses some aspects of models for MM1MS1-photons and gravitons. We discuss two concepts (the EACUNI set and the SOMMUL set) for sets of solutions that might correlate with MM1MS1-photons and gravitons. We show INTERN LADDER solutions that correlate with MM1MS1-photons and INTERN LADDER solutions that correlate with gravitons. We discuss aspects of mappings from an energy-momentum space that correlates with INTERN LADDER solutions to space-time coordinates.

Table 3.2.1 EACUNI-set models, SOMMUL-set models, and MM1MS1-photon representations

Table 3.2.2 022G2& modes and their ground-state INTERN LADDER solutions

Table 3.2.3 022G2& modes and their Nth excited states

Table 3.2.4 022G2& ground-state INTERN LADDER solution

Table 3.2.5 Coordinates relevant for 022G2&

Table 3.2.6 QE-like coordinates and QP-like coordinates, for a specific 022G2&

Table 3.2.7 QE-like coordinates and QP-like coordinates, for a specific 022G2& and a specific observer

Table 3.2.8 EACUNI-set models, SOMMUL-set models, and graviton representations

Table 3.2.9 244G4& modes and their ground-state INTERN LADDER solutions

Section 3.3 Elementary particles correlating with the G-family

Section 3.3 describes the G-family of elementary particles. People might say that G-family particles include all zero-mass elementary bosons except for particles (including gluons) correlating with Y-family solutions. We list two possible sets of G-family elementary particles. For %68both models, we list 12 particles. For %68even models, we list 7 particles. We discuss models for excitations of G-family bosons. We note that G-family particles provide for interactions with spatial dependences of r^{-2}, r^{-4}, r^{-6}, and r^{-8}. People might say that we point to a G-family member, 2G24&, that interacts with elementary-particle magnetic dipole moments. People might say that we anticipate possible correlations between Standard Model photons and a pair consisting of MM1MS1-photons and 2G24&. People might say that we anticipate interactions, intermediated by 2G24& between ordinary matter and dark matter. People might say that we anticipate the interactions intermediated by 2G68& couple to spin and do not couple to charge or to energy. We discuss a possibility for mechanisms related to G-family boson excitations and channels. We correlate zero-mass for each G-family elementary particle with work above in this monograph.

Table 3.3.1 Ground-state INTERN LADDER solutions for the G-family
Table 3.3.2 Notation, for G-family solutions, for the number of QP-like harmonic oscillator pairs for which excitement is possible
Table 3.3.3 Elementary particles correlating with G-family INTERN LADDER solutions (EACUNI-set models)
Table 3.3.4 Possible G-family elementary particles
Table 3.3.5 042G24& modes and their ground-state INTERN LADDER solutions
Table 3.3.6 042G24& modes and their first excited states (N(mode) = 1)
Table 3.3.7 Raising operator for G-family modes
Table 3.3.8 INTERN LADDER solutions for N = 0 and N = 1 for 042G2R4L
Table 3.3.9 Alternative representation of N = 0 and N = 1 for 042G2R4L
Table 3.3.10 Concepts regarding spatial dependences of interactions mediated by the G-family
Table 3.3.11 Spatial dependences of interactions mediated by the G-family bosons
Table 3.3.12 Aspects that might correlate with a MM1MS1 elementary boson that interacts with elementary-fermion magnetic dipole moments
Table 3.3.13 Correlating G-family interactions with electromagnetism
Table 3.3.14 Some possibilities for detecting dark matter and/or cross-ensemble interactions meditated by 2G24& (some ENS48 models and some ENS6 models)
Table 3.3.15 Interpretation of 2G24&-mediated interactions and particle properties related to those interactions
Table 3.3.16 Interpretation of 2G68&-intermediated interactions and particle properties related to those interactions
Table 3.3.17 Possible mechanics related to channels (for EACUNI-set models and not for SOMMUL-set models)

Section 3.4 Elementary particles correlating with the WHO-families

Section 3.4 discusses all non-zero-mass elementary bosons. We show solutions correlating with the known non-zero-mass elementary bosons. We show solutions correlating with possible non-zero-mass elementary bosons for which $\sigma = -1$. We point to work below regarding masses of WHO-family elementary particles.

Table 3.4.1 Ground-state INTERN LADDER solutions for non-zero-mass elementary bosons
Table 3.4.2 W- and H-family particles
Table 3.4.3 2W% states related to a specific interaction vertex

Section 3.5 Elementary particles correlating with the CN- and QIRD-families

Section 3.5 discusses fermion elementary particles. We show FERTRA LADDER solutions correlating with all known fermion elementary particles. We show

FERTRA LADDER solutions correlating with possible fermion elementary particles. People might say that FERTRA LADDER solutions correlate with abilities of elementary fermions to interact with elementary bosons correlating with W-family and the 2O- and 4O-subfamilies. People might say that MM1MS1-neutrinos are zero-mass Dirac fermions. People might say that we show a model correlating with handedness for leptons. We point to work below regarding relative masses of 1C-subfamily and 1Q-subfamily elementary particles.

Table 3.5.1 Types of C-, Q-, and I-family particles

Table 3.5.2 FERTRA LADDER solutions for CN-family particles and QIRD-family particles

Table 3.5.3 2W-subfamily, 2O-subfamily, and 4O-subfamily absorption interactions for elementary fermions

Table 3.5.4 Aspects regarding an interaction between an extant elementary fermion and a WO-family boson

Table 3.5.5 Concepts pertaining to fermion destruction operators and fermion creation operators correlating with interaction vertices that involve elementary fermions

Table 3.5.6 Possible opportunities for research regarding neutrinos and the possible 2O0 boson

Table 3.5.7 Handedness for leptons

Table 3.5.8 Terminology regarding 1N (or, MM1MS1-neutrinos) and 3N elementary particles

Section 3.6 Elementary particles correlating with the Y-family

Section 3.6 discusses the Y-family. We show Y-family solutions. We note that 2Y-subfamily solutions do not correlate directly with gluons. We note that 2Y-subfamily solutions correlate with a basis for 8 gluons. We show that the Y-family provides a basis for possible spin-2 zero-mass bosons. We note that Y-family solutions correlate with 3 color charges for 1QR-subfamily elementary fermions and with 5 color charges for 3QIRD-subfamily elementary fermions. We note that 4Y-subfamily solutions correlate with a basis for 24 gluon-like particles. We discuss the notion that work above correlates with zero-mass for each gluon or gluon-like elementary particle.

Table 3.6.1 Ground-state INTERN LADDER solutions for 2Y

Table 3.6.2 Characteristics of each 2Y solution

Table 3.6.3 A component for a gluon

Table 3.6.4 A restatement of a component for a gluon

Table 3.6.5 Two gluons

Table 3.6.6 Another gluon

Table 3.6.7 A correlation between the r, b, g representation for color charge and the CC2, CC1, and CC0 representation for color charge

Table 3.6.8 Some ground-state INTERN LADDER solutions for 4Y

Section 3.7 W- and H-family masses and 2O-subfamily charges

Section 3.7 discusses masses for W- and H-family particles and charges for 2O-subfamily particles. People might say that MM1MS1 models correlate with the 2O2 boson having charge $-(1/3)|q_e|$ and the 2O1 boson having charge $+(1/3)|q_e|$. We show that ratios of integers correlate approximately with ratios of squares of rest masses for H- and W-family bosons. We show that the ratios $(m'(Higgs))^2 : (m'(Z))^2 : (m'(W))^2$ are approximately $17 : 9 : \approx 7\times(1 - (0.75)\times\alpha/(2\pi))$.

Table 3.7.1 Computation of charges for 2O bosons

Table 3.7.2 Charges for 2O-subfamily bosons

Table 3.7.3 Experimental masses and first-approximation calculated masses for H- and W-family bosons

Table 3.7.4 Hypotheses regarding relative masses of H- and W-family bosons

Table 3.7.5 Estimate for an improved calculation of m'(W)

Table 3.7.6 θ_W (the weak mixing angle) and the θ_W value some work above estimates

Section 3.8 Possibilities regarding O-family masses

Section 3.8 discusses models that produce algebraic expressions that might correlate with approximate masses of O-family particles. We show a process for developing models for masses of O-family particles. We show two such models. People might say that, based on one of the models we show, a candidate mass for the spin-0 (or, 0O0) O-family boson is 155 GeV/c^2. People might say that candidate spin-1 O-family bosons and candidate values for their masses (in units of GeV/c^2) are 2O0 with a value of 182 and 2O2 and 2O1 with values of 177. People might say that an O-family particle cannot be produced unless it is part of a pair (or triplet or ...) of O-family particles. People might say that we show candidate lower bounds for energies needed to produce O-family particles. We contrast some possibilities regarding O-family masses and information regarding masses of hypothetical leptoquarks.

Table 3.8.1 A process for developing candidate masses for O-family bosons.

Table 3.8.2 H-family mass, approximate W-family masses, and possible approximate O-family masses

Table 3.8.3 Possible minimum energies to create W-, H-, and O-family bosons, assuming O-family bosons are created as parts of composite particles

Table 3.8.4 Alternative possible minimum energies to create W-, H-, and O-family bosons, assuming O-family bosons are created as parts of composite particles

Table 3.8.5 Estimated lower bounds on rest masses for leptoquarks

Section 3.9 Masses and charges of C-family and 1Q-subfamily particles

Section 3.9 discusses a formula that approximates masses of six quarks and three charged leptons. People might say that a ratio of strengths of electromagnetism and gravity correlates with the ratio of masses of the tauon and the electron. People might say that the mass of a tauon may be $1.77685(6)\times10^3$ MeV/c^2. Here, the standard deviation is less than half the standard deviation correlating with data from experiments. People might say that this result provides impetus to improve the accuracies of measurements of the tauon mass and the gravitational constant. People might say that the formula that approximates masses of six quarks and three charged leptons involves no more than 6 parameters.

Table 3.9.1 A mass-centric array for 1C- and 1Q- elementary particles
Table 3.9.2 Notation and some data regarding a relationship among the strength of electromagnetism, the strength of gravity, the mass of a tauon, and the mass of an electron
Table 3.9.3 Numeric results for and an assumed equality between β and β'
Table 3.9.4 Vertex strengths for electromagnetism and gravity
Table 3.9.5 Possibly predicted rest mass and experimental rest mass for tauons
Table 3.9.6 Possible opportunities for research, based on m'(tauon) and G_N
Table 3.9.7 Integers L and M pertaining to 1C and 1Q rest masses
Table 3.9.8 Integers L\` pertaining to 1C and 1Q rest masses
Table 3.9.9 Preparation for developing a formula that approximates the rest masses of charged leptons and of quarks
Table 3.9.10 Steps that develop a formula that approximates the rest masses of charged leptons and of quarks
Table 3.9.11 A possible hint toward additional significance related to d" and some applications of trigonometric functions
Table 3.9.12 Experimental and calculated masses for charged leptons
Table 3.9.13 Experimental and calculated masses for quarks
Table 3.9.14 Alternative sets, each of six numbers, sufficient to approximate the rest masses of the three charged leptons and six quarks
Table 3.9.15 Relationships between m\` and some results from a formula that approximates the rest masses of three charged leptons and the six quarks
Table 3.9.16 An approximate relationship between |M'| and m(M" = 0, M')
Table 3.9.17 Possible relationships between |M'| and values of D + 2ν correlating with D* = 2 and ν = −1/2
Table 3.9.18 An approximate relationship between square roots of masses and masses
Table 3.9.19 The Koide formula and results this monograph calculates
Table 3.9.20 Some G-family interactions and their relative strengths
Table 3.9.21 Some information about a pattern pertaining to masses of charged leptons

Section 3.10 Some interactions between elementary bosons and fermions

Section 3.10 provides perspective regarding interactions between elementary fermions and elementary bosons. We list elementary bosons for which conservation of fermion generation pertains regarding FRERAN interactions between spin-1/2 fermions and elementary bosons. We discuss aspects pertaining to COMPAR interactions.

Table 3.10.1 Elementary bosons for which #"G = 1 (assuming that G-family %68odd solutions do not pertain)

Table 3.10.2 Possible relationships between R_0 (and similar lengths) and some strengths of interactions

Section 3.11 MM1MS1-neutrino oscillations

Section 3.11 correlates MM1MS1-neutrino oscillations (or, flavor mixing) with various phenomena. People might say that models for MM1MS1-neutrinos correlate with the passing through ordinary matter of MM1MS1-neutrinos contributing to MM1MS1-neutrino oscillations. People might say that gravity does not directly contribute to MM1MS1-neutrino mixing. We show the possibility that 2O0 bosons can contribute to MM1MS1-neutrino mixing.

Table 3.11.1 Some phenomena that correlate with MM1MS1-neutrino oscillations (assuming that the %68even set pertains)

Section 3.12 The fine-structure constant

Section 3.12 lists traditional and possible appearances, in physics models, of the fine-structure constant.

Table 3.12.1 Possible appearances of the fine-structure constant, α, in physics models

Section 3.13 Other aspects regarding particles, properties, and interactions

Section 3.13 discusses results that may be coincidences and/or may be pointers to opportunities for further research. We show a basis for a series that, for the electron, features charge (or, q_e), mass (or, m_e), and nominal magnetic dipole moment (or, $(g_S)\hbar/2$ or $2S\hbar$). We show a possible approximate relationship between the mass of a Higgs boson and the mass of an electron. We discuss some possible significance of β regarding ratios of G-family vertex strengths. We discuss possible significance for the notion that $\beta^{1/3} \approx e^e$.

Table 3.13.1 Notes regarding R_j for which a particle property has (in Table 2.9.3) an exponent $k = 0$

Table 3.13.2 The series (for electrons) charge, mass, and magnetic dipole moment

Table 3.13.3 A possible link between masses of elementary bosons and the strength of electromagnetism

Table 3.13.4 Possible ratios of G-family vertex strengths, beyond for just 2G2&, 4G4&, and interactions involving ordinary matter

Table 3.13.5 A possible link between β and a number (or between various masses and interaction strengths)

Table 3.13.6 Possible consequences, to the extent m* has physics-relevance

Chapter 4 From particles to cosmology and astrophysics

Section 4.1 Introduction

Section 4.1 discusses correlations between values of #ENS and aspects of cosmology and astrophysics.

Table 4.1.1 Correlations between #ENS and aspects of cosmology and astrophysics

Section 4.2 The rate of expansion of the universe

Section 4.2 discusses interactions governing the rate of expansion of the universe. People might say that interactions mediated by G-family particles provide for changes in the rate of expansion of the universe. People might say that possible observations or theories that some people correlate with term inflationary epoch would (to the extent the observations or theories pertain to nature) correlate with phenomena related to 3QIRD or 1R fermions.

Table 4.2.1 Eras pertaining to interactions between two clumps and relevant %68even particles

Table 4.2.2 Some redshifts and the related times after the big bang

Table 4.2.3 Three eras relevant to the rate of expansion of the DIU

Table 4.2.4 Model for the rate of expansion of the DIU

Table 4.2.5 G-family-mediated interactions that govern the rate of expansion of the DIU (assuming %68odd particles do not pertain)

Table 4.2.6 An example of possible opportunities for research based (at least in part) on observations of large-scale phenomena

Section 4.3 Dark energy and dark matter

Section 4.3 discusses two uses of the term dark energy. People might say that some G-family particles correlate with a force-like (or, pressure-like) use of the term. We provide perspective regarding models that may correlate with densities of the DIU (or, the directly inferable universe) for ordinary matter, dark matter, and dark-energy stuff. People might say that, in ENS48 models, dark-energy stuff features 7 peers of the stuff people might associate with the combination of ordinary (or, baryonic) matter and dark matter. People might say that, for ENS48 models and ENS6 models, this work provides an explanation for the inferred ratio of density of dark matter to density of ordinary matter being equal to or greater than 5 : 1. People might say that, for ENS48 models, this work provides an explanation for the inferred ratio of density of dark-energy stuff to density of dark matter plus ordinary (or, baryonic) matter being less than 7 : 1. People might say that ENS48 models and ENS6 models point to possibly observable phenomena that would arise from clustering within a unit of stuff that features ordinary matter and dark matter, clustering within each of seven units of dark-energy stuff (ENS48 models, but not ENS6 models), anti-clustering between each pair among the just mentioned eight units (ENS48 models, but not ENS6 models), anti-clustering between ordinary matter and one of five units of dark matter, anti-clustering between two of the other four units of dark matter, anti-clustering between the remaining two units of dark matter, and clustering within each of six units - one of ordinary matter and five of dark matter. We discuss ENS1 models. We discuss possible correlations between concepts people have proposed regarding dark matter and results correlating with models this monograph discusses. We discuss concepts for experiments or observations to possibly attempt regarding the nature of dark matter and/or the nature or dark-energy stuff. People might say that we discuss, assuming ENS48 models correlate with nature, a way to envision dark matter and dark-energy stuff. People might say that we discuss, assuming ENS6 models correlate with nature, a way to envision dark matter.

Table 4.3.1 Perspective regarding stuff-centric aspects of dark energy and regarding aspects of dark matter

Table 4.3.2 The relative densities of DEE, DME, OME, and other stuff and the relative densities of dark-energy stuff, dark matter, and ordinary matter (ENS48 models)

Table 4.3.3 Concepts pertaining to inferred OMDMS and DES densities of the DIU

Table 4.3.4 Candidate reasons why the ratio of inferred DES to inferred OMDMS need not approximate or exceed 7 (ENS48 models)

Table 4.3.5 Other factors that might affect the ratio of inferred DES to inferred OMDMS

Table 4.3.6 Inferences pertaining to dark-matter and ordinary-matter densities of the universe

Table 4.3.7 Candidate reasons why the ratio of inferred DMS to inferred OMS need not equal 5 (ENS48 models and ENS6 models)

Table 4.3.8 Process for envisioning dark-energy stuff and dark matter (ENS48 models)

Table 4.3.9 Spatial dependences and spans for interactions affecting throughout much of the evolution of the universe (present-day) large-scale objects (assuming ENS48" models and %68even models)

Table 4.3.10 Possible clustering and anti-clustering of dark-energy stuff and of dark matter (ENS48" models)

Table 4.3.11 Observations to possibly attempt regarding the distributions of ordinary matter, dark matter, and dark-energy stuff

Table 4.3.12 Experiments or observations possibly to attempt regarding the nature of dark matter and/or the nature of dark-energy stuff

Table 4.3.13 Possible examples of SEDES (single-ensemble dark-energy stuff)

Table 4.3.14 Possible examples of SEDMS (single-ensemble dark-matter stuff)

Table 4.3.15 Some candidates that people discuss for dark matter

Section 4.4 Objects containing ordinary matter and dark matter

Section 4.4 discusses objects that might contain significant amounts of each of ordinary matter and dark matter. People might say that ENS48 models and ENS6 models correlate with clustering and anti-clustering phenomena that correlate with possibilities that a galaxy or a globular cluster can contain a somewhat (but statistically not completely) arbitrary ratio of ordinary matter to dark matter.

Section 4.5 Baryon asymmetry

Section 4.5 discusses the possibility that spin-1 0-family bosons catalyzed the currently existing matter/antimatter imbalance.

Table 4.5.1 Possibility for experiments or inference regarding balance between matter neutrinos and antimatter neutrinos

Section 4.6 Phenomena that people model via general relativity

Section 4.6 discusses quantum-based models for phenomena that people model via general relativity. People might say that MM1MS1 models correlate with gravitational redshift and blueshift, gravitational bending of trajectories of light, and shifts of perihelia. We discuss a means by which black holes can lose energy.

Table 4.6.1 Physics of paths of light and of redshifts and blueshifts

Table 4.6.2 Possible research opportunities related to general relativity and alternative models

Section 4.7 The galaxy rotation problem

Section 4.7 discusses the galaxy rotation problem. People might say that, depending on models people use, a gravitational somewhat analog to electromagnetism's magnetic field may point to an overlooked phenomenon that helps contain material within galaxies.

Section 4.8 The spacecraft flyby anomaly

Section 4.8 discusses the spacecraft flyby anomaly. People might say that, depending on models people use, this work may point to overlooked phenomena that would help resolve the anomaly.

Table 4.8.1 Possible explanations for the spacecraft flyby anomaly

Table 4.8.2 A research opportunity related to the spacecraft flyby anomaly

Section 4.9 Quasars

Section 4.9 discusses the formation of quasars and quasar jets. People might say that at least one of the r^{-4} and r^{-8} spatial dependence G-family-mediated interactions produces quasars.

Table 4.9.1 Chronology and phenomenology of a quasar

Section 4.10 Cosmic microwave background cooling

Section 4.10 discusses the cooling (or redshift) over time of CMB (cosmic microwave background radiation). People might say that models for CMB cooling can correlate with Doppler shifts.

Section 4.11 Quark-based plasmas

Section 4.11 discusses concepts possibly related to quark-based plasmas. People might say that we suggest possible bases for phases for which traditional research would not include the bases or the phases.

Table 4.11.1 Possible research topics regarding quark-based plasmas and related physics

Chapter 5 From MM1MS1 models to traditional models

Section 5.1 The Standard Model

Section 5.1 discusses blending aspects of work in this monograph with aspects of the Standard Model. We show an approximate value for the weak mixing angle. People might say that this result fills a gap within the current scope of the Standard Model. We note the possibility that a lack of inclusion of gravity in the early twenty-first century Standard Model may correlate with notions that non-zero mass must pertain for at least one neutrino flavor. We discuss processes for blending MM1MS1 models and results and Standard Model techniques and results. We list candidate particles people might add to the Standard Model set of elementary particles and composite particles. Candidates correlate with the MM1 meta-model and, therefore, pertain each of ENS48 models, ENS6 models, and ENS1 models. We suggest steps for harmonizing and extending the Standard Model and MM1MS1 models.

Table 5.1.1 Possible paths for improving models regarding physics related to elementary particles

Table 5.1.2 Standard-Model and possible beyond-the-Standard-Model particle subfamilies

Table 5.1.3 CTCW

Table 5.1.4 Steps to harmonize and extend the Standard Model and MM1MS1 models regarding interactions involving spin-1 and spin-0 elementary bosons this monograph discusses

Table 5.1.5 Relationships between 2G2&-related models and 2G24&-related models and photons (ENS48 models and ENS6 models)

Table 5.1.6 Steps to harmonize and extend the Standard Model and MM1MS1 models regarding interactions involving spin-2 elementary bosons this monograph discusses

Table 5.1.7 Possible steps for adding (to the Standard Model) 3N-subfamily fermions this monograph discusses and for understanding models related to neutrinos

Table 5.1.8 Possible steps for adding (to the Standard Model) other aspects correlating with work in this monograph

Section 5.2 The cosmology timeline

Section 5.2 discusses additions that might pertain for narratives about cosmology. We position, in cosmological timelines, various known and possible phenomena. We note mechanisms (related to the G-family of elementary particles) correlating with the rate of expansion of the universe. We discuss a mechanism (correlating with spin-3/2 elementary fermions and spin-2 elementary bosons) that could correlate with an inflationary epoch. We note a mechanism (correlating with 2O-subfamily bosons) that could lead to matter/antimatter imbalance. We note mechanisms (correlating with electromagnetism) that could lead to varying proportions of matter and dark matter in objects such as galaxies. We discuss mechanisms (correlating clustering and anti-clustering of clumps of perhaps galaxy size with G-family bosons fostering the clustering and anti-clustering) that could lead to subtle patterns in CMB. We discuss mechanisms (correlating with G-family bosons) that could correlate with formation of quasars.

Table 5.2.1 Possible additions to narratives about the evolution of the universe

Chapter 6 From MM1MS1 models to traditional theories

Section 6.1 Classical-physics models, QMUSPR models, and MM1MS1 models

Section 6.1 discusses and contrasts some aspects of classical-physics models, QMUSPR models, and MM1MS1 models. We discuss, for elementary particles, representations for spin and concepts related to helicity, chirality, and/or handedness. We discuss magnetic dipole moments not associated with moving charges. We discuss properties that correlate with G-family physics. We discuss roles of $\hbar$. We discuss an operator-centric interpretation of relationships between work in this monograph and aspects of traditional physics. We discuss a possible conceptual derivation of the Dirac equation. We discuss relationships among quadratic operators, linear operators, and conservation laws. We discuss relationships between models correlating with conservation laws and models correlating with internal properties of elementary particles.

Table 6.1.1 Comparisons of some aspects of MM1MS1 models, QMUSPR models, and classical-physics models

Table 6.1.2 Notation for emission or absorption of charged W-family elementary particles

Table 6.1.3 Properties correlating with %68even G-family bosons

Table 6.1.4 Thoughts regarding properties correlating with G-family bosons

Table 6.1.5 Conceptual relationships between a particle and the universe and relationships between work in this monograph and aspects of traditional physics

Table 6.1.6 Linear operators, conservation laws, SPATIM aspects, quadratic operators, elementary-particle properties, INTERN aspects, and aspects related to #b' and #b"

Table 6.1.7 Some aspects, regarding some traditional models and MM1MS1 models for phenomena for which $\sigma = +1$, of invariances, symmetries, conservation laws, and elementary-particle internal properties

Section 6.2 Atomic physics

Section 6.2 discusses models possibly pertaining to objects similar to hydrogen atoms and possibly pertaining to other aspects of atomic and molecular physics. We speculate regarding correlating four spin-1 G-family bosons, respectively, with energy levels related to the simplest electromagnetic interactions between two particles comprising a hydrogen-like atom, fine-structure splitting, hyperfine splitting, and the Lamb shift. We speculate about ATOMOL symmetries. We speculate about possible correlations between ATOMOL symmetries and the group SO(4). We point toward modeling techniques (based on G-family bosons) that might generate results similar to results (involving the presence of the fine-structure constant) that people develop via traditional approaches. People might say that work in this section is speculative.

Table 6.2.1 A possibility for modeling H1-atoms

Table 6.2.2 Concepts that might underlie ATOMOL models for the H1-atom

Table 6.2.3 Speculation about concepts for ATOMOL symmetries for H1-atoms

Table 6.2.4 A possible basis for estimating quantities for which people find that the fine-structure constant provides contributions

Table 6.2.5 Possible generalizations for and/or from applications of possible ATOMOL (a`;4)< symmetries and possible ATOMOL (a`;4)> symmetries

Section 6.3 Special relativity

Section 6.3 discusses concepts which with people correlate the term special relativity and notes a possibility that people may find that work based on other concepts can prove beneficial compared to traditional work. We list concepts people correlate with the term special relativity. People might say that, for some aspects related to QCD (or, quantum chromodynamics), work based on MM1MS1 models might provide (compared to work based on traditional models) more robust results and/or results based on simpler methods.

Table 6.3.1 Concepts with which people correlate the term special relativity

Table 6.3.2 Possible research topics regarding models that correlate COMPAR models

Section 6.4 General relativity

Section 6.4 discusses relationships between aspects of curvature inherent in uses of space-time coordinates and various models, including models this monograph discusses and including models based on general relativity. We discuss the extent to which the universe may exhibit large-scale flatness. We discuss the topic of #b" for G-family bosons. We discuss the concept that ENS48 models may not be compatible with notions that elementary particles and small mass objects traverse geodesics in a space-time having non-zero curvature. People might say that some work in this section is speculative.

Table 6.4.1 Perspective about discussions, in this monograph, related to general relativity

Table 6.4.2 Types of results that might indicate against the universe being flat

Table 6.4.3 Observations pertaining to large-scale curvature of the universe

Table 6.4.4 Cosmology that may help indicate that flat coordinates can pertain regarding adequately large-scale phenomena

Table 6.4.5 Assumptions regarding values of #b"

Table 6.4.6 Possible correlations or lack of correlations between applications of general relativity and each of ENS48 models, ENS6 models, and ENS1 models

Table 6.4.7 QE-like coordinates and QP-like coordinates possibly related to a general-relativistic approach

Table 6.4.8 Possibly useful questions, correlating with general relativity regarding representations for energy-momentum space and for curvature people might correlate with space-time

Table 6.4.9 Correlations between some spin-2 G-family effects and energy-momentum-space coordinates

Chapter 7 From the MM1 meta-model to perspective

Section 7.1 The MM1 meta-model and ENS48, ENS6, and ENS1 models

Section 7.1 compares aspects of ENS48 models, ENS6 models, and ENS1 models. People might say that ENS48 models and ENS6 models possibly correlate better with some data about dark matter than do ENS1 models. People might say that ENS6 models and ENS1 models possibly correlate better with some aspects of general relativity that do ENS48 models.

Table 7.1.1 Some similarities and differences regarding ENS48 models, ENS6 models, and ENS1 models

Table 7.1.2 Possible perspective regarding applications of ENS48 models, ENS6 models, and ENS1 models

Section 7.2 Bases for tables of elementary particles

Section 7.2 discusses characteristics regarding the periodic table for elements and characteristics that this monograph discusses regarding elementary particles. We show possible bases for analogs, for elementary particles, to the periodic table for elements.

Table 7.2.1 Characteristics of the periodic table for elements

Table 7.2.2 Characteristics that tables might feature regarding known and possible elementary particles

Section 7.3 Tables showing elementary particles and related concepts

Section 7.3 discusses a table of subfamilies of elementary particles and related instances and symmetries. We show a table of all known and some possible elementary particles. We include, in the table, various properties of the particles. We note some intricacies that people might find useful regarding constructing other tables of elementary particles.

Table 7.3.1 Some contrasts between aspects of the MM1 meta-model and aspects of MM1MS1 models

Table 7.3.2 Subfamilies of possible elementary particles (assuming the %68odd set of possible G-family elementary particles does not pertain) and characteristics for those subfamilies

Table 7.3.3 Known and possible elementary particles (assuming the %68odd set of possible G-family elementary particles does not pertain)

Table 7.3.4 Aspects of intricacy regarding MM1MS1 models

Section 7.4 Necessity and sufficiency of some particles and models

Section 7.4 discusses the extent to which particles and MM1MS1 models this monograph discusses are necessary and/or sufficient to match known data about nature and/or to match some people's conjectures about nature. We list types of known data and conjectured phenomena. We provide, for each listed item, comments based on work in this monograph. We discuss sets of elementary particles that may be necessary and/or sufficient to correlate with all known data.

Table 7.4.1 Aspects of and conjectures about nature, and discussion about models this monograph discusses

Table 7.4.2 Conjectures about a smallest set of elementary particles sufficient to correlate with data known to pertain to nature or thought likely to pertain to nature (%68even)

Section 7.5 Possible MM1MS1-correlated opportunities for research

Section 7.5 discusses possible opportunities for further research, based on MM1MS1 considerations.

Table 7.5.1 Possible math-related topics and/or modeling-related topics for further exploration

Table 7.5.2 Possible physics topics and/or physic modeling topics for further exploration

Table 7.5.3 Pointers to some possible physics topics and/or physics modeling topics for further exploration

Section 7.6 Possible general opportunities

Section 7.6 discusses advances in physics modeling. We list some previous advances. We list possible opportunities for forthcoming advances. We list generic possible steps forward.

Table 7.6.1 Some advances in physics and in models for physics

Table 7.6.2 Some advances people might derive from paralleling or continuing work in this monograph

Table 7.6.3 Possible steps forward

Chapter 8 Appendices

Section 8.1 Some physics numbers

Section 8.1 shows some physics numbers this monograph uses.

Table 8.1.1 Some physics numbers

Section 8.2 Numbers of generators for groups SU(j)

Section 8.2 notes relationships between numbers of generators for groups SU(j) for various j.

Table 8.2.1 Ratios of numbers of generators for various SU(j)

Table 8.2.2 Other arithmetic relationships between numbers of generators for various SU(j)

Chapter 9 Compendia

Section 9.1 Acronyms

Section 9.1 lists acronyms this monograph uses.

Section 9.2 Summaries of sections and names of tables

Section 9.2 lists summaries of sections and names of tables.

Section 9.3 References

Section 9.3 collects a list of references from above.

Section 9.3 References

Section 9.3 collects a list of references from above.

Chapter 1 Overview

Chapter 2 From data to the MM1 meta-model and MM1MS1 models

Reference 1 Wolfram Alpha (2014a)

Wolfram Alpha, computational knowledge engine, Wolfram Alpha LLC, (2014). http://mathworld.wolfram.com/DeltaFunction.html.

Reference 2 G. T. Adylov (1977)

G. T. Adylov, et. al., A measurement of the electromagnetic size of the pion from direct elastic pion scattering data at 50 GeV/c, *Nuclear Physics B*, Volume 128, Issue 3, 3 October 1977, pages 461-505. (http://dx.doi.org/10.1016/0550-3213(77)90056-6)

Reference 3 Particle Data Group (2014)

Particle Data Group, Electroweak (web page), The Particle Adventure, Lawrence Berkeley National Laboratory, (2014), http://www.particleadventure.org/electroweak.html.

Chapter 3 From the MM1 meta-model to particles and properties

Reference 4 K.A. Olive (2014; LEPTOQUARKS)

K.A. Olive et al. (2014) (See Reference 18). Specifically, LEPTOQUARKS, Updated August 2013 by S. Rolli (US Department of Energy) and M. Tanabashi (Nagoya U.).

Reference 5 K.A. Olive et al. (2014)

K.A. Olive et al. (Particle Data Group), *Chin. Phys. C*, 38, 090001 (2014). (http://pdg.lbl.gov/2014/html/authors_2014.html)

Chapter 4 From particles to cosmology and astrophysics

Reference 6 N. Gnedin (2014)

N. Gnedin, Cosmological Calculator for the Flat Universe (2014). (http://home.fnal.gov/~gnedin/cc/)

Reference 7 N. G. Busca (2013)

N. G. Busca, et. al., Baryon Oscillations in the Lyα forest of BOSS quasars, *Astronomy & Astrophysics* 552 A96, April 2013. (arXiv:1211.2616 [astro-ph.CO])

Reference 8 A. Riess (2004)

A. Riess, et. al., Type Ia Supernova Discoveries at z > 1 from the *Hubble Space Telescope*: Evidence for Past Deceleration and Constraints on Dark Energy Evolution, *The Astrophysical Journal*, 607, 665 (2004). (doi:10.1086/383612) (http://iopscience.iop.org/0004-637X/607/2/665)

Reference 9 T. Ferris (2015)

Timothy Ferris, A First Glimpse of the Hidden Cosmos, *National Geographic*, Vol. 227, No. 1, January 2015, pp. 108-123. (http://ngm.nationalgeographic.com/2015/01/hidden-cosmos/ferris-text)

Reference 10 J. Beringer (2012)

J. Beringer et al. (Particle Data Group), *Phys. Rev. D86*, 010001 (2012). (http://pdg.lbl.gov/2012/reviews/rpp2012-rev-cosmic-microwave-background.pdf)

Reference 11 K.A. Olive (2014; 25)

K.A. Olive et al. (2014) (See Reference 18). Specifically: 25. DARKMATTER, Revised September 2013 by M. Drees (Bonn University) and G. Gerbier (Saclay, CEA). (http://pdg.lbl.gov/2014/reviews/rpp2014-rev-dark-matter.pdf)

Reference 12 Jin Koda, et. al. (2015)

Jin Koda, Masafumi Yagi, Hitomi Yamanoi, and Yutaka Komiyama, Approximately a thousand ultra diffuse galaxies in the Coma cluster, *ApJ Letters*, Volume 807, Number 1, 2015. http://iopscience.iop.org/article/10.1088/2041-8205/807/1/L2

Reference 13 Sukanya Chakrabarti, et. al. (2015)

Sukanya Chakrabarti, Roberto Saito, Alice Quillen, Felipe Gran, Christopher Klein, and Leo Blitz, Clustered Cepheid Variables 90 kiloparsec from the Galactic Center, *Astrophysical Journal Letters*, 1502, 1358 (2015).

Reference 14 C. Yozin (2015)

C. Yozin and K. Bekki, The quenching and survival of ultra diffuse galaxies in the Coma cluster, *MNRAS* 452, 937–943 (2015).

Reference 15 Elizabeth Howell (2015)

Elizabeth Howell, Dark Matter Can Interact With Itself, Galaxy Collisions Show, Space.com (April 15, 2015). http://www.space.com/29115-dark-matter-interactions-galaxy-collisions.html

Reference 16 Evan N. Kirby, et. al. (2015)

Kirby, Evan N. and Cohen, Judith G. and Simon, Joshua D. and Guhathakurta, Puragra, Triangulum II: Possibly a Very Dense Ultra-faint Dwarf Galaxy, *Astrophysical Journal*, 814 (1), Art. No. L7, 2015.

Chapter 5 From MM1MS1 models to traditional models

Chapter 6 From MM1MS1 models to traditional theories

Reference 17 NASA (2014)

NASA, http://map.gsfc.nasa.gov/universe/uni_shape.html (2014)

Chapter 7 From the MM1 meta-model to perspective

Chapter 8 Appendices

Reference 18 K.A. Olive (2014)

K.A. Olive et al. (Particle Data Group), *Chin. Phys. C*, 38, 090001 (2014). (http://pdg.lbl.gov/2014/html/authors_2014.html)

Chapter 9 Compendia

Bibliography

Cosimo Bambi and Alexandre D. Dolgov, *Introduction to Particle Cosmology: The Standard Model of Cosmology and its Open Problems*, Springer-Verlag, Berlin Heidelberg, 2015.

Matthias Bartelmann, *Theoretical Astrophysics: An Introduction*, Wiley, 2013.

Martin Bojowald, *Quantum Cosmology: A Fundamental Description of the Universe*, Springer-Verlag, New York, 2011.

Thomas J. Buckholtz, *Theory of Particles plus the Cosmos: Small Things and Vast Effects*, CreateSpace Independent Publishing Platform, North Charleston, SC, USA, 2015.

Duncan Carlsmith, *Particle Physics*, Pearson, 2012.

Bradley Carroll and Dale Ostlie, *An Introduction to Modern Astrophysics*, 2nd Edition, Pearson, 2013.

Ghanashyam Date, *General Relativity: Basics and Beyond*, CRC Press, 2014.

Mauro D'Onofrio, et. al., *Fifty Years of Quasars: From Early Observations and Ideas to Future Research*, Springer-Verlag, Berlin Heidelberg, 2012.

John M. Davis, *Introduction to Applied Partial Differential Equations*, Macmillan, 2013.

Ulrich Ellwanger, *From the Universe to the Elementary Particles: A First Introduction to Cosmology and the Fundamental Interactions*, Springer-Verlag, Berlin Heidelberg, 2012.

E. Escultura and V. Lakshmikantham, *The Hybrid Grand Unified Theory*, Atlantis, 2009.

Rhodri Evans, *The Cosmic Microwave Background: How It Changed Our Understanding of the Universe*, Springer International Publishing, 2015.

Jane Gleeson-White, *Double Entry: How the Merchants of Venice Created Modern Finance*, Norton, 2012.

Alan Guth, *The Inflationary Universe*, Basic Books, 1998.

Pankaj Jain, *An Introduction to Astronomy and Astrophysics*, CRC Press, 2015.

Paul Langacker, *The Standard Model and Beyond*, CRC Press, 2009.

Andrew Liddle, *An Introduction to Modern Cosmology*, 3rd Edition, Wiley, 2015.

Dierck-Ekkehard Liebscher, *Cosmology*, Springer-Verlag, Berlin Heidelberg, 2005.

Debasish Majumdar, *Dark Matter: An Introduction*, CRC Press, 2014.

Franz Mandl and Graham Shaw, *Quantum Field Theory, 2nd Edition*, Wiley, 2010.

Robert Mann, *An Introduction to Particle Physics and the Standard Model*, CRC Press, 2009.

Brian R. Martin and Graham Shaw, *Particle Physics, 3rd Edition*, Wiley, 2008.

Jon Mathews and Robert Walker, *Mathematical Methods of Physics, 2nd Edition*, Pearson, 2000.

Charles W. Misner, Kip S. Thorne, and John Archibald Wheeler, *Gravitation*, Macmillan, 1973.

T.J. Buckholtz, *Models for Physics of the Very Small and Very Large*,
Atlantis Studies in Mathematics for Engineering and Science 14,
DOI 10.2991/978-94-6239-166-6

Yorikiyo Nagashima, *Beyond the Standard Model of Elementary Particle Physics*, Wiley, 2014.

Yorikiyo Nagashima, *Elementary Particle Physics: Foundations of the Standard Model, V2*, Wiley, 2013.

Chris Quigg, *Gauge Theories of the Strong, Weak, and Electromagnetic Interactions: Second Edition*, Princeton University Press, 2013.

James Rich, *Fundamentals of Cosmology, 2nd Edition*, Springer-Verlag, Berlin Heidelberg, 2010.

Matthew Robinson, *Symmetry and the Standard Model: Mathematics and Particle Physics*, Springer-Verlag, New York, 2011.

Matts Roos, *Introduction to Cosmology, 4th Edition*, Wiley, 2015.

Barbara Ryden, *Introduction to Cosmology*, Pearson, 2013.

Barbara Ryden and Bradley Peterson, *Foundations of Astrophysics*, Pearson, 2009.

Abraham Seiden, *Particle Physics: A Comprehensive Introduction*, Pearson, 2004.

Mark Srednicki, *Quantum Field Theory*, Cambridge University Press, 2007.

Edwin F. Taylor and John Archibald Wheeler, *Spacetime Physics*, Macmillan, 1992.

Kip Thorne, *Black Holes & Time Warps*, Norton, 1995.

Steven Weinberg, *Gravitation and Cosmology: Principles and Applications of the General Theory of Relativity*, Wiley, 2008.

Georg Wolschin, *Lectures on Cosmology: Accelerated Expansion of the Universe*, Springer-Verlag, Berlin Heidelberg, 2010.

Index

T.J. Buckholtz, *Models for Physics of the Very Small and Very Large*,
Atlantis Studies in Mathematics for Engineering and Science 14,
DOI 10.2991/978-94-6239-166-6

Printed in the United States
By Bookmasters